T0350233

A GRADUATE
COURSE IN
ALGEBRA

A GRADUATE COURSE IN ALGEBRA

VOLUME 1

Ioannis Farmakis
Department of Mathematics, Brooklyn College
City University of New York, USA

Martin Moskowitz
Ph.D. Program in Mathematics, CUNY Graduate Center
City University of New York, USA

 **World Scientific**

NEW JERSEY · LONDON · SINGAPORE · BEIJING · SHANGHAI · HONG KONG · TAIPEI · CHENNAI · TOKYO

Published by

World Scientific Publishing Co. Pte. Ltd.

5 Toh Tuck Link, Singapore 596224

USA office: 27 Warren Street, Suite 401-402, Hackensack, NJ 07601

UK office: 57 Shelton Street, Covent Garden, London WC2H 9HE

Library of Congress Cataloging-in-Publication Data
Names: Farmakis, Ioannis. | Moskowitz, Martin A.
Title: A graduate course in algebra : in 2 volumes / by Ioannis Farmakis
  (City University of New York, USA), Martin Moskowitz (City University of New York, USA).
Description: New Jersey : World Scientific, 2017– |
  Includes bibliographical references and index.
Identifiers: LCCN 2017001101| ISBN 9789813142626 (hardcover : alk. paper : v. 1) |
  ISBN 9789813142633 (pbk : alk. paper : v. 1) | 9789813142664 (hardcover : alk. paper : v. 2) |
  ISBN 9789813142671 (pbk : alk. paper : v. 2) | ISBN 9789813142602 (set : alk. paper) |
  ISBN 9789813142619 (pbk set : alk. paper)
Subjects: LCSH: Algebra--Textbooks. | Algebra--Study and teaching (Higher)
Classification: LCC QA154.3 .F37 2017 | DDC 512--dc23
LC record available at https://lccn.loc.gov/2017001101

British Library Cataloguing-in-Publication Data
A catalogue record for this book is available from the British Library.

Printed in Singapore

Contents

Preface and Acknowledgments

This book is a two volume graduate text in Algebra. The first volume consisting of six chapters contains basic material to prepare the reader for the (graduate) Qualifying Exam in Algebra; with its numerous exercises it is also suitable for self study. The second volume consisting of an additional seven chapters is designed for a second year topics course in Algebra. Our book was written over the course of the past five years, the second author having taught the subject many times. In our view Algebra is an integral and organic part of mathematics and not to be placed in a bubble. For this reason, the reader will find a number of topics represented here not often found in an algebra text, but which closely relate to nearby subjects such as Lie groups, Geometry, Topology, Number theory and even Analysis.

As a convenience to the instructor using our book we have also created a pamphlet containing a large number of exercises together with their solutions. The pamphlet also includes some exercises which are not in our book.

Chapter 0 (Introduction)

Among many things this introductory chapter contains three proofs that *there are infinitely many primes*: Euclid's proof, a topological proof due to H. Fürstenberg, and a proof by Erdös based on the divergence of the Euler series. We give a short proof of *the Euler formula and the Riemann Zeta function* due to Calabi. We also address *Pythagorean Triples*.

Chapter 1 (Groups)

In addition to the usual material here we give an extensive treatment of the *Quaternion* and *Dihedral* groups as well as *Modular Arithmetic*. We also present two proofs of *Fermat's Little Theorem* (the classic one, and one based on fixed points) and study *direct and inverse limits of groups* and the *profinite completion* of a group. Turning to Abelian groups, we give a detailed treatment of *divisible* and *Prüfer groups*.

Chapter 2 (Further Topics in Group Theory)

In this chapter we give an alternative proof of the Fundamental Theorem of Arithmetic based on the Jordan-Hölder Theorem and give an extensive treatment of group actions, especially the transitive ones and present various important homogeneous spaces: *the real, complex and quaternionic spheres*, the *Poincaré upper half-plane*, the *real and complex projective spaces*, the *Grassmann and flag manifolds* and study their topological properties.

We then turn to the concept of a *doubly transitive action*, and prove $\mathrm{PSL}(n, k)$ acts doubly transitively on projective space. We then prove *Iwasawa's double transitivity theorem* and use it to prove the simplicity of the projective group, $\mathrm{PSL}(n, k)$. Then we discuss *imprimitive actions* and conclude this chapter with the Sylow theorems. Here we give a second proof of Cauchy's theorem based on the fact that the action of the cyclic group $\mathbb{Z}_p$ on the space of finite sequences has a fixed point.

Chapter 3 (Vector Spaces)

In addition to the standard material, here we give two important applications of diagonalizability of linear operators : *the Fibonacci sequence*, and Liouville's theorem in *matrix differential equations*.

Chapter 4 (Inner Product Spaces)

In this chapter we study both real and complex inner product spaces and their normal operators. As an application of Gram Schmidt orthogonalization, we study the Legendre polynomials. We also do the Cayley transform. We then turn to the Iwasawa and Bruhat decomposition theorems in the case of $\mathrm{GL}(n)$. We show that $|\det|$ preserves volume in $\mathbb{R}^n$. Finally, we treat Gramians and Schur's theorems on triangularization and eigenvalue estimates.

Chapter 5 (Rings, Fields and Algebras)

Here we present the standard material, and in addition study non-negative polynomials and also show, given a finite sequence, that there is no such thing as a pattern. We then turn to *Wedderburn's little theorem* and present an application of it to projective geometry. As an application of unique factorization in the *Gaussian integers*, we solve a problem in Diophantine algebraic geometry due to Mordell. Finally, we turn to the *fundamental theorem of symmetric polynomials*.

Chapter 6 (R-Modules)

In addition to the standard material, here we do *Schur's lemma* and *simple modules*, the fundamental theorem of finitely generated modules over a Euclidean ring, the Jordan canonical form, the Jordan decomposition and the Krull-Schmidt theorem.

VOLUME II

Here we list the topics covered in Volume II.

Chapter 7 (Multilinear Algebra)

In this chapter we study the important notion of *orientation*, proving among other things that real square matrices are continuously deformable into one another if and only if the signs of their determinants are conserved. We then turn to Riesz's theorem for a separable Hilbert space and use it to define the *Hodge star* ($*$) *operator* which is especially important in differential geometry and physics. We prove *Liouville's formula* making use of the fact that diagonalizable matrices are *Zariski dense* in $M(n, \mathbb{C})$. Finally, we study Grassmannians and their relationship to the exterior product.

Chapter 8 (Symplectic Geometry)

After dealing with the standard material we show that the (real or complex) symplectic group $\mathrm{Sp}(V)$ is generated by transvections. We then prove the Williamson normal form which states that any positive definite symmetric real $2n \times 2n$ matrix is diagonalizable by a symplectic matrix. This leads to the well known *Gromov non-squeezing theorem* which that the two-dimensional shadow of a symplectic ball of radius r in $\mathbb{R}^{2n}$ has area at least πr^2. Finally, we discuss the curious relationship between Gromov's non-squeezing theorem and the Heisenberg *uncertainty principle* of quantum mechanics.

Chapter 9 (Commutative Rings with Identity)

In this chapter we systematically find all *principal ideal domains* of a certain type which are not *Euclidean rings*. We present Kaplansky's characterization of a Unique Factorization Domain (UFD), the *power series ring* $R[[x]]$, the *prime avoidance lemma* and the *Hopkins-Levitsky theorem*. Then we turn to *Nakayama's lemma*, where we give a number of applications as well as several of its variants. We then define the *Zariski topology* and derive some of its important properties (as well as giving a comparison with the *Euclidean topology*). Finally, we discuss the *completion of a ring* and prove the completion of $R[x]$ is $R[[x]]$.

Chapter 10 (Valuations and *p*-adic Numbers)

Here we study the topology of $\mathbb{Q}_p$ and its curiosities such as all triangles are isosceles and every point in a sphere is a center. Also we prove no three points in $\mathbb{Q}_p$ are collinear and right triangles do not exist. After proving Ostrowski's theorem, we study both $\mathbb{Z}_p$ and $\mathbb{Q}_p$ as topological groups and rings. As an application, we present *Monsky's theorem* that if a square is cut into triangles of equal area, the number of triangles must be even.

Chapter 11 (Galois Theory)

In addition to the standard features of Galois theory, we prove the *theorem of the primitive element* and emphasize the role of fields of characteristic zero. We also give an algebraic proof of the *fundamental theorem of algebra*.

Chapter 12 (Group Representations)

First we deal with the basic results of representations and characters of a finite group G over $\mathbb{C}$, including divisibility properties. For example, $\deg \rho$ divides the index of the center of G for every irreducible representation ρ of G. We then prove *Burnside's p, q theorem* which is an important criterion for solvability and Clifford's theorem on the decomposition of the restriction of an irreducible representation to an invariant subgroup. We then turn to infinite groups where we study linear groups over a field of characteristic zero, real representations and their relationship to calculating homotopy groups and preserving quadratic forms, and finally return to Pythagorean triples of Volume I.

Chapter 13 (Representations of Associative Algebras)

In this final chapter we prove Burnside's irreducibility theorem, the Double Commutant theorem, the *big Wedderburn theorem* concerning the structure of finite dimensional semi-simple algebras over an algebraically closed field of characteristic zero. We then give two proofs of *Frobenius' theorem* (the second one using algebraic topology) which classifies the finite dimensional division algebras over $\mathbb{R}$. Finally, using quaternions, we prove Lagrange's 4-square theorem. This involves the Hürwitz integers and leads very naturally to the study of the geometry of the unique 24-cell polyhedron in $\mathbb{R}^4$ and its exceptional properties, which are of particular interest in physics.

A significant bifurcation in all of mathematics, and not just in algebra, is the distinction between the commutative and the non-commutative. In our book this distinction is ever present. We encounter it first in Volume I in the case of groups in Chapters 1 and 2, then in rings and algebras in Chapter 5 and modules in Chapter 6, where the early sections deal with the commutative, while the later ones address the non-commutative. Chapter 9 is devoted to commutative algebra while Chapters 12 and 13 to non-commutative algebra, both being of great importance. We hope our enthusiasm for the material is communicated to our readers. We have endeavored to make this book useful to both beginning and advanced graduate students by illustrating important concepts and techniques, offering numerous exercises and providing historical notes that may be of interest to our readers. We also hope the book will be of use as a reference work to professional mathematicians specializing in areas of mathematics other than algebra. The endgame of our endeavor has turned out to be a family affair! We wish to heartily thank Andre Moskowitz for reading the text and making numerous useful suggestions for its improvement and Oleg Farmakis for creating all the graphics. We also thank our wives, Dina and Anita, for their forbearance during the long and challenging process of bringing this book to completion.

August 2016

Ioannis Farmakis   Martin Moskowitz
Brooklyn College-CUNY   The Graduate Center-CUNY

INTERDEPENCE OF THE CHAPTERS

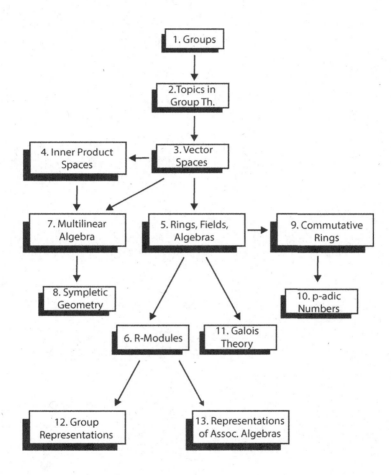

Chapter 0

Introduction

0.1 Sets

Here we take the idea of a *set* as a primitive and intuitive notion. Sets will be denoted by capital letters such as X, and their elements by x. When x is an element of X we write $x \in X$. Examples of sets are $\mathbb{Z}$, the integers; $\mathbb{Q}$, the rational numbers; $\mathbb{R}$, the real numbers; and $\mathbb{C}$, the complex numbers. The set consisting of a single element x is written $\{x\}$ and is called a *singleton*. When one wishes to define a set by its elements one writes

$$X = \{x : x \quad \text{has a certain property}\}.$$

For example, the odd integers are $\{2n + 1 : n \in \mathbb{Z}\}$ and the positive integers, $\mathbb{Z}^{+} = \{x \in \mathbb{Z} : x \geq 1\}$. The set without any elements (called the *null set*) is written $\varnothing$. A subset Y of a set X is indicated by $Y \subseteq X$. So for example $\varnothing \subseteq X$ for any set X and also $X \subseteq X$. When two sets X and Y have exactly the same elements we write $X = Y$, that is, $X \subseteq Y$ and $Y \subseteq X$. If $X \subseteq Y$ and $Y \subseteq Z$, then $X \subseteq Z$. The set of elements common to both X and Y is $X \cap Y$, while those in either is $X \cup Y$. If $X \cap Y = \varnothing$, we say X and Y are *disjoint*. If X_i is a family of sets indexed by the set I, then the set of elements common to all X_i is denoted by $\cap_{i \in I} X_i$ and is called the intersection, while the set of elements in some X_i is $\cup_{i \in I} X_i$ which is called the union. $|X|$ indicates

1

the number of elements, or *cardinality* of X (which may be finite or infinite). An important construction of sets is *Cartesian product*. An *ordered pair* (x, y) is one in which we consider the elements x and y, where it is understood that x comes first and y second. For example, latitude and longitude on a map of the earth is such a pair. We consider $(x_1, y_1) = (x_2, y_2)$ if and only if $x_1 = x_2$ and $y_1 = y_2$. If X and Y are sets, then the Cartesian product is $X \times Y = \{(x, y) : x \in X, y \in Y\}$. More generally, one can consider a finite number n, sets $X_1, \ldots, X_n$ and ordered n-tuples, $(x_1, \ldots, x_n)$ and the Cartesian product, $X_1 \times \cdots \times X_n$. For example, Euclidean 3-space is the situation in which $n = 3$ and all the $X_i = \mathbb{R}$.

We now turn to the notion of a *function* or map $f : X \to Y$. This is a rule which assigns a definite member called $f(x) \in Y$ to each $x \in X$. This is sometimes denoted by $x \mapsto f(x)$. X is called the *domain* of f and Y the *co-domain*. For example, as above $f(n) = 2n+1$ is a function from the integers to the odd integers.

For a subset $A \subseteq X$ we write $f(A) = \{f(x) : x \in A\}$. $f(X)$ is called the *range* or the *image* of f, and we write $\text{Im}(f) = f(X)$.

f is called *surjective* (or *onto*) if $f(X) = Y$. We say f is *injective* (or 1:1) if whenever $f(x_1) = f(x_2)$, where x_1 and $x_2 \in X$, then x_1 must equal x_2. If a function is both injective and surjective we say it is *bijective*. Two sets have the same cardinality if there is a bijective map between them. For example, $\mathbb{Q}$ and $\mathbb{Z}$ each have the same cardinality as $\mathbb{Z}^+$, while $\mathbb{R}$ has a larger cardinality. If $B \subseteq Y$, $f^{-1}(B)$ stands for $\{x \in X : f(x) \in B\}$. The *graph* of f is $\{(x, f(x)) : x \in X\}$. For example, a bijective correspondence of the graph of f with X is given by $x \mapsto (x, f(x))$.

The Pigeonhole Principle.

This is a convenient point to state the *Pigeonhole Principle*. Namely, if n items are to be put into m containers, with $n > m$, then at least one container must contain at least two items. So for example given three gloves, there must be at least two left or two right gloves.

0.2   Cartesian Product

We have seen that if $X_1, ..., X_n$ are n non-empty sets we define the *Cartesian product* as

$$\prod_{i=1}^{n} X_i = \{(x_1, ..., x_n) \mid x_i \in X_i, \; i = 1, ..., n\}.$$

Now, if I is an arbitrary, non-empty set, and $\{X_i\}_{i \in I}$ a family of non-empty sets their Cartesian product is defined as

$$\prod_{i \in I} X_i = \{f : I \longrightarrow \bigcup_{i \in I} X_i, \; : \; f(i) \in X_i\}.$$

Sometimes, we write the element f as $(x_i)_{i \in I}$, so if the set I is the finite set $\{1, 2, ..., n\}$, we get $(x_i)_{i \in I} = (x_1, ..., x_n)$. Here, we notice that the existence of such a map f is essentially one form of the so-called *axiom of choice*.

0.3   Relations

Closely related to (but more general) than a function is the notion of a *relation*. This is just like the definition of a function, but here one relaxes the requirement of uniqueness. That is, for each $x \in X$ there are associated elements of Y, but perhaps more than one. Formally, a relation R on X is just a subset $R \subseteq X \times X$.

Definition 0.3.1. If $X \neq \varnothing$, then a *partition* of X is a collection of non-void subsets $(X_i)_{i \in I}$ such that for each i, j, $X_i \cap X_j = \varnothing$ and $\bigcup_{i \in I} X_i = X$.

For example, we leave as an exercise to the reader to check that if $f : X \to Y$ is a surjective map and we denote $f^{-1}(y)$ by S_y, then the set of all S_y, $y \in Y$ gives a partition of X.

The relations we are particularly interested in here are *equivalence relations*. Let X be a non-empty set ($X \neq \varnothing$), then

Definition 0.3.2. An *equivalence relation* on X, is a relation $\sim$ with the following properties:

1. If $x \in X$, then $x \sim x$ (reflexive property).

2. If $x \sim y$, then $y \sim x$ (symmetric property).

3. If $x \sim y$ and $y \sim z$, then $x \sim z$ (transitive property).

Of course, equality is an equivalence relation. So what we are doing here is taking a more general view of equality.

Example 0.3.3. Let n be a fixed positive integer $n \geq 2$ and define a relation on $X = \mathbb{Z}$ by:

$$x \sim y, \text{ if and only if, } x - y \text{ is a multiple of n.}$$

Then $\sim$ is an equivalence relation.

Proof. Indeed,
1. For all $x \in \mathbb{Z}$, $x \sim x$, because, $x - x = 0 = 0 \times n$.
2. If $x \sim y$ then $y \sim x$, because, $x - y = nk \Rightarrow y - x = -nk$.
3. If $x \sim y$ and $y \sim z$, then $x - y = nk_1$, and $y - z = nk_2$. Therefore, $x - z = x - y + y - z = nk_1 + nk_2 = n(k_1 + k_2)$. $\qquad\qquad\qquad\square$

The previous example was arithmetic. We now give two geometric examples. We ask the reader to check that these are indeed equivalence relations.

Consider all triangles in the plane. We say two triangles are congruent if one can be placed upon the other and they match up perfectly. Alternatively, we say two triangles are similar if they have the same corresponding angles (or the corresponding sides are proportional). Congruence and similarity are both equivalence relations.

Definition 0.3.4. If $\sim$ is an equivalence relation on X and $a \in X$, then we define the *equivalence class* containing a by

$$[a] = \{x \in X : x \sim a\}.$$

Thus each element belongs to some equivalence class.

Proposition 0.3.5. *If $b \in [a]$, then $[b] = [a]$.*

Proof. First we show that $[b] \subseteq [a]$. Indeed, since $b \in [a]$, $b \sim a$. So, if $x \in [b]$ then $x \sim b$ and since $b \sim a$ it follows from property 3 above that $x \sim a$ and hence $x \in [a]$. Thus $[b] \subseteq [a]$. Moreover, $[a] \subseteq [b]$ since $a \sim b$, it follows that $\forall\, x \sim a \Rightarrow x \sim b$, so $x \in [b]$. $\qquad\square$

Thus, we may speak of a subset of X being an equivalence class, with no mention of any element contained in it. More precisely, we have the following proposition whose proof we leave as an exercise. It follows from the proposition below that each element of X belongs to one and only one equivalence class and thus we get a partition of X into equivalence classes.

Proposition 0.3.6. *If U, $V \subseteq X$ are two equivalence classes, and $U \cap V$ is not empty, then $U = V$.*

Conversely, a partition of X gives rise to an equivalence relation. We also leave the proof of this next proposition to the reader.

Proposition 0.3.7. *Let $X \neq \varnothing$ and $(X_i)_{i \in I}$ be a partition of X. Define the relation $a \sim b$ to mean a and b belong to the same subset of the given partition. Then $\sim$ is an equivalence relation, and the equivalence classes are just the subsets of the given partition.*

Thus there are two ways of viewing an equivalence relation. One is as a relation satisfying the conditions 1, 2 and 3 of the definition, and the other is as a partition of X into non-empty disjoint subsets. Thus equivalence relations and partitions amount to the same thing. In the example above the set of all these classes is denoted by $\mathbb{Z}_n$ and we write $x \in [y]$ as $x \cong y \bmod n$. The class of $[0]$ is denoted $n\mathbb{Z}$. There are exactly n equivalence classes. When $n = 2$ this is just the partition of $\mathbb{Z}$ into even and odd numbers.

0.4   Partially Ordered Sets

We conclude this chapter with a summary of what we shall need for Induction and Transfinite Induction as well as for other purposes. To do so we define a *partially ordered set*. Let X and $\leq$ be just as in a relation *except that two elements may or may not be comparable*. We assume properties 1, 2 and 3 as above. Here we write $(X, \leq)$. Notice that any subset of a partially ordered set is itself partially ordered by restriction of the relation.

A set $(X, \leq)$ is *totally ordered* if it is partially ordered and any two elements x and y are comparable. It is *well ordered* if each non-empty subset has a least element. The set of positive integers with its usual ordering is evidently well ordered, but the set of all integers is not although it is totally ordered.

The natural ordering on $\mathbb{R}$ and therefore on *any* of its subsets, is a total ordering. Notice that the total ordering on $\mathbb{Z}^+$ has the special feature, *not* shared by $\mathbb{R}^+$, or even $\mathbb{Q}^+$. That is, *any non-empty subset has a least element.*

An example of a partially ordered set the reader should think of is the power set, $\mathcal{P}(X)$ of X. This is the set of all subsets of a fixed set X with $\leq$ given by set inclusion, $\subseteq$. Here some subsets are not comparable with others, but when they are they satisfy conditions 1, 2 and 3 above. Thus $\mathcal{P}(X)$ under inclusion is a partially ordered, but not totally ordered set.

We note that if $|X| = n$ (a finite set), then $|\mathcal{P}(X)| = 2^n$. This follows immediately from the Binomial Theorem 0.4.2, since

$$(1+1)^n = \sum_{r=0}^{n} \binom{n}{r}.$$

0.4.1   The Principle of Induction

Let $P(n)$ be some property, depending on $n \in \mathbb{Z}^+$. If

1. $P(1)$ is true and,

2. if for any $k \in \mathbb{Z}^+$, $P(k)$ implies $P(k+1)$,   .

then, $P(n)$ is true for all $n \in \mathbb{Z}^+$. (2 is called the *inductive hypothesis*.)

It is easy to see why the Principle of Induction is valid. Let

$$X = \{n \in \mathbb{Z}^+ : P(n) \text{ is false}\}.$$

We want to show $X = \varnothing$. Suppose to the contrary that X is non-empty. Let x_0 be its *least element*. So $P(x_0)$ is false. Since by assumption $P(1)$ is true, $1 \neq x_0$ so $x_0 > 1$. But $x_0 - 1 < x_0$. Hence $P(x_0 - 1)$ is true. But by inductive hypothesis $P(x_0)$ must also be true, a contradiction.

As an example of the use of the Principle of Induction we now prove the binomial theorem.

The *binomial theorem* is a formula expressing $(a + b)^n$, for any $n \in \mathbb{Z}^+$, as a "polynomial" in a and b with certain integer coefficients. Here we will take a and $b \in \mathbb{Z}$. But, as we shall see from the argument, this important fact only depends on a and b commuting ($ab = ba$).

Before stating the binomial theorem, we make an observation concerning the coefficients. Suppose r and $n \in \mathbb{Z}^+$ with $r \leq n$. We define $\binom{n}{r} = \frac{n!}{r!(n-r)!}$. The reader will notice that $\binom{n}{r}$ is exactly the number of subsets consisting of r elements that can be chosen from $\{1, \ldots, n\}$. This is because there are n ways to fill the first spot. Having done so, there are then $n - 1$ ways to fill the second spot ... and finally, there are r ways to fill the last spot. Thus, the number of possible sets of this type is exactly $n(n - 1)(n - 2) \ldots r$, if we distinguish the order of selection. But a permutation of the elements of a set does not change the set and as we saw there are exactly $r!$ such permutations. Therefore, the number of possible sets of this type (without regard to their order) is exactly $\frac{n \cdot (n-1) \cdot (n-2) \ldots r}{r!} = \binom{n}{r}$. In particular, $\binom{n}{r}$ is always an integer. Another consequence of this interpretation as the number of possible sets of r elements is $\binom{n}{r} = \binom{n}{(n-r)}$.

Lemma 0.4.1 below follows directly from the definitions and is left to the reader as an exercise. From it we see that the $\binom{n}{r}$ can be generated in a very efficient and striking way by constructing *Pascal's triangle*.

$$
\begin{array}{ccccccccccccc}
&&&&&&1\\
&&&&&1&&1\\
&&&&1&&2&&1\\
&&&1&&3&&3&&1\\
&&1&&4&&6&&4&&1\\
&1&&5&&10&&10&&5&&1\\
1&&6&&15&&20&&15&&6&&1\\
1&7&21&&35&&35&&21&7&1
\end{array}
$$

$$\vdots \quad \vdots \quad \vdots \quad \vdots$$

Lemma 0.4.1. *For all r and $n \in \mathbb{Z}^+$ with $0 \le r \le n$, $\binom{n+1}{r} = \binom{n}{r} + \binom{n}{r-1}$.*

We now come to the binomial theorem itself.

Theorem 0.4.2. *For each $n \in \mathbb{Z}^+$ and $a, b \in \mathbb{Z}$ (or indeed for any commuting a and b)*

$$(a+b)^n = \sum_{r=0}^{n} \binom{n}{r} a^r b^{n-r}. \tag{1}$$

Proof. The proof will be by induction on n. Clearly 0.4.2 holds for $n = 1$. Assuming it holds for n, multiply (0.4.2) by $a + b$ and get

$$(a+b)^{n+1} = \sum_{r=0}^{n} \binom{n}{r} a^{r+1} b^{n-r} + \sum_{r=0}^{n} \binom{n}{r} a^r b^{n+1-r}.$$

Letting $s = r+1$ in the first summand gives $\sum_{s=1}^{n+1} \binom{n}{s-1} a^s b^{n+1-s}$. Now renaming s as r, adding this last term to the second summand and making use of 0.4.1 proves (0.4.2) for $n+1$. $\square$

Letting $a = 1$ in the binomial theorem yields the following inequality:

Corollary 0.4.3. *If $b \ge 0$ then $(1+b)^n \ge 1 + nb$ for all $n \ge 1$.*

There is also another principle of induction, called *strong induction* which is stated just below.

The Principle of Strong Induction: Let $P(n)$ be a proposition for each positive integer n. The validity of $P(k)$ for all $k < n$ implies the validity of $P(n)$.

The proof is similar to the other form of induction above, namely that $\mathbb{Z}^+$ is well ordered. That $P(1)$ is true holds vacuously. The rest of the proof is left to the reader.

0.4.2   Transfinite Induction

We now turn to Zorn's lemma. For this we shall need some vocabulary.

Given a partially ordered set $(X, \leq)$, a totally ordered subset C of X is called a *chain*. If C is a chain in X we say $x \in X$ is a cap of C if $x \geq c$ for all $c \in C$. Finally, $m \in X$ is called a *maximal element* of C if whenever $c \in X$ and $c \geq m$, then c must equal m. (Notice here we are not saying $m \geq c$ for all $c \in C$.)

Here is the statement of Zorn's lemma,

Lemma 0.4.4. *Zorn's lemma.* *Let $(X, \leq)$ be a partially ordered set. If every chain has a cap, then X has a maximal element.*

We shall take Zorn's lemma as an axiom of set theory.[1]

0.4.3   Permutations

Let $f : X \longrightarrow Y$ and $g : Y \longrightarrow Z$ be two functions.

Definition 0.4.5. We define as the *composition* of f and g, as the function $g \circ f : X \longrightarrow Z$ defined by $(g \circ f)(x) = g(f(x))$.

Definition 0.4.6. For any set X, we call *identity map*, the map $1_X : X \longrightarrow X$ defined by $x \mapsto 1_X(x) := x$.

---

[1]Readers who do not fully understand Zorn's lemma can wait until the first time it is used to better understand it. However, it is essential that the reader understand Mathematical Induction before proceeding further.

Using the identity map and the composition of two maps gives us a useful way of describing surjective and injective maps.

Proposition 0.4.7. *Let $f : X \longrightarrow Y$ be a function. Then,*

1. *The map f is surjective if and only if there is a map $g : Y \longrightarrow X$ such that $g \circ f = 1_Y$. Similarly,*

2. *The map f is injective if and only if $f \circ g = 1_X$.*

Proof. Assume such a function g exists. Then, if $y \in Y$, $y = 1_Y(y) = (g \circ f)(y) = g(f(y))$. Therefore, y is the image of $g(y)$ under f, and since $g(y) \in X$ for every $y \in Y$, this shows that the map f is surjective. Conversely, if f is surjective, then for each $y \in Y$, there is an $x \in X$ such that $f(x) = y$. Now, define the map $g : Y \longrightarrow X$ by $g(y) = x$. Then $g \circ f = 1_Y$.
The proof of 2 is similar and is left as an exercise.   □

Definition 0.4.8. If $f : X \longrightarrow Y$ is a bijective map then, we call the map g the *inverse* of f, and we write it as f^{-1}.

Definition 0.4.9. Let X be any non-empty set. We call a *permutation* of X any bijective map $f : X \longrightarrow X$. The set of all permutations of X is denoted by $S(X)$ and is called the *symmetric group* of X. In the case where X is a finite set of n elements, instead of $S(X)$ we will write S_n.

Proposition 0.4.10. *If $|X| = n$, then $|S_n| = n!$.*

Proof. To find the number of bijective maps of $X = \{1, 2, ..., n\}$ we reason as follows: The element 1 can have as its image any of the n elements of X. The element 2 can have as its image any of the $n - 1$ remaining elements, and so on up to the last element n, the image of which is then uniquely determined. Hence, $|X| = n \cdot (n - 1) \cdots 2 \cdot 1 = n!$.   □

Proposition 0.4.11. *Let X be a non-empty set. Then for any permutations (functions) f, g, h:*

  1. $f \circ (g \circ h) = (f \circ g) \circ h.$

  2. $1_X \circ f = f \circ 1_X = f.$

  3. *For any* $f \in S(X)$, $f \circ f^{-1} = f^{-1} \circ f = 1_X.$

We leave the simple proof of this as an exercise.

0.5   The set $(\mathbb{Z}, +, \times)$

Here we begin our study of algebra where it began, namely, as an abstraction of features of the set of integers, $\mathbb{Z}$, and the operations of "plus" and "times". Let a and b be any two integers.

Definition 0.5.1. We say that b *divides* a, and we write $b|a$, if there is an integer c such that $a = bc$. We call b a *divisor* of a, and we say that a is a *multiple* of b. If b does not divide a, we write $b \nmid a$.

   The most important algebraic feature of $\mathbb{Z}$, is the following *Euclidean algorithm*.

Theorem 0.5.2. (*Euclidean Algorithm*) *For any two integers a and b, with $b > 0$, there are* unique *integers q and r such that*

$$a = bq + r, \quad with \ \ 0 \le r < b.$$

   The integer q is called the *quotient*, and r the *remainder* of the division of a by b.

Proof. Let $S = \{a - bn \mid n \in \mathbb{Z}\}$. The set S contains positive and negative integers (and possibly 0). Let r be the least non-negative integer in S. Then, r has the form $r = a - qb$ for some $q \in \mathbb{Z}$. We have to show that $0 \le r < b$. Otherwise, $r \ge b$. Thus $a - bq \ge b$. Hence $a - b(q + 1) \ge 0$. But, $a - b(q + 1) < a - bq = r$. Therefore $0 \le a - b(q + 1) < r$, and since $a - b(q + 1) \in S$, we get a contradiction. Thus $0 \le r < b$.

We now turn to the uniqueness of q and r. Suppose there also exist q_1 and r_1 in $\mathbb{Z}$ such that $a = q_1 b + r_1$, with $0 \leq r_1 < b$. Then $b(q - q_1) = r_1 - r$. Since $r_1 \neq r$, we may assume without loss of generality that $r_1 > r$. Therefore, since $b > 0$, $q > q_1$. But because both r and r_1 are smaller than b, $r_1 - r < b$, it follows that $r_1 - r$ can not be a multiple of b. We conclude $r_1 = r$, and hence also $q = q_1$.   □

Definition 0.5.3. We call the *greatest common divisor* (gcd) of a and b, the positive integer d such that:

1. $d \mid a$ and $d \mid b$, and

2. if $c \mid a$ and $c \mid b$, then $c \mid d$.

The greatest common divisor d is unique. Indeed, if $d_1 > 0$ were also a greatest common divisor, then $d \mid d_1$ and $d_1 \mid d$. Therefore (since both are positive integers) $d = d_1$.

Definition 0.5.4. Two integers a and b are called *relatively prime* (or *co-prime*) if $\gcd(a, b) = 1$. More generally, a finite set $a_1, ..., a_n$ is said to be *pairwise* relatively prime if each choice a_i, a_j is relatively prime. Later we shall use the same ideas in other more general contexts.

Corollary 0.5.5. *If a and b are two non-zero integers, then there are integers n and m such that $\gcd(a, b) = an + bm$.*

Proof. Consider the set $S = \{as + bt \mid s, t \in \mathbb{Z}\}$. S contains positive integers. Let d be the least positive integer in S. Therefore d is of the form $d = an + bm$ for some n and $m \in \mathbb{Z}$. Now, $d|a$. Indeed, by Theorem 0.5.2, $a = qd + r$, with $0 \leq r < d$. So, $qd = a - r = q(an + bm)$. Hence $r = a(1 - qn) - bqm$, which means $r \in S$. But unless $r = 0$ this contradicts the definition of d. Thus, $a = qd$, so $d|a$. Similarly one shows $d|b$. Moreover since $d = an + bm$, any divisor of a and b divides d. So $d = \gcd(a, b)$.   □

In particular,

Corollary 0.5.6. *If a and b are relatively prime, then there are integers n and m such that $an + bm = 1$.*

Definition 0.5.7. We say that the positive integer m is the *least common multiple* (lcm) of the two non-zero integers a and b if

1. $a \mid m$ and $b \mid m$, and

2. if $a \mid n$ and $b \mid n$ for some $n \in \mathbb{Z}$, then $m \mid n$.

Proposition 0.5.8. *The least common multiple of two integers a and b exists and is unique.*

Proof. Let S be the set of all common multiples of a and b. Since ab and $-ab$ are in S, we conclude that S must contain at least one positive integer. Let m be the least of the positive integers in S. We claim that $m = \text{lcm}(a, b)$. Indeed, since $m \in S$, $a|m$ and $b|m$. Now, suppose that there is another integer n such that $a|n$ and $b|n$. Using Theorem 0.5.2 we conclude $n = qm + r$, with $0 \le r < m$. But if $r > 0$ this is a contradiction since $r < m$ and obviously $r = (n - qm) \in S$. So $r = 0$. It follows that $m|n$.

Concerning the uniqueness of m, suppose m_1 is also a least common multiple of a and b. Then $m \mid m_1$ and $m_1 \mid m$, and since both are positive integers, $m = m_1$.   □

Exercise 0.5.9. Let a and b be integers. Prove $ab = \gcd(a, b) \cdot \text{lcm}(a, b)$.

0.5.1   The Fundamental Theorem of Arithmetic

Here we study unique factorization of an integer into primes and the infinitude of the prime numbers themselves.

Definition 0.5.10. A positive integer $p > 1$ is called a *prime*, if it can not be factored into a product of two or more numbers in $\mathbb{Z}^+$ (except for the trivial factorization $p = 1 \cdot p$). If p is not a prime, then it is called a *composite number*.

Proposition 0.5.11. *Let p be a prime and suppose $p \mid ab$. Then $p \mid a$, or $p \mid b$.*

Proof. Assume $p \nmid a$. Since a and p are relatively prime their gcd $= 1$. Therefore, there are integers s and t such that $as + pt = 1$. Multiplying both sides of this equation by b we get $abs + bpt = b$, and since p divides abs and also bpt, it must divide b. □

We now come to the **Fundamental Theorem of Arithmetic**.

Theorem 0.5.12. *(The Fundamental Theorem of Arithmetic) Any integer in $\mathbb{Z}^+$ can be factored uniquely into primes.*

Proof. First, we will prove that the factorization exists. Indeed, if n is a prime number, this is evident. Otherwise $n = ab$, where neither a, nor b is 1. By strong induction both a and b are each a product of primes. Hence, so is n. Now for the uniqueness, suppose $n = p_1 \ldots p_r$ and also $q_1 \ldots q_s$, where the p_i and q_j are primes. Since p_1 clearly divides n it divides the product of the qs. Therefore, by Proposition 0.5.11 above and strong induction again, it divides some q_j. Renumbering this as q_1, we see $p_1 \mid q_1$, and since q_1 and p_1 are primes, $p_1 = q_1$. Canceling, $p_2 \ldots p_r = q_2 \ldots q_s$. Now we proceed by strong induction. Since this is a lower integer, $r = s$ and after reordering $p_i = q_i$ for all $i = 1 \ldots r$. □

We shall give an alternative proof of this fact in the section on Composition Series (see Theorem 2.1.6).

Theorem 0.5.13. *There are infinitely many primes.*

We give three proofs, each tailored to a different taste. The reader need not go through all of them.

The oldest and simplest proof is attributed to Euclid.

Proof. Suppose the contrary, let $\{p_1, p_2, \ldots, p_n\}$ be the full (infinite) set of all prime numbers. Let $q = p_1 p_2 \ldots p_n + 1$. Since q is greater than each of the p_i, $i = 1, \ldots, n$, it must be a composite number. By the Fundamental Theorem of Arithmetic it must have as a prime divisor one of the prime numbers p_i, say p_k. But then p_k must divide 1, since it divides both q and $p_1 p_2 \ldots p_n$, a contradiction. □

We remark that the proof of the above theorem implies that if $p_1, \ldots, p_n$ are the first n primes, the next prime, p_{n+1}, lies in the interval $p_n + 1 < p_{n+1} \leq \prod_{i=1}^n p_i + 1$.

Our second proof is due to H. Fürstenberg (see [38]).

Proof. Consider the following curious topology on the set $\mathbb{Z}$ of integers. For $a, b \in \mathbb{Z}$, with $b > 0$, let

$$U_{a,b} = \{a + nb \ : \ n \in \mathbb{Z}\}.$$

Each set $U_{a,b}$ is a two-way infinite arithmetic progression. Now, call a set $O \subset \mathbb{Z}$ *open* if either $O = \varnothing$, or if for every $a \in O$ there exists some $b > 0$ with $U_{a,b} \subset O$. Clearly, the union of open sets is open. In addition, if O_1, O_2 are two open sets, and $a \in O_1 \cap O_2$ with $U_{a,b_1} \subset O_1$ and $U_{a,b_2} \subset O_2$, then $a \in U_{a,b_1 b_2} \subset O_1 \cap O_2$. We conclude that any finite intersection of open sets is open. Thus this family of open sets gives a topology on $\mathbb{Z}$. This topology has two features:

1. Any non-empty open set is infinite.

2. The sets $U_{a,b}$ are closed.

Indeed, the first statement follows from the definition. For the second, we observe

$$U_{a,b} = \mathbb{Z} - \bigcup_{i=1}^{b-1} U_{a+i,b},$$

which proves that $U_{a,b}$ is the complement of an open set and hence is closed.

Let $\mathbb{P}$ be the set of all primes. Since any number $n \neq 1, -1$ has a prime divisor p, and hence is contained in $U_{0,p}$, it follows that

$$\mathbb{Z} - \{1, -1\} = \bigcup_{p \in \mathbb{P}} U_{0,p}.$$

If $\mathbb{P}$ were finite, then $\bigcup_{p \in \mathbb{P}} U_{0,p}$ would be a finite union of closed sets (by 2), and hence closed. Consequently, $\{1, -1\}$ would be an open set, in violation of 1. $\qquad\qquad\square$

In our third proof[2] one shows the infinitude of primes by proving the obviously stronger statement: where we write $\lfloor n \rfloor$ to mean the integer part of n.

Proposition 0.5.14. *The series*

$$\sum_{p \text{ prime}} \frac{1}{p}$$

diverges.

Proof. Let p_1, p_2, p_3,... be the sequence of primes (in increasing order), and assume that

$$\sum_{p \text{ prime}} \frac{1}{p} < +\infty.$$

Then, there must be a natural number k such that

$$\sum_{i \geq k+1} \frac{1}{p_i} < \frac{1}{2}.$$

Let us call p_1, p_2,..., p_k, the *small* primes, and p_{k+1}, p_{k+2},... the *large* primes. For an arbitrary natural number N we therefore find

$$\sum_{i \geq k+1} \frac{N}{p_i} < \frac{N}{2} \tag{1}$$

Now, let N_b be the number of positive integers $n \leq N$ which are divisible by at least one big prime, and N_s the number of positive integers $n \leq N$ which have only small prime divisors. We will show that for a suitable N

$$N_b + N_s < N,$$

which would be a contradiction, since by definition $N_b + N_s$ would have to be equal to N.

---

[2]This result is due to Euler ([35]) and was revised by Erdös, see [34] or [4].

To estimate N_b note that $\lfloor \frac{N}{p_i} \rfloor$ counts the positive integers $n \leq N$ which are multiples of p_i. Hence by (1) we obtain

$$N_b \leq \sum_{i \geq k+1} \lfloor \frac{N}{p_i} \rfloor < \frac{N}{2} \qquad (2)$$

Let us now look at N_s. We write every $n \leq N$ which has only small prime divisors in the form $n = a_n b_n^2$, where a_n is the square-free part. Every a_n is thus a product of different small primes, and we conclude that there are precisely 2^k different square-free parts. Furthermore, as $b_n \leq \sqrt{n} \leq \sqrt{N}$, we find that there are at most $\sqrt{N}$ different square parts, and so

$$N_s \leq 2^k \sqrt{N}.$$

Since (2) holds for any N, it remains to find a number N with $2^k \sqrt{N} \leq \frac{N}{2}$ or $2^{k+1} \leq \sqrt{N}$, and for this $N = 2^{2k+2}$ will do. $\qquad\square$

0.5.2 The Euler Formula and Riemann Zeta Function

For $s \in \mathbb{C}$ we define

$$\zeta(s) = \sum_{n=1}^{\infty} \frac{1}{n^s}.$$

Then

$$\sum_{n=1}^{\infty} \frac{1}{n^s} = \prod_{p} \frac{1}{1 - \frac{1}{p^s}},$$

where the product is taken over the distinct primes.

Proof. Let p be a prime and consider the geometric series

$$1 + \frac{1}{p^s} + \ldots + \frac{1}{p^{ks}} + \ldots = \frac{1}{1 - \frac{1}{p^s}}.$$

Now we multiply taking a finite number of distinct primes $p_1, \ldots, p_r$. On the left we have a series, each term of which is of the form

$$\left(\frac{1}{p_1^{k_1} \ldots p_r^{k_r}} \right)^s,$$

where $p_1, \ldots, p_r$ are the distinct primes and $k_1, \ldots, k_r$ are positive integers. But every positive integer is a product of distinct primes with positive exponents so the left hand side is exactly $\sum_{n=1}^{\infty} \frac{1}{n^s}$. On the other hand, on the right we have $\Pi_p \frac{1}{1-\frac{1}{p^s}}$, where the product is taken over the distinct primes.   □

In 1735 Euler proved the following theorem[3]:

Theorem 0.5.15. *(Euler's Theorem[4]) For any even integer $k > 0$, $\zeta(k) = a$ rational number $\times \pi^k$.*

To see this, we need the following proposition the proof of which is due to E. Calabi, see [15].

Proposition 0.5.16.

$$\zeta(2) = \frac{\pi^2}{6}.$$

---

[3]This problem is known as the Basel problem, posed in 1644 and solved by Euler in 1735 at the age of 28.

[4]Leonhard Euler, (1707-1783) Swiss mathematician and physicist, and one of the founders of modern mathematics, made decisive contributions to geometry, calculus, mechanics, the calculus of variations and number theory. Euler and Lagrange were the greatest mathematicians of the 18th century, but Euler has never been excelled by anyone in productivity or computation. He moved to Russia where he became an associate of the St. Petersburg Academy of Sciences succeeding Daniel Bernoulli to the chair of mathematics. There he wrote numerous books and memoirs throwing new light on nearly all parts of pure mathematics. Overtaxing himself, he lost the sight of one eye. Then, invited by Frederick the Great, he became a member of the Berlin Academy, where for 25 years he produced a steady stream of publications. He did for modern analytic geometry and trigonometry what the Elements of Euclid had done for ancient geometry, with the resulting tendency to render mathematics and physics in arithmetical terms continuing ever since. His investigations of complex numbers resulted in the Euler identity $e^{i\theta} = \cos(\theta) + i\sin(\theta)$. His textbook on ODE and its applications to physics have served as a prototype to the present day. After Frederick the Great became less cordial toward him, Euler accepted the invitation of Catherine II to return to Russia. Soon after his arrival at St. Petersburg, a cataract formed in his remaining good eye, and he spent his last years totally blind. Despite this, his productivity continued undiminished. His discovery of the law of quadratic reciprocity has become an essential part of modern number theory.

Proof. Expanding $(1 - x^2y^2)^{-1}$ in a geometric series we get:

$$\frac{1}{1 - x^2y^2} = 1 + x^2y^2 + x^4y^4 + \cdots,$$

where $(x, y) \in [0, 1] \times [0, 1] := D$. Integrating termwise, we obtain:

$$\iint_D (1 - x^2y^2)^{-1} dx dy = xy \Big|_0^1 + \frac{x^3y^3}{3^2} \Big|_0^1 + \frac{x^5y^5}{5^2} \Big|_0^1 + \cdots$$

$$= \frac{1}{1^2} + \frac{1}{3^2} + \frac{1}{5^2} + \cdots = 1^{-2} + 3^{-2} + 5^{-2} + \cdots$$

$$= \left(1 - \frac{1}{4}\right)\zeta(2).$$

To calculate this integral we note that, since $(x, y) \in (0, 1) \times (0, 1) = D$ (we ignore the boundary of this domain as it has zero area), we have

$$0 < \frac{1 - x^2}{1 - x^2y^2} < 1, \text{ and } 0 < \frac{1 - y^2}{1 - x^2y^2} < 1.$$

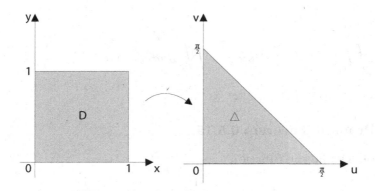

Therefore, we can find u and v such that

$$\cos u := \sqrt{\frac{1 - x^2}{1 - x^2y^2}} \text{ and } \cos v = \sqrt{\frac{1 - y^2}{1 - x^2y^2}}.$$

An easy calculation gives

$$\sin u = x\sqrt{\frac{1-y^2}{1-x^2y^2}} \text{ and } \sin v = y\sqrt{\frac{1-x^2}{1-x^2y^2}}.$$

From this we get the transformation we need. We call these equations $(\star)$.

$$x = \frac{\sin u}{\cos v} \text{ and } y = \frac{\sin v}{\cos u}.$$

The Jacobian of this transformation is:

$$\begin{vmatrix} \frac{\partial x}{\partial u} & \frac{\partial y}{\partial v} \\ \frac{\partial y}{\partial u} & \frac{\partial y}{\partial v} \end{vmatrix} = \begin{vmatrix} \frac{\cos u}{\cos v} & -\frac{\sin u}{\cos^2 v}\sin v \\ -\frac{\sin v}{\cos^2 u}\sin u & \frac{\cos v}{\cos u} \end{vmatrix} = 1 - \frac{\sin^2 u}{\cos^2 v}\frac{\sin^2 v}{\cos^2 u} = 1 - x^2y^2$$

Now, since $x, y \in (0, 1)$, from the equations $(\star)$ it follows that $0 < \sin u < \cos v$ and $0 < \sin v < \cos u$. Multiplying and adding we get $0 < \cos u \cos v - \sin u \sin v$, i.e. $\cos(u+v) > 0$, so $0 < u + v < \frac{\pi}{2}$.

This shows the transformation $(x, y) \mapsto (u, v)$ maps bijectively the open unit square D onto the open right isosceles triangle, Δ:

$$\Delta = \{(u,v) \ : \ u > 0, \ v > 0, \ u + v \le \frac{\pi}{2}\}.$$

Therefore,

$$\iint_D (1 - x^2y^2)^{-1}dxdy = \iint_\Delta dudv = \text{Area}(\Delta) = \frac{1}{2}\frac{\pi}{2}\frac{\pi}{2} = \frac{\pi^2}{8}.$$

Hence, $\frac{\pi^2}{8} = \frac{3}{4}\zeta(2)$ so, $\zeta(2) = \frac{\pi^2}{6}$.   □

Proof of Theorem 0.5.15.

Proof. Define the function

$$f(m, n) = \frac{1}{mn^3} + \frac{1}{2m^2n^2} + \frac{1}{m^3n}.$$

It follows that

$$f(m, n) - f(m + n, n) - f(m, m + n) = \frac{1}{m^2n^2}.$$

Therefore,

$$\zeta(2)^2 = \sum_{m,n>0} f(m,n) - \sum_{m>n>0} f(m,n) - \sum_{n>m>0} f(m,n)$$

$$= \sum_{n>0} f(n,n) = \frac{5}{2}\zeta(4)$$

and since $\zeta(2) = \pi^2/6$, we get

$$\zeta(4) = \frac{\pi^4}{90}.$$

For $k > 4$ set

$$f(m,n) = \frac{1}{mn^{k-1}} + \frac{1}{2}\sum_{r=2}^{k-2} \frac{1}{m^r n^{k-r}} + \frac{1}{m^{k-1}n}.$$

Hence

$$f(m,n) - f(m+n,n) - f(m,m+n) = \sum_{0<j<k,\ j=2s} \frac{1}{m^j n^{k-j}},$$

and arguing as above we get

$$\sum_{0<j<k,\ j=2s} \zeta(j)\zeta(k-j) = \frac{k+1}{2}\zeta(k),$$

where $k \geq 4$ and is even. Hence, by induction the result follows. $\square$

0.5.3   The Fermat Numbers

Definition 0.5.17. A *Fermat number* F_n is a number of the form

$$F_n = 2^{2^n} + 1, \quad n \geq 1.$$

Obviously Fermat numbers are all odd.

Proposition 0.5.18. *Any two distinct Fermat numbers are relatively prime.*

Proof. Let F_n and F_{n+k} be two distinct Fermat numbers and suppose there is an integer $m > 0$ dividing both of them. Setting $x = 2^{2^n}$, we see that $F_n = x + 1$ and

$$F_{n+k} = 2^{2^{n+k}} + 1 = x^{2^k} + 1.$$

Hence, $F_{n+k} - 2 = x^{2^k} - 1$. Therefore,

$$\frac{F_{n+k} - 2}{F_n} = \frac{x^{2^k} - 1}{x - 1} = x^{2^k - 1} - x^{2^k - 2} + \cdots - 1,$$

which implies that F_n divides $F_{n+k} - 2$. Therefore m divides $F_{n+k} - 2$, and since m also divides F_{n+k}, it must divide 2. Hence $m = 1$ or $m = 2$. But as we know all Fermat numbers are odd, so $m = 1$. $\qquad\square$

Remark 0.5.19. The first Fermat numbers

$$F_1 = 5, \quad F_2 = 17, \quad F_3 = 257, \quad F_4 = 65537,$$

are all prime, so Fermat conjectured that all F_n are prime. However, Euler proved $F_5 = 4294967297$ is not a prime $(= 641 \cdot 6700417)$, disproving this conjecture of Fermat.

Proof. We have

$$F_5 = 2^{2^5} + 1 = 2^{32} + 1 = (2 \cdot 2^7)^4 + 1.$$

Set $a = 2^7$. Then, $F_5 = (2a)^4 + 1 = 2^4 a^4 + 1$. Observe that $2^4 = 1 + 5 \cdot 3 = 1 + 5(2^7 - 5^3)$. Then,

$$F_5 = (1 + 5 \cdot a - 5^3)a^4 + 1 = (1 + 5a)\big(a^4 + (1 - 5a)(1 + a^2 5^2)\big)$$

which shows that $1 + 5a = 641$ divides F_5.[5] $\qquad\square$

---

[5]This proof is due to G.T. Bennett, in [23].

0.5.4   Pythagorean Triples

Here we will study the *Diophantine* equation $x^2 + y^2 = z^2$.

Definition 0.5.20. We call (x, y, z) a *Pythagorean triple* if x, y and z are positive integers and $x^2 + y^2 = z^2$. The *smallest* example being $3, 4, 5$.

When counting Pythagorean triples we do not consider multiples. So, for example, since we have $3, 4, 5$ we ignore $6, 8, 10$ and call the later a *duplicate*. Two immediate questions are: How does one find all Pythagorean triples and, in particular, are there infinitely many of them. To address such questions we consider the following identity in $\mathbb{Z}$ which, as we shall see (as soon as we define this) is valid in any commutative ring:

$$(a^2 - b^2)^2 + (2ab)^2 = (a^2 + b^2)^2. \qquad (2)$$

Equation 2 of subsection 0.5.4 is true because in $\mathbb{Z}$,

$$\alpha^2 - \beta^2 = (\alpha - \beta)(\alpha + \beta).$$

Hence, when $\alpha = a^2 + b^2$, $\beta = a^2 - b^2$ and $\alpha^2 - \beta^2 = (2ab)^2$.[6]

Thus, if a and b are any positive integers with $a > b$, $a^2 - b^2$, $2ab$, $a^2 + b^2$ is a Pythagorean triple. This evidently gives infinitely many of them, just by varying a and taking $b = a - 1$ because then

$$a^2 + (a - 1)^2 = (2a(a - 1))^2 + 1. \qquad (3)$$

So here the $\gcd(x, z)$ must already be 1.

Corollary 0.5.21. *For any $\epsilon > 0$ there are infinitely many non-duplicate Pythagorean triangles having an acute angle less than ϵ.*

The reader will notice that within all these non-duplicate triples, one of the two smaller numbers is even and the other is odd. This is so since $2ab$ is certainly even, while $a^2 - b^2$ must be odd because, if it

---

[6]Another approach to this can be done geometrically, or alternatively, $(\alpha + \beta)^2 - (\alpha - \beta)^2 = 4\alpha\beta$. Now the last term is not a perfect square. Taking $\alpha = a^2$ and $\beta = b^2$ we get the previous identity.

were even, $a^2 + b^2$ would also be even and therefore we would have a duplicate.

Another result along the same lines as the previous corollary is,

Corollary 0.5.22. *Although there are no isosceles Pythagorean triangles since one leg is odd and the other even, there are infinitely many non-duplicate almost isosceles Pythagorean triangles in the sense that the sides are almost equal. That is, for any $\varepsilon > 0$ both acute angles are within ε of 45 degrees.*

Proof. Let $(x, x+1, y)$ be a Pythagorean triple. Then:

$$x^2 + (x+1)^2 = y^2; \text{ in other words } (2x+1)^2 - 2y^2 = -1.$$

But this equation is of Pell type for which there are infinitely many solutions (see Proposition A.3). This shows that there are infinitely many consecutive integers x, $x+1$ which can be the legs of a right triangle. This implies that if θ is one of the acute angles,

$$\tan(\theta) = \frac{x}{x+1} = 1 - \frac{1}{x+1}.$$

Therefore, given $\varepsilon > 0$ we can always find an x such that

$$1 - \frac{1}{x+1} < \tan(\theta) < 1.$$

In other words, $\frac{\pi}{4} - \varepsilon < \theta < \frac{\pi}{4}$. $\qquad\qquad\square$

We conjecture: There are infinitely many Pythagorean triangles with prime hypothenuse. This would be yet another way to understand the infinitude of Pythagorean triangles.

We shall call the Pythagorean triple (x, y, z) *proper* if x is even and the $\gcd(x, y, z) = 1$. We denote by $\mathcal{P}$ the set of proper Pythagorean triples. As we shall see in a moment, the scheme given by equation 2 actually gives *all* proper Pythagorean triples.

Theorem 0.5.23. *Every proper Pythagorean triple comes from some appropriate a and b, as above.*

Before proving this we make a few observations about proper Pythagorean triples, (x, y, z). Since x is even both y and z must both be odd. For if one of them were even, then the other would also be even and this triple would not be proper. For this reason x must actually be divisible by 4. This is because $y = 2j + 1$ and $z = 2k + 1$ and therefore

$$x^2 = z^2 - y^2 = 4k^2 + 4k + 1 - (4j^2 + 4j + 1) = 4(k - j)(k + j + 1).$$

But $k - j$ and $k + j + 1$ can not both be odd. For if they were, their sum would have to be even and this cannot be since $k - j + k + j + 1 = 2k + 1$. Thus x^2 is divisible by 8 and so x itself must be divisible by 4. This means we may assume the a in equation 2 is itself even.

Proof of Theorem 0.5.23.

Proof. Let x, y, z be a proper Pythagorean triple and $u = \gcd(x, y)$. Then, $u^2 \mid z^2$ and hence $u \mid z$. Now $\gcd(x, y, z) = \gcd(\gcd(x, y), z)$. Therefore $u = \gcd(x, y, z)$ and by symmetry also $\gcd(x, z) = u = \gcd(y, z)$. Evidently,

$$\left(\frac{x}{u}\right)^2 + \left(\frac{y}{u}\right)^2 = \left(\frac{z}{u}\right)^2$$

and

$$\gcd\left(\frac{x}{u}, \frac{y}{u}\right) = \gcd\left(\frac{x}{u}, \frac{z}{u}\right) = \gcd\left(\frac{y}{u}, \frac{z}{u}\right) = 1.$$

This gives a new solution x', y', z' to our equation in which these elements are pairwise relatively prime $(x = ux', y = uy'$ and $z = uz')$. Thus we may assume the original x, y, z are pairwise relatively prime by adjusting the a and b we seek by u. As we saw above, we may also assume x is even and both y and z are odd. Then

$$\frac{z + y}{2} \cdot \frac{z - y}{2} = \left(\frac{x}{2}\right)^2 \tag{4}$$

and

$$\gcd\left(\frac{z + y}{2}, \frac{z - y}{2}\right) \mid \gcd\left(\frac{z + y}{2} + \frac{z - y}{2}\right) = z$$

and

$$\gcd\left(\frac{z + y}{2}, \frac{z - y}{2}\right) \mid \gcd\left(\frac{z + y}{2} - \frac{z - y}{2}\right) = y.$$

Therefore $\gcd\left(\frac{z+y}{2}, \frac{z-y}{2}\right) = 1$. This, together with equation 4, shows that $\frac{z+y}{2}$ and $\frac{z-y}{2}$ are both perfect squares, say a^2 and b^2 respectively. Hence $\gcd(a, b) = 1$ and $a > b$. Finally, $a^2 - b^2 = \frac{z+y}{2} - \frac{z-y}{2} = y$, $a^2 + b^2 = \frac{z+y}{2} + \frac{z-y}{2} = z$ and

$$2ab = 2\sqrt{\frac{z+y}{2} \cdot \frac{z-y}{2}} = \sqrt{(z+y)(z-y)} = \sqrt{x^2} = x.$$

$\square$

Here are some Pythagorean triples.

$a = 2$ and $b = 1$ gives $4, 3, 5$.
$a = 3$ and $b = 1$ gives a duplicate.
$a = 3$ and $b = 2$ gives $12, 5, 13$.
$a = 4$ and $b = 1$ gives $8, 15, 17$.
$a = 4$ and $b = 2$ gives a duplicate.
$a = 4$ and $b = 3$ gives $24, 7, 25$.
$a = 5$ and $b = 1$ gives a duplicate.
$a = 5$ and $b = 2$ gives $20, 21, 29$.
$a = 5$ and $b = 3$ gives a duplicate
$a = 5$ and $b = 4$ gives $40, 9, 41$.
$a = 6$ and $b = 1$ gives $12, 35, 37$.
$a = 6$ and $b = 2$ gives a duplicate.
$a = 6$ and $b = 3$ gives a duplicate.
$a = 6$ and $b = 4$ gives a duplicate.
$a = 6$ and $b = 5$ gives $60, 11, 61$.
$a = 7$ and $b = 1$ gives a duplicate.
$a = 7$ and $b = 2$ gives $28, 45, 53$.
$a = 7$ and $b = 3$ gives a duplicate.
$a = 7$ and $b = 4$ gives $56, 33, 65$.
$a = 7$ and $b = 5$ gives $35, 12, 37$.
$a = 7$ and $b = 6$ gives $84, 13, 85$.
$a = 8$ and $b = 1$ gives $16, 63, 65$.
$a = 8$ and $b = 2$ gives a duplicate.
$a = 8$ and $b = 3$ gives $48, 55, 73$.
$a = 8$ and $b = 4$ gives a duplicate.

Chapter 1

Groups

1.1 The Concept of a Group

Let G be a set, and denote by $\circ$ a binary operation

$$\circ : G \times G \to G.$$

Sometimes we will write $g \cdot h$, or simply even gh.

Definition 1.1.1. We shall call $(G, \circ)$ a *group* if

1. $(gh)k = g(hk)$ for all g, h, k in G (*associativity*).

2. There is $e \in G$ such that $eg = ge = g$, $\forall\, g \in G$ (*identity element*).

3. If $g \in G$, then there is a $g^{-1} \in G$ such that $gg^{-1} = g^{-1}g = e$ (*inverse element*).

Sometimes it will be more convenient to write 1_G or simply 1 instead of e.

In addition, if $gh = hg$ for every g and h, we will say that G is an *Abelian*[1] or a *commutative* group. As the reader will see, being Abelian

---

[1]Niels Henrik Abel (1802-1829) was a Norwegian mathematician who, despite being born into poverty, made pioneering contributions in a variety of area. His most famous single result is the first complete proof demonstrating the impossibility

or non-Abelian is a major distinction between groups. In the Abelian case, it is sometimes more convenient to write the operation as $+$, in which case the neutral element e will be written as 0.

Exercise 1.1.2. Check that, in a group, e is unique and that any element g has a unique inverse g^{-1}.

Exercise 1.1.3. For any g, h in G show that $(gh)^{-1} = h^{-1}g^{-1}$.[2]

Definition 1.1.4. As above, if G is finite we denote its *order* that is, the number of its elements by $|G|$. If G is not finite, we simply say that G has *infinite order*.

Exercise 1.1.5. Let X be a set of n elements e.g. $X = \{1, 2, \ldots, n\}$ and S_n be the set of all permutations of X. Then under composition S_n is a group called the *symmetric* group and, as we saw in Chapter 0, $|S_n| = n!$. Show that even if we take X infinite, $S(X)$ is still a group.

Definition 1.1.6. If G is a group and $H \subseteq G$, we say H is a *subgroup* of G if, under the restricted operations from G, H is a group in its own right.

---

of solving the general quintic equation in radicals. One of the outstanding open problems of his day, this question had been unresolved for 250 years (see Chapter 11). Abel sent his proof to Gauss, who dismissed it, failing to recognize that the famous problem had indeed been settled. In 1826 Abel went to Paris, then the world center for mathematics, where he called on the foremost mathematicians and completed a major paper on the theory of integrals of algebraic functions. However, the French would not give him the time of day and the memoir he submitted to the French Academy of Sciences was lost. Abel returned to Norway heavily in debt and suffering from tuberculosis. He subsisted by tutoring, supplemented by a small grant from the University of Christiania (Oslo) and, beginning in 1828, by a temporary teaching position. His poverty and ill health did not decrease his productivity; he wrote a great number of papers during this period rapidly developed the theory of elliptic functions in competition with the German, Carl Gustav Jacobi. By this time Abel's fame had spread to all mathematical centers, and strong efforts were made to secure a suitable position, but by the fall of 1828 Abel had succumbed to tuberculosis and died the next year. The French Academy published his memoir in 1841. The Abel Prize in mathematics, originally proposed in 1899 to complement the Nobel Prizes, is named in his honor.

[2]Coxeter pointed out that the reversal order becomes clear if we think of the operations of putting on our shoes and socks.

Since all but two of these conditions are hereditary (i.e. derive from G), what this amounts to is just closure within H and also for $h \in H$ also $h^{-1} \in H$. The reader should check that an intersection of subgroups is again a subgroup.

Evidently G and $\{e\}$ are always subgroups of any G. The following are examples of subgroups:

If $G = S_n$ and $X \subseteq \{1, \ldots, n\}$ then $H_X = \cap_{x \in X} \{g \in G : g(x) = x\}$, is a subgroup of S_n.

$G = \mathbb{Z}$ is evidently an infinite Abelian group (more on this in a moment). If $n > 1$, the multiples of n form a subgroup of G.

Definition 1.1.7. A *homomorphism* f from a group G to a group H is a function $f : G \longrightarrow H$ such that $f(xy) = f(x)f(y)$ for all x, y in G. If f is such a homomorphism, then, the *kernel* of f is defined as

$$\mathrm{Ker}(f) = \{g \in G : f(g) = 1_H\}.$$

If f is injective, it is called a *monomorphism* and if it is surjective, it is called an *epimorphism*. If f is both injective and surjective, it is called an *isomorphism*. In this case we say that G and H are *isomorphic* groups and we write $G \cong H$. Notice that when two groups are isomorphic, they share all group theoretic properties.

Exercise 1.1.8. Let $\mathbb{R}$ be the additive group of real numbers and $\mathbb{R}^+$ be the multiplicative group of positive real numbers. Prove the exponential map is a group isomorphism $\mathbb{R} \to \mathbb{R}^+$. What happens if we consider the same question, but this time $\exp : \mathbb{C} \to \mathbb{C}^\times$, where $\mathbb{C}^\times$ is the multiplicative group of non-zero complex numbers (see just below)?

Definition 1.1.9. A homomorphism from G to G is called an *endomorphism* of G, and an isomorphism of G with itself is called an *automorphism* of G. We denote the set of all automorphisms of a group G by $\mathrm{Aut}(G)$.

Definition 1.1.10. An important example of an automorphism is provided by the map α_g of $G \to G$ given by $x \mapsto gxg^{-1}$ called the *inner automorphism determined by g*. The set of all these maps is called $\mathrm{Int}(G)$.

Exercise 1.1.11. Show that $\alpha_{gh} = \alpha_g \alpha_h$ and therefore $\alpha_g \alpha_{g^{-1}} = \alpha_1 = I$.

Proposition 1.1.12. α_g *is indeed an automorphism. Its inverse is* $\alpha_{g^{-1}}$.

Proof. The map α_g is a homomorphism because

$$\alpha_g(xy) = gxyg^{-1} = (gxg^{-1})(gyg^{-1}) = \alpha_g(x)\alpha_g(y).$$

In addition, α_g is bijective because by the exercise just above, $\alpha_{g^{-1}} = (\alpha_g)^{-1}$.   □

Definition 1.1.13. An automorphism of this form is called an *inner automorphism*, and the remaining automorphisms are said to be *outer automorphisms*.

Exercise 1.1.14. Let G be a group. Show that $\mathrm{Aut}(G)$ is a group under the composition of functions and $\mathrm{Int}(G)$ is a subgroup of $\mathrm{Aut}(G)$.

Exercise 1.1.15. Let G be a group. Show $g \mapsto \alpha_g$ is a surjective homomorphism $G \to \mathrm{Int}(G)$. Show that the kernel of this map is

$$\mathrm{Ker}\, g = \{g \in G \,:\, gh = hg \text{ for all } h \in G\}.$$

This is called *the center* of G and is denoted by $\mathcal{Z}(G)$.

Given a group, especially a finite group, if we arrange the elements in some particular order and consider all possible products of elements, we arrive at a square $n \times n$ array which is called its *multiplication table*.

Definition 1.1.16. A subset S of a group G is called a *generating set* if every element of G can be written as a product of elements of S (repetitions are allowed) and in this case we shall write $G = \langle S \rangle$. If G is generated by a single element we say it is *cyclic*.

Exercise 1.1.17. Prove any cyclic group is Abelian.

Definition 1.1.18. Let G be a group and $g \in G$. It may happen that for some positive integer k

$$g^k = 1.$$

If so, then we take the least such k and call it the *order* of g and we write $|g| = k$. If there is no such k we say $|g| = \infty$. If a group has all its elements of finite order, we call it a *torsion* group. If all the elements have order, some power of a fixed prime p, we call it a *p-group*. If all the elements other than the identity have infinite order, we say G is *torsion free*.

Exercise 1.1.19. Prove that if a and b are two commuting elements of G and have relatively prime order, then,

$$|ab| = |a||b|.$$

Is it true that if a and b commute, then

$$|ab| = \text{lcm}(|a|, |b|)?$$

1.2   Examples of Groups

Example 1.2.1. $\mathbb{Z}$, $\mathbb{Q}$, $\mathbb{R}$, or $\mathbb{C}$ are groups with the usual operation of addition. Also, the set $\mathbb{Q}^\star = \mathbb{Q} - \{0\}$ of non-zero rational numbers is a group under multiplication. Similarly, this is true for $\mathbb{R}^\times = \mathbb{R} - \{0\}$, and $\mathbb{C}^\times = \mathbb{C} - \{0\}$, *but not for* $\mathbb{Z} - \{0\}$. Why?

Example 1.2.2. The set $\pm 1 := \{1, -1\}$ is a group under the usual multiplication of integers.

Example 1.2.3. Let n be an integer > 1. The group $\mathbb{Z}_n$ is defined as follows: If $a, b \in \mathbb{Z}$, and n divides $a - b$, we will say that a is *congruent* to b *modulo* n, and we will write $a \equiv b \pmod{n}$. One easily checks that " $\equiv$ " is an equivalence relation and therefore we consider the set of all equivalence classes here called residue classes.

$$\mathbb{Z}_n := = \{[a], \, a \in \mathbb{Z}\}.$$

It is clear that in $\mathbb{Z}_n$ there are exactly n distinct *residue classes* $[a]$, since every integer is congruent modulo n to precisely one of the integers $0, 1, \ldots, (n-1)$.

Now, we define in $\mathbb{Z}$ an operation by setting

$$[a] + [b] = [a + b].$$

We leave it to the reader to check that this operation does not depend on the choice of the representatives of the residue classes $[a]$ and $[b]$, and that under this operation $\mathbb{Z}_n$ is a group.

Exercise 1.2.4. Show that $\mathbb{Z}$ and $\mathbb{Z}_n$ are cyclic. Show that the map $x \mapsto [x]$ taking $\mathbb{Z} \to \mathbb{Z}_n$ is a surjective group homomorphism. Does $\mathbb{Z}$ have any elements of finite order? Is there an element of $\mathbb{Z}_n$ of order n?

Example 1.2.5. The additive group $\mathbb{Q}$ of rational numbers has some interesting subgroups. For example, let p be a prime number and consider the set of all rational numbers $\frac{m}{p^n}$, with $m \in \mathbb{Z}$ and $n \in \mathbb{N}$. This is a subgroup that we denote by $\frac{1}{p^\infty}\mathbb{Z}$.

Except for the symmetric group, so far all our examples have been Abelian. But group theory is mostly concerned with non-Abelian groups.

Example 1.2.6. The following is an example from linear algebra (see Chapter 4). Let V be a finite dimension vector space over a field k. Consider the set

$$\mathrm{GL}(V) = \{g : V \to V, \quad g \text{ invertible k-linear map}\}$$

Then $\mathrm{GL}(V)$, with operation the composition of maps, is a group called the *general linear group*. In the case where $V = k^n$ we shall write $\mathrm{GL}(n, k)$.

The reader should check that this group is non-Abelian.

Other important examples are subgroups of $\mathrm{GL}(n, k)$.

Example 1.2.7. Let k be a field. The *affine group of k* is defined as

$$G = \left\{ \begin{pmatrix} a & b \\ 0 & 1 \end{pmatrix}, \, a \neq 0, \, a, b \in k \right\}$$

under the operation of matrix multiplication.

More generally, we can consider the *affine group of k^n* which contains $\mathrm{GL}(n, k)$, and is defined as:

$$G = \left\{ \begin{pmatrix} A & \vec{b} \\ 0 & 1 \end{pmatrix}, \, A \in \mathrm{GL}(n, k), \, \vec{b} \in k^n \right\}$$

also under the operation of matrix multiplication.

Example 1.2.8. The *Klein Four Group*, K which is the group

$$K = \mathbb{Z}_2 \oplus \mathbb{Z}_2 = \{(x, y) \mid x = 0, 1, \, y = 0, 1\},$$

and has the following multiplication table:

$\mathbb{Z}_2 \oplus \mathbb{Z}_2$	$(0,0)$	$(0,1)$	$(1,0)$	$(1,1)$
$(0,0)$	$(0,0)$	$(0,1)$	$(1,0)$	$(1,1)$
$(0,1)$	$(0,1)$	$(0,0)$	$(1,1)$	$(1,0)$
$(1,0)$	$(1,0)$	$(1,1)$	$(0,0)$	$(0,1)$
$(1,1)$	$(1,1)$	$(1,0)$	$(0,1)$	$(0,0)$

1.2.1 The Quaternion Group

An important finite group is the *quaternion group* Q_8. This is usually defined as the subgroup of the general linear group $GL(2, \mathbb{C})$ generated by the matrices

$$\mathbf{1} = \begin{pmatrix} 1 & 0 \\ 0 & 1 \end{pmatrix}, \quad \mathbf{i} = \begin{pmatrix} i & 0 \\ 0 & -i \end{pmatrix}, \quad \mathbf{j} = \begin{pmatrix} 0 & 1 \\ -1 & 0 \end{pmatrix}, \quad \mathbf{k} = \begin{pmatrix} 0 & i \\ i & 0 \end{pmatrix}.$$

The quaternion group is

$$Q_8 = \{\pm\mathbf{1}, \, \pm\mathbf{i}, \, \pm\mathbf{j}, \, \pm\mathbf{k}\}$$

under the matrix multiplication,

$$\mathbf{i}^2 = \mathbf{j}^2 = \mathbf{k}^2 = -1, \quad \mathbf{ij} = -\mathbf{ji} = \mathbf{k}, \quad \mathbf{jk} = -\mathbf{kj} = \mathbf{i}, \quad \mathbf{ki} = -\mathbf{ik} = \mathbf{j}.$$

Moreover, $\mathbf{1}$ is the identity of Q_8 and $-\mathbf{1}$ commutes with all elements of Q_8. Here $\mathbf{i}$, $\mathbf{j}$, and $\mathbf{k}$ have order 4 and any two of them generate the entire group. This gives rise to what is called a *presentation* of a group.

A presentation of Q_8 is

$$Q_8 = \{a,\ b \mid a^4 = 1,\ a^2 = b^2,\ b^{-1}ab = a^{-1}\}.$$

Take, for instance, $\mathbf{i} = a$, $\mathbf{j} = b$ and $\mathbf{k} = ab$.

Another concept is the *subgroup lattice*, namely the set of all subgroups of a given group. Here the subgroup lattice $L(Q_8)$ consists of Q_8 itself and the cyclic subgroups $\langle \mathbf{1} \rangle$, $\langle -\mathbf{1} \rangle$, $\langle \mathbf{i} \rangle$, $\langle \mathbf{j} \rangle$ and $\langle \mathbf{k} \rangle$. Hence Q_8 is a non-Abelian group all of whose subgroups are normal (see definition 1.6.1 below).

Evidently every subgroup of an Abelian group is normal. Although the quaternion group is not Abelian it also has this property.

Definition 1.2.9. A non-Abelian group G all of whose subgroups are normal is called a *Hamiltonian* group.

One can show that any Hamiltonian group G can be written as

$$G = Q_8 \times \mathbb{Z}_2^k \times H,$$

where H is an Abelian group with all its elements of odd order. (For a proof see [53] p. 190.)

Therefore,

$$|G| = 2^3 \cdot 2^k \cdot |H| = 2^{3+k} \cdot |H|.$$

Thus, Q_8 is the Hamiltonian group of lowest order.

Other basic properties of the subgroups of Q_8 are the following: These give us some simple, but useful characterizations of the group of quaternions (see [109]).

- $\langle i \rangle$, $\langle j \rangle$ and $\langle k \rangle$ have the property that none of them is contained in the union of the other two, and they determine a group which maps onto Q_8, that is

$$Q_8 = \langle i \rangle \cup \langle j \rangle \cup \langle k \rangle.$$

- Q_8 is the unique non-Abelian p-group all of whose proper subgroups are cyclic.

- Q_8 is the unique non-Abelian group that can be covered by any three irredundant[3] proper subgroups, respectively.

1.2.2   The Dihedral Group

The *dihedral group* D_n is the group of symmetries of a regular planar polygon with n vertices. We think of this polygon as having vertices on the unit circle, with vertices labeled $0, 1, \dots, n-1$ starting at $(1,0)$ and proceeding counterclockwise at angles in multiples of $360/n$ degrees, that is, $2\pi/n$ radians. There are two types of symmetries of the n-gon, each giving rise to n elements in the group D_n:

1. Rotations $R_0, R_1, \dots, R_{n-1}$, where R_k is a rotation of angle $2\pi k/n$.

2. Reflections $S_0, S_1, \dots, S_{n-1}$, where S_k is a reflection about the line through the origin making an angle of $\pi k/n$ with the horizontal axis.

The group operation is given by composition of symmetries: if a and b are two elements in D_n, then $a \cdot b = b \circ a$. That is to say, $a \cdot b$ is the symmetry obtained by first applying a and then b.

The elements of D_n can be thought of as linear transformations of the plane, leaving the given n-gon invariant. This lets us represent the elements of D_n as 2×2 matrices, with the group operation corresponding to matrix multiplication. Specifically,

$$R_k = \begin{pmatrix} \cos(2\pi k/n) & -\sin(2\pi k/n) \\ \sin(2\pi k/n) & \cos(2\pi k/n) \end{pmatrix}$$

---

[3]If G is a finite group and $\mathcal{H}$ a family of proper subgroups such that $G = \bigcup_{H \in \mathcal{H}} H$, we say that $\mathcal{H}$ is *irredundant* if no proper subfamily of $\mathcal{H}$ covers G.

$$S_k = \begin{pmatrix} \cos(2\pi k/n) & \sin(2\pi k/n) \\ \sin(2\pi k/n) & -\cos(2\pi k/n) \end{pmatrix}.$$

It is now a simple matter to verify that the following relations hold in D_n:

$$R_i \cdot R_j = R_{i+j},$$

$$R_i \cdot S_j = S_{i+j},$$

$$S_i \cdot R_j = S_{i-j},$$

$$S_i \cdot S_j = R_{i-j},$$

where $0 \le i, j \le n-1$, and both $i+j$ and $i-j$ are computed modulo n. The multiplication table (or Cayley table) for D_n can be readily gotten from the above relations. In particular, we see that R_0 is the identity, $R_i^{-1} = R_{n-i}$, and $S_i^{-1} = S_i$.

In particular, **The group** D_3. This is the symmetry group of the equilateral triangle, with vertices on the unit circle, at angles 0, $2\pi/3$, and $4\pi/3$. The matrix representation is given by

$$R_0 = \begin{pmatrix} 1 & 0 \\ 0 & 1 \end{pmatrix}, \quad R_1 = \begin{pmatrix} -\frac{1}{2} & -\frac{\sqrt{3}}{2} \\ \frac{\sqrt{3}}{2} & -\frac{1}{2} \end{pmatrix}, \quad R_2 = \begin{pmatrix} -\frac{1}{2} & \frac{\sqrt{3}}{2} \\ -\frac{\sqrt{3}}{2} & -\frac{1}{2} \end{pmatrix}$$

$$S_0 = \begin{pmatrix} 1 & 0 \\ 0 & -1 \end{pmatrix}, \quad S_1 = \begin{pmatrix} -\frac{1}{2} & \frac{\sqrt{3}}{2} \\ \frac{\sqrt{3}}{2} & \frac{1}{2} \end{pmatrix}, \quad S_2 = \begin{pmatrix} -\frac{1}{2} & -\frac{\sqrt{3}}{2} \\ -\frac{\sqrt{3}}{2} & \frac{1}{2} \end{pmatrix}.$$

The group D_4. This is the symmetry group of the square with vertices on the unit circle, at angles 0, $\pi/2$, π, and $3\pi/2$. The matrix representation is given by

$$R_0 = \begin{pmatrix} 1 & 0 \\ 0 & 1 \end{pmatrix}, \quad R_1 = \begin{pmatrix} 0 & -1 \\ 1 & 0 \end{pmatrix}, \quad R_2 = \begin{pmatrix} -1 & 0 \\ 0 & -1 \end{pmatrix}, \quad R_3 = \begin{pmatrix} 0 & 1 \\ -1 & 0 \end{pmatrix}$$

$$S_0 = \begin{pmatrix} 1 & 0 \\ 0 & -1 \end{pmatrix}, \quad S_1 = \begin{pmatrix} 0 & 1 \\ 1 & 0 \end{pmatrix}, \quad S_2 = \begin{pmatrix} -1 & 0 \\ 0 & 1 \end{pmatrix}, \quad S_3 = \begin{pmatrix} 0 & -1 \\ -1 & 0 \end{pmatrix}$$

Remark 1.2.10. One can also view the dihedral group D_{2n} as the set of the following bijective maps

$$\sigma : \mathbb{Z}/n\mathbb{Z} \longrightarrow \mathbb{Z}/n\mathbb{Z} \ : \ k \mapsto \sigma(k) = ak + b,$$

where $a = \pm 1$ and $b \in \mathbb{Z}/n\mathbb{Z}$, with group multiplication the composition of maps. Now define the two elements σ and $\tau \in D_{2n}$ by $\sigma(k) = k + 1$ and $\tau(k) = -k$. Then

$$D_{2n} = \{\sigma^m, \ \tau\sigma^m \ | \ 0 \le m \le n\}.$$

Indeed, both sets have $2n$ elements and therefore

$$D_{2n} \subset \{\sigma^m, \ \tau\sigma^m \ | \ 0 \le m \le n\}.$$

At the same time we observe that

$$\sigma^b(k) = k + b, \ \text{and} \ \tau\sigma^{-b}(k) = -k + b,$$

so both are in D_{2n}. Hence,

$$D_{2n} \supset \{\sigma^m, \ \tau\sigma^m \ | \ 0 \le m \le n\}.$$

Thus these two sets are equal. The elements σ and τ are the *generating elements* of D_{2n}. Here σ^m is a *rotation* and $\tau\sigma^m$ a *reflection* for any $0 \le m \le n$.

Exercise 1.2.11. Find the center, $\mathcal{Z}(D_{2n})$, when $n \ge 1$.

Exercise 1.2.12.

1. Prove that D_3 is the symmetry group of an equilateral triangle.

2. Prove that D_4 is the symmetry group of a square.

1.3   Subgroups

Example 1.3.1. Let S_n be the symmetric group. A transposition is a permutation which leaves all but two elements fixed and permutes these two elements. It is easy to see that any permutation is a product of transpositions.

Exercise 1.3.2. Prove that any permutation is a product of transpositions. Also prove that when a permutation is written as such a product, the number of ways to do this is either always even, or always odd. We then say the permutation is even or odd, respectively.

Define the map $h : S_n \to \pm 1$ by $h(even) = 1$ and $h(odd) = -1$. Check to be sure that h is a homomorphism. Its kernel (the subgroup of even permutations) is denoted by A_n and is called the *alternating group*. It has order $\frac{n!}{2}$.

Exercise 1.3.3. Prove the following:

1. $\mathbb{Z}$ is a subgroup of $\mathbb{Q}$ and of $\mathbb{R}$.

2. ± 1 is a subgroup of $\mathbb{R}^\star$.

3. The set of complex numbers z with $|z| = 1$ is a subgroup of $\mathbb{C}^\star$ (the non-zero complex numbers under multiplication).

4. If the reader would like to look at the definition of the quaternions in Chapter 5 it might be interesting to try to extend the idea in the two items above to the quaternions.

Exercise 1.3.4. Verify that the following are subgroups of $\mathrm{GL}(n, \mathbb{R})$. Here A^t denotes the transpose of the matrix A.

1. $\mathrm{SL}(n, \mathbb{R}) = \{g : g \in \mathrm{GL}(n, \mathbb{R}), \text{ with } \det(g) = 1\}$ (*the special linear group*).

2. The (real) *the orthogonal group* $\mathrm{O}(n, \mathbb{R})$, and the (real) *the special orthogonal group* $\mathrm{SO}(n, \mathbb{R})$, are respectively defined as:

$$\mathrm{O}(n, \mathbb{R}) = \{A \in GL(n, \mathbb{R}) : AA^t = A^t A = I\}$$

and
$$SO(n, \mathbb{R}) = O(n, \mathbb{R}) \cap SL(n, \mathbb{R}).$$

Exercise 1.3.5. Check that the *unitary group* $U(n, \mathbb{C})$, and its subgroup, the *special unitary group* $SU(n, \mathbb{C})$, defined as

$$U(n, \mathbb{C}) = \{A \in GL(n, \mathbb{C}) \ A\bar{A}^t = \bar{A}^t A = I\}$$

and
$$SU(n, \mathbb{C}) = U(n, \mathbb{C}) \cap SL(n, \mathbb{C})$$

where $\bar{A}^t$ is the conjugate transpose of A, are subgroups of $GL(n, \mathbb{C})$.

Here we see that the subgroups of $\mathbb{Z}$ that we know, namely, $k\mathbb{Z}$ where, $k \geq 0$, $k \in \mathbb{Z}$ are the only ones.

Proposition 1.3.6. *Suppose that G is a subgroup of $\mathbb{Z}$. Then, there is a non-negative integer k such that $G = k\mathbb{Z}$.*

Proof. Since G is a subgroup of $\mathbb{Z}$ we must have $0 \in G$. For the same reason, if g_1 and g_2 are in G, then $n_1 g_1 + n_2 g_2 \in G$.
Now, if $G \neq \{0\}$, it must contain a positive integer. Let k be the smallest positive integer in G. Take any $g \in G$. Then, for this g we have $g = nk + r$ with $0 \leq r < k$. But $nk \in G$ and therefore $r \in G$. But $r < k$, so r must be 0. Hence, $G = k\mathbb{Z}$. $\qquad \square$

Proposition 1.3.7. *In a finite group G each $g \in G$ has finite order.*

Proof. Let $g \neq e$. Since G is finite, the set of all g^n as $n \geq 1$ varies, cannot result in distinct elements. Therefore $g^i = g^j$ for some $i \neq j$ and we may assume $i > j$. Hence $g^{i-j} = g^i g^{-j} = e$. $\qquad \square$

It may happen that in a group G there are elements x and y of finite order, but their product xy is of infinite order. An example of this is the following:
Consider the group $SL(2, \mathbb{Z})$, of all 2×2 integer matrices of determinant 1. Let

$$A = \begin{pmatrix} 0 & -1 \\ 1 & 0 \end{pmatrix}, \text{ and } B = \begin{pmatrix} 0 & 1 \\ -1 & -1 \end{pmatrix}.$$

One checks that $A^4 = B^3 = I$, so A has order 4 and B has order 3. Now, their product is

$$AB = \begin{pmatrix} 1 & 1 \\ 0 & 1 \end{pmatrix}$$

and one sees easily that,

$$(AB)^n = \begin{pmatrix} 1 & n \\ 0 & 1 \end{pmatrix}$$

for each $n \in \mathbb{Z}^+$. Therefore AB has infinite order.

Exercise 1.3.8. Show if x and y have finite order in an Abelian group, then their product xy must also have finite order.

1.3.0.1   Exercises

Exercise 1.3.9. If G is a group such that $(a \cdot b)^2 = a^2 \cdot b^2$ for all $a,\ b \in G$, show that G must be Abelian.

Exercise 1.3.10. If G is a group in which $(a \cdot b)^i = a^i \cdot b^i$ for three consecutive integers i for all $a,\ b \in G$, show G is Abelian.

Exercise 1.3.11. Show that the above conclusion does not follow if we merely assume the relation $(a \cdot b)^i = a^i \cdot b^i$ for just two consecutive integers.

Exercise 1.3.12. Let G be a group in which $(ab)^3 = a^3b^3$ and $(ab)^5 = a^5b^5$ for all a, b in G. Show that G is Abelian.

Exercise 1.3.13. Let G be a group and $Z_G(g)$ denote the centralizer of $g \in G$, that is all the elements of G which commute with g. Prove $Z_G(g)$ is a subgroup. Prove $\mathcal{Z}(G) = \bigcap_{g \in G} Z_G(g)$. In addition, show $g \in \mathcal{Z}(G)$ if and only if $Z_G(g) = G$.

Exercise 1.3.14.   1. If the group G has order three, must it be Abelian? Why?

2. Are groups of order four, or five. Abelian? How many such non-isomorphic groups are there?

3. Give an example of a group of order 6 which is not Abelian.

Exercise 1.3.15. Let

$$G = \left\{ \begin{pmatrix} a & b \\ -b & a \end{pmatrix} \ \middle|\ a,\, b \in \mathbb{R},\ a^2 + b^2 \neq 0 \right\}.$$

Show G is an Abelian group under matrix multiplication. Can you think of a group that this is isomorphic to?

Exercise 1.3.16. Suppose that the finite group G has even order. Prove that the number of elements in G of order 2 is odd.

1.4   Quotient Groups

1.4.1   Cosets

Let G be a group and H be a subgroup of G.

Definition 1.4.1. The *right coset* of H determined by $g \in G$, is the set

$$Hg = \{hg, h \in H\}.$$

The *left coset* of H determined by $g \in G$, is the set

$$gH = \{gh, h \in H\}.$$

Since $1 \in H$, gH and Hg always contain g, so they are non-empty.

Proposition 1.4.2. *The set of left (or right) cosets constitute a partition of G. In particular, if G is a group and H is a subgroup, then G is the union of its distinct left (or right) H cosets.*

Proof. We will restrict ourselves to left cosets, the right cosets can be handled similarly. We shall show that two left cosets either coincide or else are disjoint. Therefore (see 0.3.1), since each left coset is non-empty, they constitute a partition of G.

First, we will prove that $g_1 H = g_2 H$ if and only if $g_2^{-1} g_1 \in H$. Indeed,

$\Rightarrow$. Let $g_1 H = g_2 H$. Then, an element $g_1 h_1 \in g_1 H$ must be equal to one $g_2 h_2 \in g_2 H$. In other words $g_1 h_1 = g_2 h_2$, and so $g_2^{-1} g_1 = h_2 h_1 \in H$.

$\Leftarrow$. Let $g_2^{-1} g_1 \in H$. We will first show that $g_1 H \subseteq g_2 H$. Take an element $g_1 h \in g_1 H$. By assumption $g_2^{-1} g_1 \in H$, therefore, $g_1 \in g_2 H$, which implies that there is an element $\tilde{h} \in H$ such that $g_1 = g_2 \tilde{h}$. Hence since H is a subgroup, $g_1 h = g_2 \tilde{h} h \in g_2 H$.

Now, to see that $g_2 H \subseteq g_1 H$ take a $g_2 h \in g_2 H$. Since $g_2^{-1} g_1 \in H$ and H is a subgroup of G, we get $(g_2^{-1} g_1)^{-1} \in H$ and therefore $g_1^{-1} g_2 \in H$, so there is an $\tilde{h}$ such that $g_{-1} g_2 = \tilde{h}$ and therefore $g_2 = g_1 \tilde{h}$. Hence $g_2 h = g_1 \tilde{h} h \in g_1 H$.

Secondly, since for every $g \in G$, $g = g.1 \in gH$; therefore,

$$\bigcup_{g \in G} gH = G.$$

Finally, we show that if $g_1 H \cap g_2 H \neq \varnothing$, then $g_1 H = g_2 H$. To do this, take an element in the intersection, i.e. $g_1 h_1 = g_2 h_2$. Then, $g_2^{-1} g_1 = h_2 h_1^{-1} \in H$, and so as above, $g_1 H = g_2 H$. $\square$

Let G be a group and H be a subgroup. Consider the equivalence relation $g_1 \sim g_2 \Leftrightarrow g_2^{-1} g_1 \in H$. Then, because of Proposition 1.4.2, we can make the following definition:

Definition 1.4.3. We define the *quotient space of left cosets* of G by H, the set

$$G/H = \{gH, g \in G\}.$$

Similarly, we define the space of right cosets

$$H \backslash G = \{Hg \mid g \in G\}.$$

The quotient G/H can not in general be given a group structure compatible with G. Nonetheless, we have the map

$$\pi : G \to G/H \text{ defined by } \pi(g) := gH$$

that we call the *natural projection*. Obviously, π is surjective.

Definition 1.4.4. The number of distinct cosets in G/H is called the *index of H in G* and is denoted by $[G : H]$.

Corollary 1.4.5. *(Lagrange's theorem) If G is a finite group and H is a subgroup, then* $|G| = [G : H]|H|$. *In particular, the order of H divides the order of G.*

Proof. This is because all the cosets gH have the same cardinality. Indeed, for each $g \in G$, the map $h \mapsto gh$ is a bijection $H \to gH$. □

Corollary 1.4.6. *In a finite group the order of any element divides the order of the group.*

Proof. Let $g \in G$. Then, as we know, g has finite order, say k. But the cyclic subgroup H generated by g has order k. Hence k divides $|G|$. □

Exercise 1.4.7. If $H \subseteq F$ are subgroups of G, then

$$|G : H| = |G : F| \cdot |F : H|.$$

Proof. Since $|G : H| = \frac{|G|}{|H|}$, $|G : F| = \frac{|G|}{|F|}$, and $|F : H| = \frac{|F|}{|H|}$, the result follows. □

Let G be a group and g_1 an element of G. The map

$$L_{g_1} : G \to G, \text{where} g \mapsto g_1 g$$

is called the *left translation* (i.e. left multiplication by g_1).

Proposition 1.4.8. *The map L_g is bijective for each $g \in G$.*

Proof. We first prove that L_g is injective. For if $L_g(g_1) = L_g(g_2)$, then $gg_1 = gg_2$, and so $g_1 = g_2$.
L_g is also surjective. Indeed, for any g_1 in G, $L_g(g^{-1}g_1) = gg^{-1}g_1 = g_1$. □

Proposition 1.4.9. *(Cayley's theorem) Any group G is isomorphic to a subgroup of S(G), the set of all permutations of G. In particular, a finite group of order n can be identified with a subgroup of the permutations, S_n.*

Proof. Since L_g is a bijection for any $g \in G$, $L_g \in S(G)$. Now, consider the map

$$\phi : G \to S(G), \text{ such that } g \mapsto L_g.$$

We will show that this map is a homomorphism. Indeed, for any $g, g' \in G$, $L_g L_{g'} = L_{gg'}$. This is because $L_g L_{g'}(g_1) = L_g(g'g_1) = gg'g_1 = L_{gg'}(g_1)$. Also, $L_g L_{g^{-1}} = L_{gg^{-1}} = Id$ and so $L_{g^{-1}} = (L_g)^{-1}$. Therefore $G \approx \text{Im}\phi(G) \subseteq S(G)$. $\qquad\qquad\qquad\qquad\qquad\qquad\qquad\qquad\qquad\qquad\square$

Unfortunately, when G has large order n, S_n is too large to be manageable. But, as we shall see later, G can often be embedded in a permutation group of much smaller order than $n!$.

1.5   Modular Arithmetic

Gauss[4] in his textbook *Disquisitiones Arithmeticae*[5] (Arithmetic Investigations) ([40]) introduced the following notion:

Definition 1.5.1. Let m be an integer. For any a, $b \in \mathbb{Z}$, we write

$$a \equiv b \pmod{m}$$

and say *a is congruent to b modulo m*, if m divides $a - b$. The integer m is called the *modulus*.

Note that the following conditions are equivalent:

1. $a \equiv b \pmod{m}$.

2. $a = b + km$ for some integer k.

---

[4]Carl Friedrich Gauss (1777-1855) German mathematician worked in a wide variety of fields in both mathematics and physics including number theory, analysis, differential geometry, geodesy, magnetism, astronomy and optics. His work is to be found everywhere and it would be both a daunting and absurd task to try to summarize his achievements; Gauss' influence is immense.

[5]Disquisitiones Arithmeticae is a textbook of number theory written in Latin by Carl Friedrich Gauss in 1798 when he was 21, and first published in 1801. Here Gauss brings together results in number theory obtained by mathematicians such as Fermat, Euler, Lagrange and Legendre and adds important new results of his own.

3. a and b have the same remainder when divided by m.

Since $a \equiv b \pmod{m}$ means $b = a + mk$ for some $k \in \mathbb{Z}$, adjusting an integer modulo m is the same as adding (or subtracting) multiples of m to it. Thus, if we want to find a positive integer congruent to -18 (mod 5), we can add a multiple of 5 to -18 until it becomes positive. Adding 20 does the trick: $-18 + 20 = 2$, so $-18 \equiv 2 \pmod{5}$ (check!).

Lemma 1.5.2. *If $\gcd(a, n) = 1$, then the equation $ax \equiv b \pmod{n}$ has a solution, and that solution is unique modulo n.*

Proof. If $\gcd(a, n) = 1$, then the map

$$\mathbb{Z}/n\mathbb{Z} \longrightarrow \mathbb{Z}/n\mathbb{Z},$$

given by left multiplication by a is a bijection. $\square$

For example, $2x \equiv 1 \pmod{4}$ has no solution, but $2x \equiv 2 \pmod{4}$ does, and in fact it has more than one since ($x \equiv 1 \pmod{4}$ and $x \equiv 3 \pmod{4}$).

Lemma 1.5.3. *The equation $ax \equiv b \pmod{n}$ has one solution if and only if $d = \gcd(a, n)$ divides b.*

Proof. Let $d = \gcd(a, n)$. If there is a solution x to the equation $ax \equiv b \pmod{n}$, then $n \mid (ax - b)$. Since $d \mid n$ and $d \mid a$, it follows that $d \mid b$.

Conversely, suppose that $d \mid b$. Then $d \mid (ax - b)$ if and only if

$$\frac{n}{d} \mid \left(\frac{a}{d} x - \frac{b}{d} \right).$$

Thus $ax \equiv b \pmod{n}$ has a solution if and only if

$$\frac{a}{d} x \equiv \frac{b}{d} \pmod{\frac{n}{d}},$$

has a solution. Since $\gcd(a/d, n/d) = 1$, Lemma 1.5.2 implies the above equation does have a solution. $\square$

Remark 1.5.4. There is a useful analogy between integers $\pmod{m}$ and angle measurements (in radians). In both cases, the objects involved admit different representations, i.e., the angles 0, 2π, and -4π are the same, just as $2 \equiv 12 \equiv -13 \pmod{5}$. Every angle can be put in *standard form* as a real number in the interval $[0, 2\pi)$. There is a similar convention for the standard representation of an integer $\pmod{m}$ using remainders, as follows.

Proposition 1.5.5. *(Euclidean Algorithm) Let $m \in \mathbb{Z}$ be a non-zero integer. For each $a \in \mathbb{Z}$, there is a unique r with $a \equiv r \pmod{m}$ and $0 \le r < |m|$.*

Proof. As we know, using division with a remainder in $\mathbb{Z}$, there are q and r in $\mathbb{Z}$ such that $a = mq + r$, with $0 \le r < |m|$. Then $m \mid a - r$, and therefore $a \equiv r \pmod{m}$. To show r is the unique number in the range $0, 1, ..., |m| - 1$ that is congruent to $a \pmod{m}$, suppose two numbers in this range work:

$$a \equiv r \pmod{m}, \quad a \equiv r_1 \pmod{m},$$

where $0 \le rr_1 < |m|$. Then

$$a = r + mk, \quad a = r_+ ml$$

for some k and $l \in \mathbb{Z}$, so the remainders r and r_1 have a difference $r - r_1 = m(l - k)$. This is a multiple of m, and the bounds on r and r_1 tell us $| r - r_1 | j < |m|$. But a multiple of m has an absolute value less than $|m|$ only if it is 0, so $r - r_1 = 0$, i.e. $r = r_1$. $\square$

Lemma 1.5.6. *If $a^2 \equiv 1 \pmod{p}$ and $\gcd(a, p) = 1$ then $a \equiv 1 \pmod{p}$ or $a \equiv p - 1 \pmod{p}$.*

Lemma 1.5.7. *If p is prime and $0 < j < p$, then $p \mid \binom{p}{j}$.*

Proof. Note that $\binom{p}{j} = \frac{p!}{j!(p-j)!}$. Since $0 < j < p$, there is no p in the denominator of $\binom{p}{j}$, but there is a factor of p in the numerator. Thus

$$\binom{p}{j} \equiv 0 \pmod{p}$$

which implies the statement.   □

Definition 1.5.8. We call the integers $\{0, 1, 2, ..., |m| - 1\}$ the *standard representatives* for integers modulo m.

Exercise 1.5.9. Check that $mk \equiv 0 \pmod{m}$, for any $k \in \mathbb{Z}$.

As we saw earlier,

Theorem 1.5.10. *The relation* $a \equiv b \pmod{m}$ *is an equivalence relation in* $\mathbb{Z}$.

Proposition 1.5.11. *If* $a \equiv b \pmod{m}$ *and* $c \equiv d \pmod{m}$, *then*

  1. $(a + c) \equiv (b + d) \pmod{m}$,

  2. $ac \equiv bd \pmod{m}$.

Proof. We want to show that $(a + c) - (b + d)$ and $ac - bd$ are multiples of m. Write $a = b + mk$ and $c = d + ml$ for some k and $l \in \mathbb{Z}$. Then

$$(a + c) - (b + d) = (a - b) + (c - d) = m(k + l),$$

hence $(a + c) \equiv (b + d) \pmod{m}$.

  For the second equation,

$$ac - bd = (b + mk)(d + ml) - bd = m(kd + bl + mk),$$

so $ac \equiv bd \pmod{m}$.   □

Definition 1.5.12. The Euler ϕ-function is the function ϕ defined on positive integers, where $\phi(n)$ is the number of integers in $\{1, 2, ..., n - 1\}$ which are relatively prime to n.

Example 1.5.13. $\phi(24) = 8$ because there are eight positive integers less than 24 which are relatively prime to 24: $1, 5, 7, 11, 13, 17, 19, 23$. On the other hand, $\phi(11) = 10$ because all of the numbers in $\{1, ..., 10\}$ are relatively prime to 11.

Proposition 1.5.14. *The following are properties of* ϕ.

1. *If p is prime, $\phi(p) = p - 1$.*

2. *If p is prime and $n \geq 1$, then $\phi(p^n) = p^n - p^{n-1}$.*

3. *$\phi(n)$ counts the elements in $\{1, 2, ..., n - 1\}$ which are invertible (mod n).*

Proof. 1. If p is prime, then all of the numbers $\{1, ..., p - 1\}$ are relatively prime to p. Hence, $\phi(p) = p - 1$. 2. There are p^n elements in $\{1, 2, ..., p^n\}$. An element of this set is not relatively prime to p if and only if it is divisible by p. The elements of this set which are divisible by p are

$$1.p, \quad 2.p, \quad 3.p, \quad ..., \quad p^{n-1}.p.$$

(Note that $p^{n-1}.p = p^n$ is the last element of the set.) Thus, there are p^{n-1} elements of the set which are divisible by p, i.e. p^{n-1} elements of the set which are not relatively prime to p. Hence, there are $p^n - p^{n-1}$ elements of the set which are relatively prime to p. The definition of $\phi(p^n)$ applies to the set $\{1, 2, ..., p^{n-1}\}$, whereas we just counted the numbers from 1 to p^n. But this is not a problem, because we counted p^n in the set, but then subtracted it off since it was not relatively prime to p. 3. $(a, n) = 1$ if and only if $ax = 1$ (mod n) for some x, so a is relatively prime to n if and only if a is invertible (mod n). Now, $\phi(n)$ is the number of elements in $\{1, 2, ..., n - 1\}$ which are relatively prime to n, so $\phi(n)$ is also the number of elements in $\{1, 2, ..., n - 1\}$ which are invertible (mod n). $\qquad \square$

1.5.1 Chinese Remainder Theorem

The Chinese Remainder Theorem is a result from elementary number theory about the solution of systems of simultaneous congruences. The Chinese mathematician Sun-Tsï wrote about the theorem in the first century C.E.[6] This theorem also has important applications to the de-

---

[6]Oysteen Ore in [95] mentions a puzzle from *Brahma-Sphuta-Siddharta* (Brahma's Correct System) by Brahmagupta (born 598 C.E.):

An old woman goes to market and a horse steps on her basket and crushes the eggs. The rider offers to pay for the damages and asks her how many eggs she had brought.

sign of software for parallel processors. Here we present it in a simple form. For the general case see 5.5.14.

Proposition 1.5.15. *Let* m *and* n *be positive integers such that* $gcd(m, n) = 1$. *Then for* $a, b \in \mathbb{Z}$ *the system*

$$x \equiv a \pmod{m}$$
$$x \equiv b \pmod{n}$$

has a solution. If x_1 *and* x_2 *are two solutions of the system, then*

$$x_1 \equiv x_2 \pmod{mn}.$$

Proof. It is clear that the equation $x \equiv a \pmod{m}$ has at least one solution since $a + km$ satisfies the equation for all $k \in \mathbb{Z}$. What we have to show is that we can find an integer k_1 such that

$$a + k_1 m \equiv b \pmod{n},$$

which is equivalent to finding a solution for

$$k_1 m \equiv (b - a) \pmod{n}.$$

Since m and n are relatively prime, there exist integers s and t such that $ms + nt = 1$, so $(b - a)ms = (b - a) - (b - a)nt$, or

$$[(b - a)s]m \equiv (b - a) \pmod{n}.$$

Set $k_1 = (b - a)s$. To show that any two solutions are congruent $\pmod{mn}$, let c_1 and c_2 be two solutions of the system. That is,

$$c_i \equiv a \pmod{m}$$
$$c_i \equiv b \pmod{n}$$

She does not remember the exact number, but when she had taken them out two at a time, there was one egg left. The same happened when she picked them out three, four, five, and six at a time, but when she took them out seven at a time they came out even. What is the smallest number of eggs she could have had?

for $i = 1,\ 2$. Then

$$c_2 \equiv c_1 \pmod{m}$$
$$c_2 \equiv c_1 \pmod{n}.$$

Thus, both m and n divide $c_1 - c_2$, so, $c_2 \equiv c_1 \pmod{mn}$.   □

1.5.2   Fermat's Little Theorem

Theorem 1.5.16. (Euler-Fermat Theorem[7]) *If* $(a, n) = 1$, *then* $a^{\varphi(n)} \equiv 1 \pmod{n}$.

Proof. Let $\{a_1, a_2, ..., a_{\varphi(n)}\}$ be the set of all numbers less than n and relatively prime to it and $c \in \{a_1, a_2, ..., a_{\varphi(n)}\}$. It follows immediately from the definition that since c and a_i are themselves relatively prime to n, so is their product. Because c is relatively prime to n we can cancel. Hence if $ca_i \equiv ca_j \pmod{n}$, then $a_i = a_j$. Hence, the set $\{ca_1, ca_2, ..., ca_{\varphi(n)}\}$ is just a permutation of the set $\{a_1, a_2, ..., a_{\varphi(n)}\}$ and so,

$$\prod_{k=1}^{\varphi(n)} ca_k = \prod_{k=1}^{\varphi(n)} a_k \pmod{n}.$$

Therefore,

$$c^{\varphi(n)} \prod_{k=1}^{\varphi(n)} a_k = \prod_{k=1}^{\varphi(n)} a_k \pmod{n}.$$

Now, $\prod_{k=1}^{\varphi(n)} a_k$ is relatively prime to n and hence we can cancel on both sides to get

$$c^{\varphi(n)} \equiv 1 \pmod{n},$$

whenever $(c, n) = 1$.   □

---

[7]Fermat stated this theorem in a letter to Fre'nicle on 18 October 1640. As usual he didn't provide a proof, this time writing "...I would send you the demonstration, if I did not fear it being too long...". Euler first published a proof in 1736, but Leibniz left almost the same proof in an unpublished manuscript sometime before 1683.

The converse of this theorem is also true. Namely, if the above congruence holds, then a and n are relatively prime.

A second even simpler proof of Euler-Fermat can be given via Lagrange's theorem:

Proof. We are given $\gcd(a, n) = 1$ and so $[a]$ is invertible in $\mathbb{Z}_n$. The order of the group $\mathbb{Z}_n$ is $\varphi(n)$. Thus by the corollary to Lagrange's theorem

$$[a]^{\varphi(n)} = [1].$$

Hence $a^{\varphi(n)} \equiv 1 \pmod{n}$.   □

Corollary 1.5.17. *If p is prime, then $a^p = a \pmod{p}$ for all integers a.*

Here is a second proof of this corollary using fixed points. First we need the following result

Proposition 1.5.18. *Let S be a finite set, and p be a prime. Suppose that the map $f : S \longrightarrow S$ has the property that $f^p(x) = x$ for any x in S, where f^p is the p-fold composition of f. Then*

$$|S| \equiv |F| \pmod{p},$$

where F is the set of fixed points of f.

Proof. Suppose $s \in S$ and let m be the smallest positive integer such that $f^m(s) = s$. By assumption $f^p(s) = s$, so $1 \leq m \leq p$. Now, suppose n is a positive integer such that $f^n(s) = s$. Then, applying the division algorithm to n, there exist integers $q \geq 0$ and r such that $n = qm + r$, where $0 \leq r < m$. Then,

$$s = f^n(s) = f^{qm+r}(s) = f^r(f^{qm}(s)) = f^r(s),$$

since $f^{qm}(s) = f^m(f^m(\cdots(f^m(s)))) = s$. Hence, $f^r(s) = s$, which contradicts the minimality of m. Hence, $r = 0$ and m divides n.

Since $f^p(s) = s$ for all s, for any $s \in S$, the smallest positive integer m such that $f^m(s) = s$ divides p and, since p is prime, $m = 1$ or $m = p$. Now, for any $n > 0$, let $n = pq + r$ where $q \geq 0$ and $0 \leq r < p$. Then,

$$f^n(s) = f^{pq+r}(s) = f^r(f^{pq}(s)) = f^r(s).$$

Now, for any s, $t \in S$, we say that $s \sim t$ if there is a positive integer k such that $f^k(s) = t$. Clearly, $\sim$ is reflexive. Also, if $s \sim t$ we can choose k so that $f^k(s) = t$ and $k < p$. Then $f^{p-k}(t) = f^{p-k}(f^k(s)) = f^p(s) = s$. Therefore $t \sim s$ and so $\sim$ is symmetric. Finally, if $s \sim t$ and $t \sim u$, there are integers k, $l < p$ such that $f^k(s) = t$ and $f^l(t) = u$. Then, $f^{k+l}(s) = f^l(t) = u$, so $s \sim u$. Thus, $\sim$ is an equivalence relation. Clearly, if $f(s) = s$, then s lies in an equivalence class of size 1. Now, if $f(s) \neq s$, then the smallest integer m such that $f^m(s) = s$ is p. Hence, if $t \sim s$, then there exists an integer k with $0 \leq k < p$ such that $f^k(s) = t$. Hence, $t \in (s, f(s), ..., f^{p-1}(s))$.

Evidently, all elements in the above p-tuple are equivalent to s. We claim that each of these elements is distinct, since if $f^i(s) = f^j(s)$ with $0 < i < j < p$, then applying f $p - j$ times to each sides gives $f^{p-j+i}(s) = s$. Since $j > i$, $0 < p - j - i < p$, a contradiction. Hence, for all $s \in S$, the equivalence class containing s has size 1 or p. Note that the number of equivalence classes of size 1 is $|F|$, where $F = \{s \in S : f(s) = s\}$. Thus, $|S| \equiv |F| \pmod{p}$, as desired. □

To complete the proof using fixed points we now turn to *Necklaces and Colors*. Consider the set N of necklaces of length n (i.e., n beads) with a color (i.e. a choice for a bead's color). Since a bead can have any color, and there are n beads in total, the total number of necklaces is $|N| = a^n$. Of these necklaces, the mono-colored necklaces are those with the same color for all beads; the others are multi-colored necklaces. Let S (for single) denote the set of mono-colored necklaces, and M (for multiple) denote the multicolored necklaces. Then $N = S \cup M$, and $S \cap M = \varnothing$, that is, S and M are disjoint, and they form a partition of the set of all necklaces N. Since there is only 1 mono-colored necklace for each color, the number of mono-colored necklaces $|S|$ is just a. Given that the two types of necklaces partition the whole set, the number of multi-colored necklaces $|M|$ is equal to $|N| - |S| = a^n - a$.

In order to show that the last expression $a^n - a$ is divisible by n when length n is prime, we need to know something more about the multi-colored necklaces, especially how an equivalence relation involving cyclic permutations partitions the set.

Theorem 1.5.19. *(**Fermat's Little Theorem**[8]) For any integer a,*

$$a^p \equiv a \pmod{p}.$$

Proof. Consider the set S of trinkets of a bracelet of p in number using trinkets of a colors. Clearly, $|S| = a^p$. For $s \in S$, define $f(s)$ to be the

---

[8]Pierre de Fermat (160?-1665) was a French lawyer at the Parlement of Toulouse and an amature mathematician who made notable contributions to analytic geometry, probability, optics and number theory. He is best known for Fermat's Last Theorem, which he stated in the margin of a copy of Diophantus' Arithmetica, saying the margin doesn't leave enough room to prove it. It was not proved in full generality until 1994 by Andrew Wiles, using techniques unavailable to Fermat. In 1630 he bought the office of Conseillé at the Parlement de Toulouse, one of the High Courts in France and held this office for the remainder of his life. Fluent in six languages: French, Latin, Occitan, classical Greek, Italian, and Spanish, Fermat was praised for his written verse in several languages, and his advice was often sought regarding the interpretation of Greek texts. He communicated most of his work in letters to friends, often with little or no proof of his theorems. This naturally led to priority disputes with contemporaries such as Descartes. Fermat developed a method for determining maxima, minima, and tangents to various curves. He was the first person known to have evaluated the integral of general power functions and reduced this evaluation to the sum of geometric series. The resulting formula was helpful to Newton, and then Leibniz, when they independently developed the fundamental theorem of calculus. In optics Fermat refined and generalized the classical principle "the angle of incidence equals the angle of reflection" to "light travels between two given points along the path of shortest time". In this way Fermat played an important role in the historical development of the fundamental *principle of least action* in physics and the calculus of variations in mathematics. Through their correspondence Fermat and Blaise Pascal helped lay the fundamental groundwork for the theory of probability. In number theory, his forte, Fermat studied Pell's equation, perfect numbers, now called Fermat numbers and discovered Fermat's little theorem. He invented the technique of infinite descent and used it to prove Fermat's Last Theorem for the case $n = 4$. Fermat developed the two-square theorem. Although he claimed to have proved all his arithmetic theorems, few records of his proofs have survived. Many mathematicians, including Gauss, doubted several of his claims, especially given the difficulty of some of the problems and the limited mathematical methods available to Fermat.

trinket obtained by shifting every trinket of S one cell to the right. It is clear that $f^p(s) = s$ for all $s \in S$. Let F be the set of trinkets fixed by f. From Lemma 2.1, it follows that $|S| \equiv |F|$ (mod p). Now, suppose t is fixed by f. Then, since if $j > i$, $f^{j-i}(t) = t$ and f^{j-i} sends cell i to cell j, the contents of these two cells must be the same. Thus, all the trinkets in t must be the same color. Conversely, if s is a trinket where all the trinkets are of the same color, then $s \in F$. Thus, $|F| = a$. It follows that $a^p \equiv a$ (mod p). $\qquad\qquad\qquad\qquad\qquad\qquad\qquad\qquad\square$

Example 1.5.20. Consider finding the last decimal digit of 7^{222}, i.e. 7^{222} (mod 10). Note that 7 and 10 are relatively prime, and $\varphi(10) = 4$. So the theorem yields $7^4 \equiv 1$ (mod 10), and we get

$$7^{222} \equiv 7^{4 \times 55 + 2} \equiv (7^4)^{55} \times 7^2 \equiv 1^{55} \times 7^2 \equiv 49 \equiv 9 \quad (\text{mod } 10).$$

Example 1.5.21. Compute 50^{250} (mod 83).

One way is to multiply out 50^{250}. A computer tells us that this is equal to

527147875260444560247265192192255725514240233239220086415170220907898754023953317101764802222264464998750268125535784702076863325972445883937922417317167855799198150634765625000.

Now just reduce (mod 83).

If you do not have a powerful computer, maybe you should use Fermat's theorem. $(83, 50) = 1$, so Fermat says

$$50^{82} = 1 \quad (\text{mod } 83).$$

Now $3 \cdot 82 = 246$, so

$$50^{250} = 50^{246} \cdot 50^4 = (50^{82})^3 \cdot 2500^2 = 1^3 \cdot 10^2 = 100 = 17 \quad (\text{mod } 83).$$

Example 1.5.22. Solve $15x = 7$ (mod 32).

Note that $(15, 32) = 1$ and $\phi(32) = 16$. Therefore, $15^{16} = 1$ (mod 32). Multiply the equation by 15^{15}: $x = 7 \cdot 15^{15}$ (mod 32).

Now $7 \cdot 15^{15} = 105 \cdot 15^{14} = 105 \cdot (15^2)^7 = 105 \cdot 225^7 = 9 \cdot 1^7 = 9$ (mod 32). So the solution is $x = 9$ (mod 32).

Exercise 1.5.23. Calculate $2^{20} + 3^{30} + 4^{40} + 5^{50} + 6^{60}$ (mod 7).

1.5.3 Wilson's Theorem

The statement of Wilson's Theorem first appeared in Meditationes Algebraicae [113] in 1770 in a work by the English mathematician Edward Waring. John Wilson, a former student of Waring, stated the theorem but provided no proof, much like Fermat did for Fermat's Little Theorem. In 1771 Lagrange provided the first proof.

Theorem 1.5.24. *(Wilson's Theorem[9]) If p is a prime, then*

$$(p-1)! \equiv 1 \pmod{p}.$$

The first proof uses the group U_p of all invertible elements of $\mathbb{Z}_p$.

Proof. In a group, each element has a unique inverse. Sometimes an element may be its own inverse: the identity element is such an element but there may be others. We shall begin by finding the elements that are their own inverse in U_p. Such an element has the property that $1 < x < p1$ and $[x]^2 = [1]$. Thus $x^2 \equiv 1 \pmod{p}$. This means that $p \mid (x+1)(x-1)$. Thus either $x \equiv 1 \pmod{p}$ or $x \equiv -1 \pmod{p}$, and so $x \equiv 1$ or $x \equiv p - 1$. Thus there are only two such elements which are self inverse. Multiply all the elements of the group U_p together. Each element that has an inverse different from itself will multiply that inverse to get the identity. We are therefore left with $p-1$ being paired. Thus $(p-1)! \equiv p - 1 \pmod{p}$. □

---

[9]John Wilson (1741-1793) English mathematician responsible for Wilson's theorem. He was later knighted, and became a Fellow of the Royal Society in 1782. He was Judge of Common Pleas from 1786 until his death in 1793.

Proof. If $p = 2$ then $(2 - 1)! = 1 \equiv 1 \pmod 2$ and if $p = 3$ then $(3 - 1)! = 2 \equiv 1 \pmod 3$. So, we can assume p is a prime > 3. Since $(p - 1) \equiv 1 \pmod p$, it suffices to show that $(p - 2)! \equiv 1 \pmod p$. By Lemma 7, for each j such that $1 \le j \le p - 1$ there exists an integer k such that $1 \le k \le p - 1$ such that $jk \equiv 1 \pmod p$. If $k = j$, then $j^2 \equiv 1 \pmod p$ and so $j = 1$ or $j = p - 1$. Thus, if $2 \le j \le p - 2$, then there exists an integer k such that $j \ne k$ and $2 \le k \le p - 2$ and $jk \equiv 1 \pmod p$. Since there are $\frac{1}{2}(p - 3)$ such pairs, multiplying them together yields $(p - 2)! \equiv 1 \pmod p$, and the conclusion.   □

The converse is also true. Indeed,

Theorem 1.5.25. *If $(p - 1)! \equiv -1 \pmod p$, then p is a prime.*

Proof. Suppose p is composite. Then p has at least a divisor d, with $1 < d < p$. By assumption p divides $(p - 1)! + 1$, hence, since $d \mid p$, d must divide $(p - 1)! + 1$, which implies that there is some integer k such that

$$(p - 1)! + 1 = k \cdot d \tag{1}$$

Now, since $1 < d < p$, d must divide $(p - 1)!$, and since it divides the right side of (1), it must divide 1, so $d = 1$, a contradiction.   □

Example 1.5.26. Consider 6! (mod 7).

$$6! \equiv 1 \cdot 2 \cdot 3 \cdot 4 \cdot 5 \cdot 6 \equiv 1 \cdot 6 \cdot (2 \cdot 4) \cdot (3 \cdot 5) \equiv 1 \cdot (-1) \cdot 1 \cdot 1 \equiv -1 \pmod 7.$$

From a direct calculation $6! + 1 = 721$ is seen to be divisible by 7.

Definition 1.5.27. An integer a is called a *quadratic residue* modulo n, if it is congruent to a perfect square modulo n, i.e. if there is an integer x such that $x^2 \equiv a \pmod n$. Otherwise, a is called *quadratic non-residue* modulo n.

Using Wilson's Theorem for any odd prime $p = 2n + 1$ we can rearrange the left side of the congruence to obtain

$$1 \cdot (p-1) \cdot 2 \cdot (p-2) \cdots n \cdot (p-n) \equiv 1 \cdot (-1) \cdot 2 \cdot (-2) \cdots n \cdot (-n) \equiv -1 \pmod p,$$

and this becomes
$$(n!)^2 \equiv (-1)^{n+1} \pmod{p}.$$

This last result yields the following famous theorem.

Theorem 1.5.28. *If p is a prime and $p \equiv 1 \pmod 4$, the number -1 is a quadratic residue $\pmod p$.*

Proof. Suppose that $p = 4k + 1$ for some k. Then, we set $n = 2k$ and the above equation gives us that $(n!)^2$ is congruent to -1. $\square$

1.5.3.1 Exercises

Exercise 1.5.29. For any two distinct primes p and q we have

$$p^q + q^p = (p + q) \bmod pq.$$

Exercise 1.5.30. Solve $x^{62} - 16 = 0$ in $\mathbb{Z}/31\mathbb{Z}$.

Exercise 1.5.31. Solve $19x - 11 = 0$ in $\mathbb{Z}/31\mathbb{Z}$.

Exercise 1.5.32. Solve $13x - 11 = 0$ in $\mathbb{Z}/31\mathbb{Z}$.

Exercise 1.5.33. Solve $21x - 24 = 0$ in $\mathbb{Z}/30\mathbb{Z}$.

Exercise 1.5.34. Choose a number from 1 to 9 (inclusive). Multiply it by 3. Add 4 to the result. Multiply this by 3 and subtract 8. Now add the digits of what you get. If the result is a two-digit number, add the digits again. Explain why this gives 4 regardless of what number you start with.

Exercise 1.5.35. Use Fermat's little theorem to do the following:

1. Find the least residue of 9^{794} modulo 73.

2. Solve $x^{86} \equiv 6 \pmod{29}$.

3. Solve $x^{39} \equiv 3 \pmod{13}$.

Exercise 1.5.36. The congruence $7^{1734250} \equiv 1660565 \pmod{1734251}$ is true. Can you conclude that 1734251 is a composite number? Why or why not?

Exercise 1.5.37. Use Euler's theorem to find the least residue of 2^{2007} modulo 15.

Exercise 1.5.38. Show that $x^5 \equiv 3 \pmod{11}$ has no solutions.

Exercise 1.5.39. Find the last two digits of 13^{1010}.

Exercise 1.5.40. Show $2^{35} + 1$ divisible by 11?

Exercise 1.5.41. Show that $2^{1194} + 1$ is divisible by 65.

Exercise 1.5.42. Solve $x^{22} + x^{11} \equiv 2 \pmod{11}$.

Exercise 1.5.43. Show that there are no integer solutions (x, y) to the equation

$$x^{12} - 11x^6 \, y^5 + y^{10} \equiv 8.$$

Exercise 1.5.44. Calculate 20! $\pmod{23}$.

1.6 Automorphisms, Characteristic and Normal Subgroups

Let G be a group and H a subgroup.

Definition 1.6.1. We say H is *normal* in G if $gHg^{-1} \subseteq H$ for every $g \in G$.

In other words, H is invariant under every inner automorphism α_g. When this happens we write $H \triangleleft G$. Of course 1 and G are normal. We leave to the reader the easy verification that $H \triangleleft G$ if and only if $gH = Hg$ for every $g \in G$. In other words, there is no distinction between left and right cosets.

Remark 1.6.2. It is possible for a subgroup H and a $g \in G$ to have $gHg^{-1} \subseteq H$, but $gHg^{-1} \neq H$.
For example, take $G = \mathrm{GL}(2, \mathbb{Q})$ and

$$H = \left\{ \begin{pmatrix} 1 & n \\ 0 & 1 \end{pmatrix} \, \middle| \, n \in \mathbb{Z} \right\}.$$

Then H is a subgroup of G isomorphic to $\mathbb{Z}$.
Take
$$g = \begin{pmatrix} 5 & 0 \\ 0 & 1 \end{pmatrix}.$$

Now,

$$g \begin{pmatrix} 1 & n \\ 0 & 1 \end{pmatrix} g^{-1} = \begin{pmatrix} 5 & 0 \\ 0 & 1 \end{pmatrix} \begin{pmatrix} 1 & n \\ 0 & 1 \end{pmatrix} \begin{pmatrix} 5 & 0 \\ 0 & 1 \end{pmatrix}^{-1}$$

$$= \begin{pmatrix} 5 & 5n \\ 0 & 1 \end{pmatrix} \frac{1}{5} \begin{pmatrix} 1 & 0 \\ 0 & 5 \end{pmatrix} = \begin{pmatrix} 5 & 5n \\ 0 & 1 \end{pmatrix} \begin{pmatrix} 5^{-1} & 0 \\ 0 & 1 \end{pmatrix} = \begin{pmatrix} 1 & 5n \\ 0 & 1 \end{pmatrix}.$$

So, $gHg^{-1} \subseteq H$, but $gHg^{-1} \neq H$. The element

$$\begin{pmatrix} 1 & 6 \\ 0 & 1 \end{pmatrix} \in H$$

does not belong to gHg^{-1}.

A typical example of a normal subgroup is given as follows.

Proposition 1.6.3. *The group* $\mathrm{SL}(n,k)$ *is a normal subgroup of* $\mathrm{GL}(n,k)$.

Proof. Since $\det(AB) = \det(A)\det(B)$ for any two $n \times n$ matrices A and B (see Volume II, Corollary 7.3.10) it follows that $\det(ABA^{-1}) = \det B$. So that if $\det(B) = 1$, then

$$\det(ABA^{-1}) = \det(A)\det(B)\det(A^{-1}) = \det(A)\det(A^{-1}) = \det(I) = 1.$$

$\square$

The extreme opposite situation is when a group has no non-trivial normal subgroups.

Definition 1.6.4. A group G is called a *simple group*[10] if it has no non-trivial normal subgroups.

---

[10] One of the great accomplishments of twentieth century mathematics is a complete classification of all *finite* simple groups. Most of these groups lie in several infinite

We now state and prove the first isomorphism theorem.

Theorem 1.6.5. *Let G and G' be groups and $f : G \to G'$ be a surjective homomorphism. Then $\mathrm{Ker}\, f$ is a normal subgroup of G, the following diagram commutes*

$$
\begin{array}{ccc}
G & \longrightarrow & G' \\
\pi \downarrow & \curvearrowright & \uparrow \tilde{f} \\
G/\mathrm{Ker}\, f & \longrightarrow & G/Kerf
\end{array}
$$

and $\tilde{f}$ is an isomorphism.

Proof. To see that $\mathrm{Ker}(f)$ is normal we argue exactly as with det in the case of $\mathrm{GL}(n, k)$. Let $f(h) = 1$. Then

$$f(ghg^{-1}) = f(g)f(h)f(g^{-1}) = f(g)f(g^{-1}) = f(gg^{-1}) = f(1) = 1.$$

Now define $\tilde{f}(g\,\mathrm{Ker}\, f) = f(g)$. This is a well defined map. Indeed, if $g_1\,\mathrm{Ker}\, f = g_2\,\mathrm{Ker}\, f$ then $g_1 g_2^{-1} \in \mathrm{Ker}\, f$, so $1_{G'} = f(g_1 g_2^{-1}) = f(g_1)f(g_2)^{-1}$. Therefore, $f(g_2) = f(g_1)$.
To prove $\tilde{f}$ is injective we need the following lemma.   □

Lemma 1.6.6. *A homomorphism $f : G \to G'$ is injective if and only if $\mathrm{Ker}\, f = \{1_G\}$.*

Proof. $\Longrightarrow$ Let $g \in \mathrm{Ker}\, f$. Then $f(g) = 1_{G'}$. But $f(1_G) = 1_{G'}$, and since f is injective, $g = 1_G$.
$\Longleftarrow$ Suppose $\mathrm{Ker}\, f = \{1_G\}$ and $f(x) = f(y)$. Then $f(x)f(y)^{-1} = 1_{G'}$ and so $f(xy^{-1}) = 1_{G'}$. Therefore $xy^{-1} \in \mathrm{Ker}\, f = \{1_G\}$, in other words $xy^{-1} = 1_G$, i.e. f is injective.   □

Returning to the proof of the theorem.

---

families: for example, the alternating groups A_n are simple for $n \geq 5$, see 2.2.14. There are also 26 simple groups which do not fit into such families; they are called the *sporadic* simple groups. This complete classification was announced in 1983 by Daniel Gorenstein, who had outlined a program for approaching the classification several years earlier. It is estimated that the complete proof takes about $30,000$ pages. However, in 1983 when the announcement was made, a missing step in the proof was discovered. This gap was only filled in 2004 by Michael Aschbacher and Stephen Smith, in two volumes totaling 1221 pages.

Proof. To see that f is injective, by the previous lemma, it is sufficient to prove that $\operatorname{Ker} \tilde{f} = 1_{G/\operatorname{Ker} f}$. Take any element in $\operatorname{Ker} \tilde{f}$. This will be of the form $g \operatorname{Ker} f$ and it will satisfy $\tilde{f}(g \operatorname{Ker} f) = 1_{G'}$. But $\tilde{f}(g \operatorname{Ker} f) = f(g)$, and so $f(g) = 1_{G'}$, i.e. $g \in \operatorname{Ker} f$. Hence $g \operatorname{Ker} f = \operatorname{Ker} f = 1_{G/\operatorname{Ker} f}$, i.e. $\operatorname{Ker} \tilde{f} = \{1_{G/\operatorname{Ker} f}\}$. $\square$

Conversely, if G is a group and H a normal subgroup then G/H is a group and the natural map $\pi : G \to G/H$ given by $\pi(g) = gH$ is a surjective group homomorphism. In particular, the normal subgroup H is $\operatorname{Ker}(\pi)$. Thus every normal subgroup is the kernel of a surjective homomorphism. We leave the routine proof of this to the reader.

Exercise 1.6.7. As we just saw $\operatorname{SL}(n, \mathbb{R})$ is a normal subgroup of $\operatorname{GL}(n, \mathbb{R})$ therefore we can form the quotient group. Can you identify this group? Is there a difference in its properties between n odd and even?

As we know, the inner automorphisms α_g form a subgroup $\operatorname{Int}(G)$ of $\operatorname{Aut}(G)$. The remaining automorphisms are said to be *outer automorphisms*. Some important features here are $\operatorname{Int}(G)$ is normal in $\operatorname{Aut}(G)$. This is because

$$\alpha \alpha_g \alpha^{-1} = \alpha_{\alpha_g}.$$

Thus we can form the group $\operatorname{Aut}(G)/\operatorname{Int}(G)$. Sometimes this is called the group of outer automorphisms. Also, since the map $g \mapsto \alpha_g$ is an epimorphism from $G \to \operatorname{Int}(G)$ whose kernel is $\mathcal{Z}(G)$ we see $G/\mathcal{Z}(G)$ is isomorphic to $\operatorname{Int}(G)$.

Exercise 1.6.8. Show a group G is Abelian if and only if the map $i : G \to G$ defined by $x \mapsto f(x) = x^{-1}$ is an automorphism.

Proposition 1.6.9. *If H is a normal, cyclic subgroup of G, then every subgroup of H is also normal in G.*

Proof. Let $H = \langle a \rangle$ be a normal cyclic subgroup of G. Every subgroup of H must be cyclic. Therefore if K is a subgroup of H, then $K = \langle a^m \rangle$ for some $m \in \mathbb{Z}^+$. Since H is normal, for any $g \in G$, $gag^{-1} = a^n$ for some $n \in \mathbb{Z}^+$. It follows that $ga^m g^{-1} = (gag^{-1})^m = (a^n)^m = (a^m)^n$. So $gKg^{-1} \subseteq K$, for every $g \in G$. Hence K is normal in G. $\square$

Definition 1.6.10. H is called a *characteristic subgroup* of G if

$$\alpha(H) \subseteq H$$

for every $\alpha \in \text{Aut}(G)$.

Of course characteristic subgroups are normal. The reader should find an example of a normal subgroup of a group which is not characteristic. For example try an Abelian group.

Proposition 1.6.11. *Let $L \subseteq H \subseteq G$ be subgroups of G. If L is characteristic in H and H is normal in G, then L is normal in G.*

Proof. G acts on itself by conjugation, i.e. $\alpha_g : G \to G$ leaving H stable. The restriction, $\alpha_g \mid H$, of this map to H is an automorphism of H. Hence $\alpha_g(L) \subseteq L$. $\square$

Exercise 1.6.12. If $H \triangleleft G$ and $K \triangleleft H$, is $K \triangleleft G$?

1.7 The Center of a Group, Commutators

Definition 1.7.1. The *center* $\mathcal{Z}(G)$ is defined as

$$\mathcal{Z}(G) = \{x \ : \ xy = yx, \ \text{for all } y \in G\}.$$

Exercise 1.7.2. The center $\mathcal{Z}(G)$ is a characteristic Abelian subgroup of G.

Exercise 1.7.3. If $G/\mathcal{Z}(G)$ is a cyclic group, then $G = \mathcal{Z}(G)$, and so G is Abelian.

Proof. Since $G/\mathcal{Z}(G)$ is cyclic, there must exist an element g in G such that its image $\pi(h)$ is a generator of the quotient. In other words any coset $g\mathcal{Z}(G)$ is a power of $\pi(h)$. Therefore, any g in G is of the form zh^k, where $z \in \mathcal{Z}(G)$, and $k \in \mathbb{Z}$. Now, we can see that any two elements g_1, g_2 of G commute, since $g_1 g_2 = z_1 h^k z_2 h^l = z_2 h^{k+l} z_1 = z_2 h^{l+k} z_1 = z_2 h^l h^k z_1 = z_2 h^l z_1 h^k = g_2 g_1$. So, $G = \mathcal{Z}(G)$ is Abelian. $\square$

Definition 1.7.4. Given x, y in G, their *commutator* is defined by $[x, y] = xyx^{-1}y^{-1}$.

The commutator measures the degree that x, y do not commute.

Definition 1.7.5. We define the *commutator* of G (or equivalently the *derived group* of G), and we denote it by $[G, G]$,

$$[G, G] = \{\text{all finite products } \prod_{i=1}^{n}[x_i, y_i] | n \in \mathbb{Z}, \ x_i, y_i \in G\}.$$

More generally, for subgroups H and K of G, $[H, K]$ stands for the all finite products of the type $[h, k]$, for $h \in H$ and $k \in K$.

Proposition 1.7.6. $[G, G]$ *is a normal subgroup of G. More generally, if H, K are normal subgroups of G, then $[H, K]$ is a normal subgroup of G.*

Proof. We have only to check inverses. $[h, k]^{-1} = (hkh^{-1}k^{-1})^{-1} = khk^{-1}h^{-1} = [k, h]$. Now suppose H and K are normal. To see $[H, K]$ is normal, it is sufficient to show $g[h, k]g^{-1} \in [H, K]$ for every g in G. Let $x = [h, k]$. Then,

$$gxg^{-1} = (ghg^{-1})(gkg^{-1})(gh^{-1}g^{-1})(gk^{-1}g^{-1}$$
$$= (ghg^{-1})(gkg^{-1})(ghg^{-1})^{-1}(gkg^{-1})^{-1} \in [H, K].$$

$\square$

1.8   The Three Isomorphism Theorems

Having already dealt with the 1st isomorphism theorem for groups, here we will prove the other two. As the reader will see, all three isomorphism theorems will remain valid in all the other contexts of this book - vector spaces, rings, modules, algebras, etc. The statements and proofs of these results will be functorial. That is to say, the maps will be the same as for groups section, but they will also have to respect the appropriate additional structures.

We now turn to the 2nd and 3rd isomorphism theorems. These important results will follow from the 1st isomorphism theorem.

Definition 1.8.1. If G is a group and H, F are subgroups, we define

$$H \cdot F = \{hf \,|\, h \in H,\, f \in F\}.$$

Proposition 1.8.2. *If H and F are subgroups of G and F is normal, then HF is a group. If H is normal, then so is HF.*

Proof. The operation is well defined and closed on HF. Indeed, since F is normal,

$$(h_1 f_1)(h_2 f_2) = h_1(f_1 h_2) f_2 = h_1(h_2' f_1) f_2 = (h_1 h_2')(f_1 f_2) \in HF.$$

For the same reason

$$(hf)^{-1} = f^{-1} h^{-1} = h^{-1} f' \in HF.$$

Now if, in addition, H is normal, then for all $g \in G$,

$$gHFg^{-1} = (gHg^{-1})(gFg^{-1}) = HF.$$

$\square$

Hence,

Corollary 1.8.3. *If H and F are subgroups of G and F is normal in G, then F is normal in HF.*

Proof. We must prove that for every hf in HF we have $hfF(hf)^{-1} = F$. Indeed, because F is normal in G,

$$hfF(hf)^{-1} = hfFf^{-1}h^{-1} = h(fFf^{-1})h^{-1} = hFh^{-1} = F.$$

$\square$

We now come to the *2nd isomorphism theorem*.

Theorem 1.8.4. *Let H be a subgroup of G and N be a normal subgroup of G. Then, HN is a subgroup of G, $H \cap N$ is a normal subgroup of H and*

$$HN/N \simeq H/(H \cap N).$$

Proof. As we already know, HN is a subgroup of G. Since N is normal in G, G/N is a group. Consider the map

$$\tilde{\pi} : H \to G/N, : h \mapsto hN.$$

This is a homomorphism and its kernel is $H \cap N$. By the 1st isomorphism theorem, $H \cap N$ is a normal subgroup of H. Applying it again we obtain following the isomorphism:

$$H/(H \cap N) \simeq \mathrm{Im}(\tilde{\pi}).$$

But $\mathrm{Im}(\tilde{\pi})$ is the set of all cosets of the form hN with $h \in H$. So, $\mathrm{Im}(\tilde{\pi}) = HN/N$. Thus

$$HN/N \cong H/(H \cap N).$$

$\square$

Example 1.8.5. Consider a general linear group $G = \mathrm{GL}(2, \mathbb{C})$, its special linear subgroup $H = \mathrm{SL}(2, \mathbb{C})$, and its center, $N = \mathbb{C}^* I_{2 \times 2}$. Then

$$HN = G, \qquad H \cap N = \{\pm I_{2 \times 2}\}.$$

The 2nd isomorphism theorem implies

$$\mathrm{GL}(2, \mathbb{C})/\mathbb{C}^* I_{2 \times 2} \xrightarrow{\simeq} \mathrm{SL}(2, \mathbb{C})/\{\pm I_{2 \times 2}\} \quad \mathbb{C}^* I_{2 \times 2} \mapsto \{\pm m\}.$$

The groups on each side of the isomorphism are the projective general and special linear groups (see also 2.4). Even though the general linear group is larger than the special linear group, the difference disappears after projectivizing:

$$\mathrm{PGL}(2, \mathbb{C}) \xrightarrow{\simeq} \mathrm{PSL}(2, \mathbb{C}).$$

Theorem 1.8.6. (3rd Isomorphism Theorem). *Let G be a group, H and L be normal subgroups of G, with $L \subseteq H$. Then, H/L is a normal subgroup of G/L, and there is a natural isomorphism*

$$(G/L)(H/L) \xrightarrow{\simeq} G/H \quad : \quad gL \cdot (H/L) \mapsto gH.$$

Proof. The map

$$G/L \longrightarrow G/H, \quad \text{such that given by} gL \mapsto gH.$$

This is well defined because if $g'L = gL$ then $g' = gl$ for some $l \in L$ and so because $L \subset H$ we have $g'KH = gH$. It is a homomorphism since

$$gL \cdot g'L = gg'L \mapsto gg'H = gH \cdot g'H.$$

The map is clearly surjective. Its kernel is H/L, showing that H/L is a normal subgroup of G/L, and the 1st isomorphism theorem gives an isomorphism

$$(G/L)/(H/L) \xrightarrow{\simeq} G/H.$$

□

Example 1.8.7. Let n and m be positive integers with $n \mid m$. Thus $m\mathbb{Z} \subset n\mathbb{Z} \subset \mathbb{Z}$, and all subgroups are normal since $\mathbb{Z}$ is Abelian. By the 3rd isomorphism theorem

$$(\mathbb{Z}/m\mathbb{Z})/(n\mathbb{Z}/m\mathbb{Z}) \longrightarrow \mathbb{Z}/n\mathbb{Z}, \quad (k + m\mathbb{Z}) + n\mathbb{Z} \mapsto k + n\mathbb{Z}.$$

An easy check shows the following diagram commutes:

$$\mathbb{Z}$$
$$\swarrow \qquad \searrow$$
$$\mathbb{Z}/m\mathbb{Z} \qquad\qquad \mathbb{Z}/n\mathbb{Z}$$
$$\searrow \qquad \swarrow$$
$$(\mathbb{Z}/m\mathbb{Z})/(n\mathbb{Z}/m\mathbb{Z})$$

All this to say that if we reduce modulo m and then we further reduce modulo n, the second reduction subsumes the first.

1.9   Groups of Low Order

In this section we will classify (up to isomorphism) all groups of order 1 through 9.

To do so, we need the following proposition:

Proposition 1.9.1. *Any group G of prime order, p, is isomorphic to $\mathbb{Z}_p$.*

Proof. We know that, if G is any group and $g \neq e$, then the order of g is the smallest positive integer n such that $g^n = e$. We write $n = o(g)$. If $\langle g \rangle$ is the group generated by a g in G, we will prove that

$$\langle g \rangle \simeq \mathbb{Z}_{o(g)}.$$

For this, let us consider the homomorphism

$$\mathbb{Z} \longrightarrow \langle g \rangle \text{, such that } n \mapsto g^n.$$

This map is onto. We also note that if $n = q.o(g) + r$,

$$g^n = g^{qo(g)} g^r = eg^r = g^r.$$

The kernel of this homomorphism is $o(g)\mathbb{Z}$. Applying the first isomorphic theorem we get $\mathbb{Z}/o(g) \simeq \langle g \rangle$. $\qquad\square$

Proposition 1.9.2. *Let m and n be positive integers. Then*

$$\mathbb{Z}_{mn} \cong \mathbb{Z}_m \times \mathbb{Z}_n$$

if and only if m and n are relatively prime.

Proof. Let $\mathbb{Z}_m = \langle x \rangle$, $\mathbb{Z}_n = \langle y \rangle$, and $d = \gcd(m, n)$.

($\Rightarrow$) Assume that $d = 1$. Consider the group homomorphism

$$\phi : \mathbb{Z} \longrightarrow \mathbb{Z}_m \times \mathbb{Z}_n, \quad \text{defined by} \quad k \mapsto (x^k, y^k).$$

We see that ϕ is surjective. Indeed, let $(x^a, y^b) \in \mathbb{Z}_m \times \mathbb{Z}_n$. Since m and n are relatively prime, there are integers r and s such that $rm + sn = 1$. Write

$$k = (b - a)rm + a = (a - b)sn + b.$$

Then $k \equiv a \pmod{m}$ and $k \equiv b \pmod{n}$, so that $\phi(k) = (x^a, y^b)$.

Now, we will find the kernel of ϕ. Suppose $k \in \mathrm{Ker}(\phi)$. Then, $(x^k, y^k) = \phi(k) = (1, 1)$. Recall that $x^k = 1$ if and only if $k \equiv 0$

$\pmod{m}$, so that $k \equiv 0 \pmod{m}$ and $k \equiv 0 \pmod{n}$. Since $(m, n) = 1$, we have $k \equiv 0 \pmod{mn}$. Hence $\text{Ker}(\phi) = mn\mathbb{Z}$ and by the 1st isomorphism theorem

$$\mathbb{Z}_{mn} \cong \mathbb{Z}/mn\mathbb{Z} = \mathbb{Z}/\text{Ker}(\phi) \cong Im(\phi) = \mathbb{Z}_m \times \mathbb{Z}_n.$$

($\Leftarrow$) Assume that $d > 1$. We show that every group homomorphism $\phi : \mathbb{Z}_{mn} \longrightarrow \mathbb{Z}_m \times \mathbb{Z}_n$ is not surjective. Write $\mathbb{Z}_{mn} = \langle z \rangle$, and assume that a surjection exists. Let $\phi(z) = (x^a, y^b)$. If ϕ were surjective, then $(x, 1) = \phi(z^{k_1}) = (x^{k_1 a}, y^{k_1 b})$ and $(1, y) = \phi(z^{k_2}) = (x^{k_2 a}, y^{k_2 b})$, for some integers k_1 and k_2. Then,

$$k_1 a \equiv 1 \pmod{m} \qquad k_1 b \equiv 0 \pmod{n}$$
$$\text{and}$$
$$k_2 a \equiv 0 \pmod{m} \qquad k_2 b \equiv 1 \pmod{n}.$$

In particular, a is relatively prime to m and b is relatively prime to n. But then k_1 is relatively prime to m yet divisible by n, so that d divides k_1. This is a contradiction since by assumption $d > 1$. $\qquad\qquad\square$

Corollary 1.9.3. *The groups of order 1,2,3,5,7 are isomorphic to* (1), $\mathbb{Z}_2$, $\mathbb{Z}_3$, $\mathbb{Z}_5$, $\mathbb{Z}_7$ *respectively.*

Proposition 1.9.4. *A group of order p^2, where p is a prime, is Abelian.*

Proof. We will use the fact (which we will prove later) that "if the order of a group is p^n, p prime, then its center is not trivial, i.e. $\mathcal{Z}(G) \neq \{1\}$". In our case, since $\mathcal{Z}(G)$ is a (normal) subgroup, by Lagrange's theorem, its order must divide p^2. So

$$|\mathcal{Z}(G)| = 1, p, \text{ or, } p^2.$$

As we just saw, $|\mathcal{Z}(G)| \neq 1$. Hence $|\mathcal{Z}(G)| = p$ or p^2.
If $|\mathcal{Z}(G)| = p^2$, the center is the whole group G, and so G is Abelian.
If $|\mathcal{Z}(G)| = p$, then $|G/\mathcal{Z}(G)| = p$, hence the group $G/\mathcal{Z}(G)$ is cyclic.
We will show that in this case G is also Abelian.
Indeed, since $G/\mathcal{Z}(G)$ is cyclic that means that this group is generated

by the powers of some element, say the element $g_0 \mathcal{Z}(G)$, i.e. $G/\mathcal{Z}(G) = \langle g_0 \mathcal{Z}(G) \rangle$, and so

$$G = \bigcup_{n \in \mathbb{Z}} g_0^n \mathcal{Z}(G).$$

Now, take g and h of G. Then, $g \in g_0^i \mathcal{Z}(G)$ and $h \in g_0^j \mathcal{Z}(G)$. So, $g = g_0^i z$ and $h = g_0^j z'$, where z, z' in $\mathcal{Z}(G)$. Hence,

$$gh = g_0^i z g_0^j z = g_0^{i+j} z z' = g_0^{j+i} z' z = g_0^j z' g_0^i z = hg,$$

and so G is Abelian. $\qquad\square$

We remark that the Abelian groups $\mathbb{Z}_p \oplus \mathbb{Z}_p$ and $\mathbb{Z}_{p^2}$ are not isomorphic even though they have the same order. This is because $\mathbb{Z}_p \oplus \mathbb{Z}_p$ has all its elements of order 1 or p, while $\mathbb{Z}_{p^2}$ has an element of order p^2. $\mathbb{Z}_2 \oplus \mathbb{Z}_2$ is called the Klein 4 group.

The following lemma is the Abelian version of Cauchy's Theorem whose general proof will be given in 2.6.1.

Lemma 1.9.5. *Cauchy's lemma.* *If G is a finite Abelian group and p is a prime divisor of $|G|$, then G has an element of order p.*

Proof. If G is cyclic, then the statement is true and does not depend on p being prime. If G is not cyclic, since it is finite, it has non-trivial cyclic subgroups. Let H be a maximal cyclic subgroup, i.e. there is no cyclic subgroup L of G such that $H \subseteq L$, except $L = H$. Since we assume that $H \neq G$, there is a $g_0 \notin H$. Let G_0 be $< g_0 >$, the cyclic subgroup generated by g_0. Since H is a normal subgroup of G because G is Abelian, it follows by 2.6.1 that, HG_0 is a subgroup of G, and since $g_0 \notin H$, it properly contains H. Since H is maximal, $HG_0 = G$. Now, we apply the second isomorphism theorem,

$$G/H = (HG_0)/H = G_0/(G_0 \cap H).$$

In particular, $|G|/|H| = |G_0|/|G_0 \cap H|$. Thus, $|G|.|G_0 \cap H| = |H|.|G_0|$. Since p divides $|G|$, it divides $|H|.|G_0|$ and since p is prime it divides either $|H|$ or $|G_0|$. If p divides $|H|$, then we finish by arguing by induction. On the other hand if p divides $|G_0|$, we also finished, because G_0 is cyclic. $\qquad\square$

Exercise 1.9.6. The Klein group $\mathbb{Z}_2 \oplus \mathbb{Z}_2$ is not isomorphic to $\mathbb{Z}_4$.

Exercise 1.9.7. Let G be a group of order 6. Then G is either isomorphic to $\mathbb{Z}_6$, or to D_3 (which is also S_3).

Exercise 1.9.8. There are exactly two distinct, non-Abelian groups of order 8 (up to isomorphism): the quaternion group Q_8, and the dihedral group D_4.

We are now left with the question of the Abelian groups of order 8. If there is an element of order 8, then $G = \mathbb{Z}_8$. Otherwise, there are only non-trivial elements of order 4 or 2. If there are only those of order 2, then $G = \mathbb{Z}_2^3$. The only remaining possibility is there are both elements of order 2 and 4, but none of order 8. In this case we get $\mathbb{Z}_2^2 \times \mathbb{Z}_4$. We leave the details of this to the reader and summarize the results in the following table:

Order	Group
1	Trivial group
2	$\mathbb{Z}_2$
3	$\mathbb{Z}_3$
4	Klein four group $\mathbb{Z}_2 \oplus \mathbb{Z}_2$, $\mathbb{Z}_4$
5	$\mathbb{Z}_5$
6	$\mathbb{Z}_6 \simeq \mathbb{Z}_3 \oplus \mathbb{Z}_2$, D_3
7	$\mathbb{Z}_7$
8	Q_8, D_4, $\mathbb{Z}_8$, $\mathbb{Z}_4 \oplus \mathbb{Z}_2$, $\mathbb{Z}_2 \oplus \mathbb{Z}_2 \oplus \mathbb{Z}_2$
9	$\mathbb{Z}_9$, $\mathbb{Z}_3 \oplus \mathbb{Z}_3$

1.9.0.1   Exercises

Exercise 1.9.9. Let G be a group and H a subgroup. Analogous to the centralizer $Z_G(H)$ we also have the normalizer, $N_G(H)$, of H in G. This is, $\{g \in G : gHg^{-1} \subseteq H\}$. Show $N_G(H)$ is a subgroup of G containing H as a normal subgroup. Show the number of H conjugates is equal to the index $[G : N_G(H)]$.

Exercise 1.9.10. Let G be a finite group, and H a proper subgroup. Prove G can never been covered by H and its conjugates.

Proof. Let
$$B = \bigcup_{g \in G} gHg^{-1}.$$

If H is a normal subgroup of G, then $B = H$, and hence $B \neq G$. If H is not a normal subgroup, then the number of its conjugates is equal to the index $[G : N(H)]$ of the normalizer $N(H)$ of H in G. Since $H \subseteq N(H)$, we get $[G : N(H)] \leq [G : H]$, and since the identity element belongs to all the conjugates, it follows that

$$|B| < [G : N(H)] \cdot |H|.$$

But
$$[G : N(H)] \cdot |H| \leq [G : H] \cdot |H| = |G|.$$

Therefore $B \neq G$.   □

Exercise 1.9.11. Let G be a finite group and A be a subset of G such that $|A| > \frac{1}{2}|G|$. If $A^2 = \{ab : a, b \in A\}$, show $A^2 = G$.

Exercise 1.9.12. Let G be the group generated by x and y with relations $x^5y^3 = x^8y^5 = 1$. Prove that G is the trivial group.

Exercise 1.9.13.

1. Let H_i be a finite family of subgroups of G each of finite index in G. Prove $\cap_i H_i$ is a subgroup of G of finite index.

2. Let G be a group and H be a subgroup of finite index. Prove that within H there is a normal subgroup of finite index in G.

1.10   Direct and Semi-direct Products

1.10.1   Direct Products

In the 1880s, Otto Hölder considered the following question: Suppose we are given two groups H and K. Of course, we can think of these as being subgroups of a larger group G if we choose $G = H \times K$ as their direct product. We wish to find all groups G such that $N \simeq H$ and $G/N \simeq K$ for some normal subgroup $N \lhd G$. Is $G = H \times K$ the only possibility? Here, we will answer this question.

Let G, H be two (multiplicative) groups, and let

$$G \times H = \{(g, h),\ g \in G,\ h \in H\}.$$

Definition 1.10.1. We call (external) *direct product*, the group, $G \times H$ equipped with pointwise multiplication:

$$(g_1, h_1).(g_2, h_2) = (g_1 g_2, h_1 h_2).$$

When G and H are Abelian groups written additively, we write $G \oplus H$, instead of $G \times H$.

The direct product of two groups is characterized by the following theorem:

Proposition 1.10.2. *Let L be a group and G and H be two subgroups. Then, the following statements are equivalent:*

1. *L, G and H satisfy the following relations: $L = GH$, $G \cap H = \{1\}$, and $[G, H] = \{1\}$.*

2. *L is the direct product of G and H, i.e. $L \simeq G \times H$.*

Proof. $\Longrightarrow$ Write

$$G = \{(g, 1),\ g \in G\}$$

and

$$H = \{(1, h),\ h \in H\}.$$

Then, obviously $L = GH$ and $G \cap H = \{1\}$. For $[G, H] = \{1\}$, we have

$$(g, 1)(1, h)(g, 1)^{-1}(1, h)^{-1} = (1, 1).$$

$\Longleftarrow$ Pick a $l \in L$. Then, since $L = GH$, we get $l = gh$. This is a unique representation of the elements of L, since, if l is also equal to $g_1 h_1$, then $gh = g_1 h_1$, thus $g_1^{-1} g = h_1 h^{-1}$, which is a common element of H and G, and since $G \cap H = \{1\}$, $g_1^{-1}g = h_1 h^{-1} = 1$. Hence $g = g_1$ and $h = h_1$. Now, consider the map

$$f : L \to G \times H \quad : l \mapsto (g, h).$$

This map is one-to-one and onto, since the decomposition is unique. It is a homomorphism. To see this, let

$$f(l) = (g, h) \text{ and } f(l_1) = (g_1, h_1).$$

We must check if $f(ll_1) = f(l)f(l_1)$. Indeed, $[G, H] = \{1\}$ implies $[g_1, h] = \{1\} = g_1 h g_1^{-1} h^{-1}$, and therefore $g_1 h = h g_1$. Hence, $g_1 h h_1 = h g_1 h_1$, which implies $g g_1 h h_1 = g h g_1 h_1$. Then, $ll_1 = g h g_1 h_1 = g g_1 h h_1$, therefore $f(ll_1) = (g g_1, h h_1) = (g, h) \cdot (g_1, h_1) = f(l) f(l_1)$. □

Remark 1.10.3. In this case, we call L the *internal direct product* of G and H.

It is also possible to take the direct product of an infinite (possibly uncountable) number of groups. This can defined as follows:

Let $(G_i)_{i \in I}$ be a family of groups. Then, the Cartesian product

$$\prod_{i \in I} G_i = \{f : I \longrightarrow \bigcup_{i \in I} G_i \; : \; i \mapsto f(i) \in G_i\}$$

becomes a group under the following pointwise operation

$$f \cdot g \text{ such that } (f \cdot g)(i) := f(i) \cdot g(i) \in G_i$$

with unit the element 1 defined by $1(i) = 1_{G_i}$ for each $i \in I$. Together with this product there is a family of surjective homomorphisms (check this)

$$\pi_j : \prod_{i \in I} G_i \longrightarrow G_j \text{ such that } \pi_j(f) := f(j),$$

for any $j \in I$. The homomorphism π_j is called the $j - th$ *canonical projection*.

The direct product of groups is characterized by the following universal property:

Theorem 1.10.4. *For any group G and any family of homomorphisms*

$$\phi_i : G \longrightarrow G_i, \quad i \in I$$

there is a unique homomorphism $\phi : G \to \prod_{i \in I} G_i$ so that the following diagram commutes for each $i \in I$.

$$
\begin{array}{ccc}
G & \overset{\phi}{\longrightarrow} & \prod_{i \in I} G_i \\
 & \phi_i \searrow & \downarrow \pi_i \\
 & & G_i
\end{array}
$$

for each $i \in I$.

Proof. For any $g \in G$ we define $\phi(g)$ as the unique element of $\prod_{i \in I} G_i$ with $\pi_i(\phi(g)) = \phi_i(g)$. One checks that ϕ is a homomorphism and $\pi_i \circ \phi = \phi_i$ for each i. To see that ϕ is unique, let ψ be another such homomorphism satisfying $\pi_i \circ \psi = \phi_i$. Then, $\pi_i(\phi(g)) = \pi_i(\psi(g))$, for any $i \in I$, so,

$$\pi_i(\phi(g)) \left(\pi_i(\psi(g)) \right)^{-1} = 1_{G_i},$$

or

$$\pi_i \left(\phi(g) \psi(g)^{-1} \right) = 1_{G_i},$$

i.e. $\phi(g)\psi(g)^{-1} = 1$. Hence $\phi(g) = \psi(g)$ for each $g in G$. $\qquad\square$

We can also consider the subgroup

$$\bigoplus_{i \in I} G_i := \{ f \in \prod_{i \in I} G_i \text{ such that } f(i) \neq 1_{G_i} \text{ all but a finite number of } i \}$$

with the restricted operation. This is called the *direct sum* of the family.

Of course if I is a finite set, these two products coincide and in general the direct sum is a proper subgroup of the direct product.

1.10.2   Semi-Direct Products

Let G and H be groups and ϕ be a homomorphism

$$\phi : G \longrightarrow \text{Aut}(H).$$

Definition 1.10.5. We call the (*external*) *semi-direct product* of the two groups H and G the set

$$H \rtimes_\phi G = \{(h, g) \mid h \in H, \ g \in G\}$$

equipped with the following multiplication:

$$(h, g) \cdot (h_1, g_1) = (h\phi_g(h_1), gg_1),$$

where $\phi_g = \phi(g) \in \text{Aut}(H)$.

In the special case where $\phi_g = 1_H$, we obtain the usual (external) direct product $H \times G$.

Proposition 1.10.6. $H \rtimes_\phi G$ *is a group.*

Proof. We leave the easy check that the multiplication is associative and that $(1, 1)$ is the identity element to the reader. For the inverse,

$$(h, g)^{-1} = \left(h\phi_g(\phi(g^{-1})(h^{-1})) \ , \ gg^{-1} \right).$$

Indeed,

$$(h, g)\left(h\phi_g(\phi(g^{-1})(h^{-1})) \ , \ gg^{-1} \right) = \left(h\phi_g(\phi_{g^{-1}})(h^{-1}), gg^{-1} \right)$$

$$= \left(h(\phi_g\phi_{g^{-1}})(h^{-1}), 1 \right)$$

$$= \left(h\phi_{gg^{-1}}(h^{-1}), 1 \right)$$

$$= (hh^{-1}, 1)$$

$$= (1, 1),$$

and

$$\left(h\phi_g(\phi(g^{-1})(h^{-1}))\,,gg^{-1}\right)(h,g) = \left(\phi_{g^{-1}}(h^{-1}1)\phi_{g^{-1}}h, g^{-1}g\right)$$

$$= \left(\phi_{g^{-1}}(h^{-1}h),1\right)$$

$$= \left(\phi_{g^{-1}}(1),1\right)$$

$$= (1,1).$$

$\square$

Now, when $G = H \rtimes_\phi K$ there are two injective homomorphisms

$$i_1 : H \longrightarrow G \quad h \mapsto i_1(h) = (h,1),$$

and

$$i_2 : K \longrightarrow G \quad h \mapsto i_2(k) = (1,k).$$

Set $H_1 = i_1(H)$ and $K_1 = i_2(K)$. Then,

1. $G = H_1 K_1$, since $(h,k) = (h,1)(1,k) \in H_1 K_1$.

2. $H_1 \cap K_1 = \{(1,1)\}$.

3. $H_1 \lhd G$ and this since

$$(h_1,k)(h,1)(h_1,k)^{-1} = (h_1,k)(h,1)\left(\phi_{k^{-1}}(h_1^{-1}), k^{-1}\right) = (\star,1),$$

where $\star$ is some element of H_1.

K_1 is not in general a normal subgroup of $G = H \rtimes_\phi K$. In fact:

Proposition 1.10.7. $K_1 \lhd G = H \rtimes_\phi K$ *if and only if ϕ is trivial. (In that case the semi-direct product is simply the direct product.)*

Proof. We will show that $\phi_k(h) = h$ for any $h \in H$ and $k \in K$. Indeed, since H_1, and by assumption K_1, are normal in G and $H_1 \cap K_1 = \{(1,1)\}$, we know $h_1 k_1 = k_1 h_1$ for all $h_1 \in H_1$ and $k_1 \in K_1$. Therefore

$$(h,k) = (h,1)(1,k) = (1,k)(h,1) = (\phi_k(h),k),$$

and so $\phi_k(h) = h$.

$\square$

Corollary 1.10.8. $G = H \rtimes_\phi K$ *is Abelian if and only if ϕ is trivial and H, K are Abelian.*

Theorem 1.10.9. *Let G be a group and H, K subgroups such that*

1. $G = HK$,

2. $H \cap K = \{1\}$,

3. $H \triangleleft G$.

Then, $\phi : K \longrightarrow \mathrm{Aut}(H)$ defined by $k \mapsto \phi_k(h) = khk^{-1}$ is a group homomorphism and $G \cong H \rtimes_\phi K$. In this case, we say that G is the internal semi-direct product of H and K.

Proof. In order to construct an isomorphism ψ between G and $H \rtimes_\phi K$ we define
$$\psi : H \rtimes_\phi K \longrightarrow G \quad : \quad \psi(h,k) = hk.$$
We check that ψ is a homomorphism. Indeed

$$
\begin{aligned}
\psi\big((h_1, k_1)(h_2, k_2)\big) &= \psi\big(h_1 \phi_{k_1}(h_2), k_1 k_2\big) \\
&= \psi\big(h_1 k_1 h_2 k_1^{-1}, k_1 k_2\big) \\
&= h_1 k_1 h_2 k_2 = \psi\big((h_1, k_1)\big)\psi\big((h_2, k_2)\big).
\end{aligned}
$$

To prove that ψ is onto, the assumption $G = HK$ implies that if $g \in G$, there are $h \in H$ and $k \in K$ such that $g = hk$, i.e. $g = \psi(h, k)$. To see that ψ is $1:1$, we show that $\mathrm{Ker}\,\psi = \{(1,1)\}$. Suppose $\psi(h, k) = 1$, then $hk = 1$ and $k = h^{-1} \in H \cap K = \{1\}$, i.e. $k = 1$. Similarly, we can prove that $h = 1$. Hence, $\psi(h, k) = 1$. $\square$

However, most can not be written as semi-direct products of smaller groups. For example, $G = Q_8$, the group of quaternions has only the subgroups $\langle i \rangle$, $\langle j \rangle$, $\langle k \rangle$ and $\langle -1 \rangle$, and each contains -1, which implies that there do not exist two non-trivial subgroups having trivial intersection. Other examples are a cyclic group or a simple group.

Exercise 1.10.10. Prove that if H is central in G then $H \rtimes_\phi K$ is a direct product.

An important example of a semi-direct product is the *affine group* of dimension n.

As defined above, this is the group of matrices,

$$G = \left\{ \begin{pmatrix} A & b \\ 0 & 1 \end{pmatrix} \quad : \quad A \in \mathrm{GL}(n,k),\ b \in k^n \right\}.$$

We will show that

$$G = k^n \rtimes_\varphi \mathrm{GL}(n,k),$$

where

$$\varphi : \mathrm{GL}(n,k) \longrightarrow \mathrm{Aut}(k^n) \quad : \quad A \mapsto \varphi(A) = \varphi_A \quad : \varphi_A(x) = Ax$$

with $x \in k^n$. Indeed, in the semi-direct product $k^n \rtimes_\varphi \mathrm{GL}(n,k)$, the product of the two elements (b, A), and (b_1, A_1) is:

$$(b, A) \cdot (b_1, A_1) = (b + \varphi_A(b_1)\,, AA_1) = (b + Ab_1\ AA_1).$$

On the other hand, if $(b, A) = \begin{pmatrix} A & b \\ 0 & 1 \end{pmatrix}$ and $(b_1, A_1) = \begin{pmatrix} A_1 & b_1 \\ 0 & 1 \end{pmatrix}$ then,

$$(b, A) \cdot (b_1, A_1) = \begin{pmatrix} AA_1 & Ab_1 + b \\ 0 & 1 \end{pmatrix} = (b + Ab_1\,, AA_1).$$

Exercise 1.10.11. Prove that the group $E(n, \mathbb{R})$ of *Euclidean motions* of $\mathbb{R}^n$, that is the group generated by translations, rotations and reflections through a hyperplane, is the semi-direct product $\mathbb{R}^n \rtimes_\phi O(n, \mathbb{R})$.

Exercise 1.10.12. Let k be a field. Prove that the 3-dimensional Heisenberg group given by

$$N = \left\{ \begin{pmatrix} 1 & x & z \\ 0 & 1 & y \\ 0 & 0 & 1 \end{pmatrix}, \quad \text{with}\quad x,\, y,\, z \in k \right\}$$

is the semi-direct product

$$N = K \rtimes_\phi H,$$

where K is the subgroup of N generated by the ys and H is the subgroup generated by (x, z)'s. This subgroup is normalized by K, and ϕ is the restriction to H of conjugation by the y's. Note that the subgroup generated by the z's is $\mathcal{Z}(N)$ (which equals $[N, N]$).

1.11   Exact Sequences of Groups

A finite or infinite sequence of groups and homomorphisms

$$\cdots \longrightarrow G \overset{\varphi}{\longrightarrow} K \overset{\chi}{\longrightarrow} L \longrightarrow \cdots$$

is called an *exact sequence*.

When there are only 3 non-trivial groups involved we call it a *short exact sequence*. In this case we have

$$1 \longrightarrow H \overset{i}{\longrightarrow} G \overset{\pi}{\longrightarrow} K \longrightarrow 1$$

where i is the inclusion map (injective) and π the projection map (surjective). Here H is a normal subgroup of G since $i(H) = \text{Ker}(\pi)$, and we shall say that G is *an extension of H by K*. The exactness also tells us that $K \cong G/H$. The map i is injective and the map π surjective.

An example of a short exact sequence is

$$1 \longrightarrow \text{SL}(n, \mathbb{R}) \overset{i}{\longrightarrow} \text{GL}(n, \mathbb{R}) \overset{\det}{\longrightarrow} \mathbb{R}^* \longrightarrow 1$$

where i is the inclusion map and det the determinant map.

Actually, a short exact sequence of groups is just another way of talking about a normal subgroup and the corresponding quotient group. However, even if H is normal in G, knowing H and G/H does not determine G. Here is an example:

$$0 \longrightarrow \mathbb{Z}/p\mathbb{Z} \to \mathbb{Z}/p\mathbb{Z} \times \mathbb{Z}/p\mathbb{Z} \longrightarrow \mathbb{Z}/p\mathbb{Z} \longrightarrow 0$$

$$0 \longrightarrow \mathbb{Z}/p\mathbb{Z} \longrightarrow \mathbb{Z}/p^2\mathbb{Z} \longrightarrow \mathbb{Z}/p\mathbb{Z} \longrightarrow 0$$

Here the first and third groups in each of the two exact sequences are the same, but the two groups in the middle, although they have the same order, are not isomorphic as we saw earlier.

If G and H are groups, their direct product gives rise to a short sequence as follows:

$$1 \longrightarrow H \xrightarrow{\ \eta\ } H \times G \xrightarrow{\ \zeta\ } G \longrightarrow 1$$

where η is the map $h \mapsto (h, 1)$ and ζ the map $(h, g) \mapsto g$.

More generally semi-direct products $H \rtimes_{\varphi} G$ where $\varphi : G \to \mathrm{Aut}(H)$ do the same. Here we have the short exact sequence

$$1 \longrightarrow H \xrightarrow{\ \eta\ } H \rtimes_{\varphi} G \xrightarrow{\ \zeta\ } G \longrightarrow 1$$

where the two maps η and ζ are as in the previous example.

Proposition 1.11.1. *For an arbitrary short exact sequence*

$$A \xrightarrow{\ \varphi\ } B \xrightarrow{\ \chi\ } C \xrightarrow{\ \psi\ } D$$

the following are equivalent:

1. φ *is surjective.*

2. χ *is trivial.*

3. ψ *is injective.*

Proof. Since φ is onto, $\mathrm{Im}\,\varphi = \mathrm{Ker}\,\chi$ implies χ is trivial. Therefore $\mathrm{Ker}\,\psi = \{1\}$, i.e. ψ is $1:1$. All these implications are reversible. $\square$

Corollary 1.11.2. *In the exact sequence*

$$A \xrightarrow{\ \varphi\ } B \xrightarrow{\ \chi\ } C \xrightarrow{\ \psi\ } D \xrightarrow{\ \tau\ } E$$

$C = \{1\}$ *if and only if φ is onto and τ is $1:1$.*

We leave the proof of this as an exercise.

Definition 1.11.3. We say that the following short exact sequence

$$\{1\} \longrightarrow N \xrightarrow{\varphi} G \xrightarrow{\chi} H \longrightarrow \{1\}$$

splits if there exists a homomorphism $\psi : H \longrightarrow G$ (called *a section or a cross section*) satisfying,

$$\chi \circ \psi = Id_H.$$

In this case, we write:

$$1 \longrightarrow N \xrightarrow{\varphi} G \overset{\chi}{\underset{\psi}{\rightleftharpoons}} H \longrightarrow 1.$$

One checks easily that the short exact sequence given by the semi-direct product just above splits. An example of one that does not split is the following:

$$\mathbb{Z} \xrightarrow{\times n} \mathbb{Z} \xrightarrow{\pi} \mathbb{Z}/n\mathbb{Z}$$

does not split for any $n \neq -1,\, 0,\, 1$.

Indeed, any homomorphism $\mathbb{Z}/n\mathbb{Z} \to \mathbb{Z}$ must send any element of order n to some element of finite order. But in $\mathbb{Z}$ the only element of finite order is 0, so there is no such a 1 : 1 homomorphism.

Definition 1.11.4. Two short exact sequences

$$1 \longrightarrow N_i \xrightarrow{\varphi_i} G_i \xrightarrow{\chi_i} H_i \longrightarrow 1, \qquad i = 1, 2$$

are *equivalent* if there exist isomorphisms,

$$f_1 : N_1 \longrightarrow N_2, \quad f_2 : G_1 \longrightarrow G_2, \quad f_3 : H_1 \longrightarrow H_2,$$

such that the following diagram commutes

$$
\begin{array}{ccccccccc}
1 & \longrightarrow & N_1 & \xrightarrow{\varphi_1} & G_1 & \xrightarrow{\chi_1} & H_1 & \longrightarrow & 1 \\
& & \downarrow{\scriptstyle f_1} & & \downarrow{\scriptstyle f_2} & & \downarrow{\scriptstyle f_3} & & \\
1 & \longrightarrow & N_2 & \xrightarrow[\varphi_2]{} & G_2 & \xrightarrow[\chi_2]{} & H_2 & \longrightarrow & 1
\end{array}
$$

The following theorem is another characterization semi-direct products:

Theorem 1.11.5. *The short exact sequence of groups,*

$$1 \longrightarrow H \xrightarrow{\ i\ } G \xrightarrow{\ j\ } K \longrightarrow 1$$

splits if and only if $G = H \rtimes_\phi K$, *for some* $\phi : K \to \mathrm{Aut}(H)$.

Proof. Given a section $s : K \longrightarrow G$, we define $\phi : K \to \mathrm{Aut}(H)$ by setting $\phi_k(h) = s(k)hs(k)^{-1}$. Then the isomorphism is

$$G \longrightarrow H \rtimes_\phi K \ : g \mapsto \Big(gs(j(g)^{-1}), j(g)\Big)$$

with inverse $(h, k) \mapsto hs(k)$.

Conversely, given an isomorphism $\psi : G \longrightarrow H \rtimes_\phi K$, we define $s : K \longrightarrow G$ by $k \mapsto \psi^{-1}(1, k)$. $\square$

Exercise 1.11.6. Does this short exact sequence split?

$$1 \longrightarrow \mathrm{SL}(n, \mathbb{R}) \xrightarrow{\ i\ } \mathrm{GL}(n, \mathbb{R}) \xrightarrow{\ \det\ } \mathbb{R}^* \longrightarrow 1$$

1.12   Direct and Inverse Limits of Groups

Let $(I, \preceq)$ be a partially ordered set (see 0.4) with the property that for any two elements i, j there is some $k \in I$ such that $i \preceq k$, $j \preceq k$.

Definition 1.12.1. A *directed system* of sets (groups, rings, modules, topological spaces, etc.) indexed by I is a collection of sets (respectively groups, rings, modules, topological spaces, etc.) X_i, $i \in I$, and maps (respectively homomorphisms, continuous maps, etc.) $f_{ij} : X_i \longrightarrow X_j$, defined whenever $i \preceq j$, satisfies the following compatibility conditions:

1. $f_{ik} = f_{jk} \circ f_{ij}$, $i \preceq j \preceq k$, and

2. $f_{ii} = \mathrm{Id}_{X_i}$.

Example 1.12.2. Let X be a set and $(X_i)_{i \in I}$ be a family of subsets. Consider the partial order

$$i \leq j \quad \text{when } X_i \subset X_j,$$

and $f_{ij} : X_i \hookrightarrow X_j$ as the inclusion maps. Then, (X_i, f_{ij}) is a direct system.

Definition 1.12.3. A *direct* (or *inductive*) *limit*[11] [12] of the directed system of groups G_i, is defined by giving a group G and a homomorphism

$$\phi_i : G_i \longrightarrow G$$

for any $i \in I$ such that

1. if $i < j$ then $\phi_i = \phi_j \circ \phi_{ij}$,

2. if (G', ϕ'_i) is another group and family of homomorphisms with the same property then there exists a unique homomorphism

$$\phi : G \to G'$$

   such that $\phi_i = \phi \circ \phi'_i$ for any $i \in I$.

It follows that a direct limit, if it exists, is unique up to a unique isomorphism. We may therefore speak of *the direct limit* of the directed system G_i. We denote this limit (if it exists) by

$$G = \varinjlim_{i \in I} G_i.$$

Remark 1.12.4. The limit depends on the category in which it is taken. For example, the underlying set of a direct limit of groups is not in general the direct limit of the underlying sets. Also, the direct limit of Abelian groups in the category of Abelian groups is different from the direct limit in the category of all groups.

---

[11]We first note that what topologists call the *limit*, algebraists call the *inverse limit* and denote it by $\lim_{\leftarrow}$. Likewise, what topologists call the *colimit*, algebraists call the *direct limit* and denote $\lim_{\rightarrow}$.

[12]Inverse and direct limits were first studied as such in the 1930s, in connection with topological concepts as Čech cohomology. The general concept of the limit was introduced in 1958 by D. M. Kan [64].

The direct limit can be constructed as the quotient of the subgroup $\bigoplus_i G_i$ of $\prod_i G_i$ as follows:

$$\varinjlim_{i \in I} G_i = \bigoplus_i G_i / \sim$$

where, $x_i \in G_i$ is equivalent to $x_j \in G_j$ if and only if

$$\phi_{ik}(x_i) = \phi_{jk}(x_j)$$

for some $k \in I$, with the additional structure that the product of $x_i \in G_i$, $x_j \in G_j$ is defined as the product of their images in G_k.

Definition 1.12.5. An *inverse system* of groups (respectively topological groups, rings, modules, topological spaces, etc.) is a family G_i, $i \in I$, of groups, together with a family of homomorphisms $f_{ij} : G_i \to G_j$, $i \succeq j$ such that:

1. $f_{ii} = 1_{G_i}$.

2. Whenever $i \succeq j \succeq k$, then $f_{ik} = f_{jk} \circ f_{ij}$.

Definition 1.12.6. The *inverse* (or *projective*) *limit* of the inverse system of groups G_i, is defined by giving a group G and a homomorphism

$$\phi_i : G \longrightarrow G_i$$

for each $i \in I$ satisfying

1. If $i \succeq j$, then $\phi_j = f_{ij} \circ \phi_i$, i.e. the homomorphisms ϕ_i are compatible with the inverse system.

2. Given another group H, and homomorphisms $\psi_i : G_i \to H$ compatible with the inverse system, then there is a unique homomorphism $\psi : H \to G$ such that $\psi_i = \phi_i \circ \psi$ for any $i \in I$.

We write

$$G = \varprojlim_{i \in I} G_i.$$

In the categories of sets, groups, rings etc., inverse limits always exist and admit the following description:

$$\varprojlim_{i \in I} G_i = \{(g_i) \in \prod_{i \in I} G_i \; : \; f_{ji}(g_j) = g_i \text{ for all } i \preceq j\}.$$

Now, we give some important examples of direct and inverse limits.

Example 1.12.7. Consider the following family of groups and homomorphisms:

$$\cdots \xrightarrow{\phi_4} G_3 \xrightarrow{\phi_3} G_2 \xrightarrow{\phi_2} G_1 \xrightarrow{\phi_1} G_0.$$

Then, the inverse limit of this family is

$$\varprojlim_{i \in I} G_i = \{(g_0, g_1, \cdots) \in \prod_{i=0}^{\infty} G_i \mid \phi_i(g_i) = g_{i-1}\}.$$

An important application is the following inverse system:

$$\cdots \xrightarrow{\phi_4} \mathbb{Z}/m^3\mathbb{Z} \xrightarrow{\phi_3} \mathbb{Z}/m^2\mathbb{Z}. \xrightarrow{\phi_2} \mathbb{Z}/m\mathbb{Z}.$$

Taking $m = 10$, the elements of the inverse limit will be sequences $(n_1, n_2, \ldots)$, where each n_i is an i digit decimal whose last $i - 1$ digits give n_{i-1}, as for example $(3, 13, 013, 7013, 57013...)$. This can be viewed as an infinite decimal going to the left:

$$...57013.$$

As we shall see in Chapter 10 this is related to the p-adic numbers.

Example 1.12.8. Let G be a group and let $(N_i)_{i \in I}$ be a family of normal subgroups of G which have finite index in G. Define a partial order in I by

$$i \leq j \quad \text{if and only if } N_i \subset N_j.$$

Also, define whenever $i \leq j$, the group homomorphism

$$f_{ij} : G/N_j \longrightarrow G/N_i \quad \text{such that } f_{ij}(gN_j) = gN_i.$$

Then, one sees that $(G/N_i, f_{ij})$ is an inverse system in the category of finite groups. The inverse limit

$$\varprojlim_{i \in I} G/N_i$$

is called the *profinite completion* of G.

Example 1.12.9. *Double arrow.* Let G and H be groups, and let f_1, f_2 be homomorphisms as in the following diagram

$$
\begin{array}{c}
G \\
\downarrow {\scriptstyle f_1} \\
G \xrightarrow[\;f_2\;]{} H
\end{array}
$$

The limit L has the property that there are maps $p_1 : L \to G$, and $p_2 : L \to H$ such that the following diagram commutes

$$
\begin{array}{ccc}
L & \xrightarrow{\;p_1\;} & G \\
{\scriptstyle p_1} \downarrow & {\scriptstyle p_2} \searrow & \downarrow {\scriptstyle f_1} \\
G & \xrightarrow[\;f_2\;]{} & H
\end{array}
$$

This implies that $f_1 \circ p_1 = f_2 \circ p_1$. In other words, the map $p_1 : L \to G$ has as image the subgroup

$$K = \{g \in G \mid f_1(g) = f_2(g)\}.$$

Since K sits in G, the map p_1 is unique. K is called the *equalizer* of the two arrows f_1 and f_2.

Example 1.12.10. *Pull-back.* The standard construction of the *pull-back* is the limit of the following diagram:

$$
\begin{array}{c}
G \\
\downarrow {\scriptstyle f_1} \\
H \xrightarrow[\;f_2\;]{} K
\end{array}
$$

Definition 1.12.11. Let $f_1 : G \to K$, and $f_2 : H \to K$ be two group homomorphisms. Then, the *pull-back* P is defined as the subgroup of the product $G \times H$ given by

$$P = \{(g, h) \in G \times H \mid f_1(g) = f_2(h)\},$$

together with the projection homomorphisms $p_1 : P \to G$, and $p_2 : P \to H$.

Proposition 1.12.12. *The pull-back is the limit of diagram 1.12.10 above.*

Proof. Suppose that there is another group N with homomorphisms $n_1 : N \to G$, and $n_2 : N \to H$ making the following diagram commute.

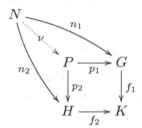

Then, $f_1\big(n_1(x)\big) = f_2\big(n_2(x)\big)$ for every $x \in N$. This is the same as saying that $\nu(x) = \big(p_1(x), p_2(x)\big) \in P$ which means that ν exists and is unique. $\square$

Example 1.12.13. We consider the category of Abelian groups, and let (A_i), (f_{ij}), $i, j \in I$ be a direct system of Abelian groups. Then,

$$\varinjlim_{i \in I} A_i = \bigoplus_{i \in I} A_i / H$$

where H is the subgroup of $\bigoplus A_i$ generated by the set

$$\{f_{ij}(x_i) - x_i \quad \text{where } i \preceq j, \ x_i \in A_i\}.$$

Example 1.12.14. Let $(X_i)_{i \in I}$ be a family of subsets of a set X and let (X_i, f_{ij}) be the direct system as in Example 1.12.2. Then,

$$\varinjlim_{i \in I} X_i = \bigcup_{i \in I} X_i.$$

Proof. Let

$$\mathcal{U} = \bigcup_{i \in I} X_i$$

and let $f_i : X_i \hookrightarrow \mathcal{U}$ be the inclusion map for every $i \in I$. If $x \in X_i$ and $i \leq j$, then $f_j \circ f_{ij}(x) = f_j(x) = x = f_i(x)$. Hence, $f_j \circ f_{ij} = f_i$.

Now, suppose that there is a set Y and maps $g_i : X_i \longrightarrow Y$ such that $g_j \circ f_{ij} = g_i$ for all $i \leq j$. We must find a map $f : \mathcal{U} \longrightarrow Y$ such that $f \circ f_i = g_i$ for all $i \in I$. Assume we have it. Then, if $u \in \mathcal{U}$ and $u \in X_i$ for some i, this implies $f(u) = f \circ f_i(u) = g_i(u)$. If, u is also in some X_j, then we can choose a $k \in I$ such that $i \leq k$ and $j \leq k$ and we get

$$g_i(u) = g_k \circ f_{ik}(u) = g_k(u) \quad \text{and} \quad g_j \circ f_{jk}(u) = g_k(u).$$

Therefore, $g_i(u) = g_j(u)$, which means that our map f must be defined by setting $f(u) = g_i(u)$ whenever $u \in X_i$. We also proved that f is well defined, and also that $f \circ f_i = g_i$ for each $i \in I$. $\qquad\qquad\square$

An important example of an inverse limit is the following:

Example 1.12.15. Let $I = \mathbb{N}$, with order given by divisibility, i.e.

$$n \preceq m \quad \text{if and only if } n \mid m$$

let $X_n = \mathbb{Z}/n\mathbb{Z}$, and $f_{n,m}$ for $n \mid m$ be the reduction modulo m. Then the inverse limit

$$\overline{\mathbb{Z}} := \varprojlim_{n \in \mathbb{N}} \mathbb{Z}/n\mathbb{Z}$$

is the *Prüfer group*.

An important example of a direct limit is the group $\mathbb{Z}(p^\infty)$ which is usually defined as,

$$\mathbb{Z}(p^\infty) = \varinjlim_{k \in \mathbb{N}} \mathbb{Z}/(p^k).$$

Here the maps $f_{k,k+1} : \mathbb{Z}/(p^k) \to \mathbb{Z}/(p^{k+1})$ are multiplication by p. This is the quotient of the direct sum $\oplus \mathbb{Z}/(p^k)$ by the equivalence relation

$$\mathbb{Z}/(p^k) \ni \sim p^j n \in \mathbb{Z}/(p^{j+k}).$$

Another way to say this is, $\mathbb{Z}(p^\infty)$ is generated by elements x_0, x_1, x_2, x_3,... which are related by

$$x_k = p x_{k+1} \quad \text{and} \quad x_0 = 0.$$

Since x_k has order p^k it is divisible by any number relatively prime to p and is divisible by any power of p since $x_k = p^j x_{j+k}$.

Yet another description of $\mathbb{Z}(p^\infty)$ is the multiplicative group of all p-power roots of unity. These are complex numbers of the form $e^{2\pi i n/p^k}$ where n is an integer. An isomorphism φ with the additive version above is given by

$$\varphi(x_k) = e^{2\pi i/p^k}.$$

1.13   Free Groups

For completeness here we define the concept of a *free group*. As we shall see in Chapter 3, every vector space admits special generating sets: namely those generating sets that are as free as possible (meaning having as few linear relations among them as possible), i.e. the linearly independent ones. Also, in the setting of group theory, we can formulate what it means to be a free generating set. However, as we shall see, most groups do not admit free generating sets. This is one of the reasons why group theory is much more complicated than linear algebra.

Definition 1.13.1. (Universal Property of Free Groups) Let S be a set. We say that the group F is *freely generated* by S if F has the following universal property: For any group G and any map $\phi : S \to G$ there is a unique group homomorphism $\psi : F \to G$ such that the following diagram (where i is the inclusion map) commutes.

We say that the group F is *free* if it contains a free generating set.

The term *universal property* means that objects having this property are unique (up to an isomorphism). As we shall see, for every set S there exists a freely generated group (generated by S) as follows.

Proposition 1.13.2. (Uniqueness of the Free Groups) *Let S be a set. Then, up to canonical isomorphism, there is at most one group freely generated by S.*

For any set S, we will define a group $F(S)$, known as the free group generated by S. Informally, one should view S as an *alphabet*, and elements of $F(S)$ as *words* in this alphabet. So, for example, if $S = \{a, b\}$, then the words, ab and ba are elements of $F(S)$. The group operation is *concatenation*, in other words, to compose two words in $F(S)$, we simply write one and then follow it by the other. For example, the product of ab and ba is $abba$. This description is slightly oversimplified because it does not take account of the fact that since groups have inverses there is the possibility of cancelation. The words aa^{-1} and 1 represent the same element of the $F(S)$. In this way, elements of $F(S)$ are not words in the alphabet S, but rather are equivalence classes of words.

We are now ready to give the formal definitions. Throughout, S is some set, known as the *alphabet*. From this set, create a new set S^{-1}. This is a copy of the set S, but for each element x of S, we denote the corresponding element of S^{-1} by x^{-1} and require $S \cap S^{-1} = \varnothing$, $x^{-1} \in S^{-1}$, and $(x^{-1})^{-1} = x$.

Definition 1.13.3. A *word* w is a finite sequence $x_1, ..., x_m$, where $m = 0, 1, 2, ...$ and each $x_i \in S \cup S^{-1}$. We write w as $x_1 x_2 ... x_m$. Note that the empty sequence, where $m = 0$, is allowed as a word. We denote it by $\varnothing$.

Definition 1.13.4. The *concatenation* of two words $x_1 x_2 \ldots x_m$ and $y_1 y_2 \ldots y_n$ is the word $x_1 x_2 \ldots x_m y_1 y_2 \ldots y_n$.

Definition 1.13.5. A word w_0 is an *elementary contraction* of a word w, if $w = y_1 x x^{-1} y_2$ and $w_0 = y_1 y_2$, for words y_1 and y_2, and some $x \in S \cup S^{-1}$.

Definition 1.13.6. Two words w_0 and w are *equivalent*, written $w \sim w_0$, if there are words $w_1 \ldots w_n$, where $w = w_1$ and $w_0 = w_n$, and for each i, w_i is an elementary contraction of the word w_{i+1}. The equivalence class of a word w is denoted $[w]$.

Definition 1.13.7. The free group $F(S)$ generated by the set S consists of equivalence classes of words in the alphabet S. The composition of two elements $[w]$ and $[w_0]$ is the class $[w w_0]$. The identity element is $[\varnothing]$, and is denoted e. The inverse of an element $[x_1 x_2 \cdots x_n]$ is $[x_n^{-1} \cdots x_2^{-1} x_1^{-1}]$.

One checks easily that composition is well defined. That is, if $w_1 \sim \bar{w}_1$ and $w_2 \sim \bar{w}_2$, then $w_1 w_2 \sim \bar{w}_1 \bar{w}_2$. Evidently $F(S)$ is a group.

Remark 1.13.8. If $|S| = \{a\}$, then $F(S) \cong \mathbb{Z}$ via the isomorphism $n \mapsto a^n$. However, when $S = \{a, b\}$, then F_2 is the free group on 2 generators and things become much more complicated. For example, the derived subgroup of F_2 is isomorphic to the free group of countable rank.

Another interesting way of understanding the free group F_2 is this: Let $\Gamma(2)$ be the group of integer matrices

$$\begin{pmatrix} a & b \\ c & d \end{pmatrix}$$

where $a, d = 1 \pmod 2$, $b, c = 0 \pmod 2$, and T be the index 2 subgroup of $\Gamma(2)$ generated by the matrices, where $a = d = -1$ and $b = c = 0$. Then, $F_2 = \Gamma(2)/T$.

Proposition 1.13.9. *Every group G is a quotient of a free group.*

Proof. Consider a set $X = \{x_g, \ g \in G\}$ such that the map $\xi : X \to G$ defined by $\xi(x_g) = g$ is bijective. If $F(X)$ is the free group generated by X, there is a homomorphism $\phi : F(X) \to G$ extending ξ which must also be surjective, since ξ is. Therefore, by the first isomorphism theorem,

$$G \cong F(X)/\operatorname{Ker}(\phi).$$

$\square$

1.14   Some Features of Abelian Groups

In this section A will always stand for an Abelian group (written additively) although we may sometimes also use G.

The most important result in Abelian groups is the *Fundamental Theorem of Abelian Groups* which states that a finitely generated Abelian group is a direct sum of a finite number of cyclic groups (with certain additional features) (see Corollary 6.4.5). This obviously applies as well to finite Abelian groups, but is not valid if the group is not finitely generated, or if it is not Abelian. This result is of great importance in both algebra and topology and a number of other areas of mathematics. For technical reasons we defer the proof to Chapter 6.

1.14.1   Torsion Groups

As we have seen, the elements of A of finite order form a subgroup.

Definition 1.14.1. The *torsion subgroup* of A is defined as

$$\operatorname{Tor}(A) = \{a \in A \ \mid \ na = 0 \ \text{ for some } \ n \in \mathbb{N}\}.$$

If $A = \operatorname{Tor}(A)$, then we say that A is a *torsion group*. If $\operatorname{Tor}(A) = \{0\}$ we say A is *torsion-free*.

Proposition 1.14.2. *Let A be an Abelian group. Then $\operatorname{Tor}(A)$ is a subgroup of A and $A/\operatorname{Tor}(A)$ is torsion-free.*

Proof. We have already established that $\text{Tor}(A)$ is a subgroup of A. To see that $A/\text{Tor}(A)$ is torsion-free, suppose $a + \text{Tor}(A)$ that has finite order n. Then $n(a + \text{Tor}(A)) = \text{Tor}(A)$. Hence $na \in \text{Tor}(A)$, so there is an $m \in \mathbb{N}$ so that $m(na) = (mn)a = 0$. Hence $a \in \text{Tor}(A)$ and so $a + \text{Tor}(A) = \bar{0}$ in $A/\text{Tor}(A)$. $\qquad\square$

Definition 1.14.3. Let G be a torsion group and p be a prime. G is called a *p primary* group if every element of G has order some power of p. The p-primary components of G are denoted by $\text{Tor}_p(G)$.

Theorem 1.14.4. *A torsion group G is a direct sum of its primary subgroups. These summands are all uniquely determined.*

Proof. As defined,

$$\text{Tor}_p(G) = \{g \in G \ : \ p^r g = 0 \ \text{ for some } \ r\}.$$

Since $0 \in \text{Tor}_p(G)$, $\text{Tor}_p(G) \neq \varnothing$ for any p. Each $\text{Tor}_p(G)$ is a subgroup of G. Indeed, if a and b are in $\text{Tor}_p(G)$ then, $p^n a = p^m b = 0$ for some n, m in $\mathbb{N}$. Hence

$$p^{\max(n,m)}(a - b) = 0,$$

and therefore $a - b$ is in $\text{Tor}_p(G)$.
    We claim that
$$G = \bigoplus_{p \text{ prime}} \text{Tor}_p(G).$$

To prove this, we have to show that the collection of all $\text{Tor}_p(G)$'s generate G and, that the various $\text{Tor}_p(G)$ have only (0) in common.
    Let $g \in G$ and let n be its order, i.e. $ng = 0$. By the Fundamental Theorem of Arithmetic (see 0.5.12)

$$n = p_1^{n_1} p_2^{n_2} \cdots p_k^{n_k},$$

where all the p_i are prime. Now, set

$$n_i = \frac{n}{p_i^{n_i}}.$$

These n_i are relatively prime, i.e. $(n_1, ..., n_k) = 1$. Hence by Corollary 0.5.6 there are integers $m_1, ..., m_k$ such that

$$m_1 n_1 + m_2 n_2 + \cdots + m_k n_k = 1.$$

Therefore,

$$g = m_1 n_1 g + \cdots + m_k n_k g. \tag{1}$$

Notice that $p_i^{n_i} m_i n_i g = m_i n g = 0$, which means that $m_i n_i g \in G_{p_i}$ for each i.

Now, any element in $\text{Tor}_{p_1}(G) + \text{Tor}_{p_2}(G) + \cdots + \text{Tor}_{p_k}(G)$ is annihilated by a product of powers of $p_1, ..., p_k$. Hence

$$\text{Tor}_p(G) \cap [\text{Tor}_{p_1}(G) + \cdots + \text{Tor}_{p_k}(G)] = \{0\},$$

for $p \neq p_1, ..., p_k$. To see that this forces the decomposition to be unique suppose

$$g = d_1 + \cdots + d_k = e_1 + \cdots e_k,$$

with d_i, e_i lying in the same $\text{Tor}_{p_i}(G)$. Consider the equation

$$d_1 - e_1 = (d_2 + \cdots + d_k) - (e_2 + \cdots + e_k) \tag{2}$$

Then $d_1 - e_1$ has order a power of p_1. On the other hand the right side of (2) is an element whose power is a product of the powers of $p_2, ..., p_k$, which is possible only if $d_1 - e_1 = 0$. Hence, $d_1 = e_1$. Continuing in this way we see $d_i = e_i$ for all i. $\square$

What really lies behind this theorem is the following corollary.

Corollary 1.14.5. *If the integer n has prime factorization $p_1^{e_1} \cdots p_k^{e_k}$, then $\mathbb{Z}_n$ is the direct sum of the cyclic groups,*

$$\mathbb{Z}_n = \mathbb{Z}_{p_1^{e_1}} + \cdots + \mathbb{Z}_{p_k^{e_k}}.$$

Here is a more robust example of an Abelian torsion group. We leave the details to be checked by the reader.

Example 1.14.6. The group $\mathbb{Q}/\mathbb{Z}$ is a torsion group. Its p-primary decomposition is

$$\bigoplus_p \mathbb{Z}(p^\infty).$$

Because of the Kronecker approximation theorem B.3, $\mathbb{Q}/\mathbb{Z}$ is easily seen to be the torsion subgroup of $\mathbb{R}/\mathbb{Z}$.

The reader might like to check the additive group $\mathbb{R}/\mathbb{Z}$ is isomorphic to the (multiplicative) circle group of complex numbers z, with $|z| = 1$. This is true both algebraically and topologically. The reason for this is that the first isomorphism theorem also works for continuous homomorphisms. The map $t \mapsto e^{2\pi i t}$ is a continuous surjective homomorphism from $\mathbb{R}$ to $\mathbb{T}$ whose kernel is $\mathbb{Z}$. The induced map is continuous and injective and hence a homeomorphism because the domain is compact.

Definition 1.14.7. When a torsion group is finite we call a p primary component an *elementary p-group*.

The following result requires knowledge of Chapters 3 and 5. The method of dealing with questions concerning p groups by turning to vector spaces over $\mathbb{Z}_p$ is very typical and effective. Notice that 1.14.8 fails if G is an infinite p group. In Chapter 6 we will get both a stronger and more general result by a different method.

Theorem 1.14.8. *An elementary p-group is the direct sum of cyclic groups of order p^n for various n. Therefore using the primary decomposition, any finite Abelian group is the direct sum of cyclic groups of order p^n for various p and n.*

Proof. To prove this, we will show that an elementary p-group G is in a natural way, a vector space over the finite field $\mathbb{Z}_p$ (in Chapter 5 we shall prove $\mathbb{Z}_p$ is a field). Indeed, $pg = 0$ for any $g \in G$, which shows that for $m, n \in \mathbb{Z}_p$, we have $mg = ng$ if and only if $n \equiv m \pmod{p}$. In other words $[n] = [m]$ in $\mathbb{Z}_p$. We leave to the reader to check the vector space axioms. Therefore, G, as a vector space over the field $\mathbb{Z}_p$, has a basis, say $(g_i)_{i \in I}$. Hence, $G = \bigoplus_{i \in I} \langle g_i \rangle$. $\square$

1.14.2   Divisible and Injective Groups

Divisible groups are a generalization of infinite Abelian groups such as $\mathbb{Q}$ and $\mathbb{R}$.

Definition 1.14.9. An Abelian group G (written additively) is called a *divisible group* if for every $g \in G$ and each integer $n \neq 0$ there is an $h \in G$ so that $g = nh$.

We note that a non-trivial divisible group G must be infinite. For if $|G|$ were finite, then as we know, each element has finite order. Let $g \neq 0 \in G$, $n > 1$ be its order. Then there would be an $h \in G$ with $g = nh$, and so $h \neq g$ and $n^2 h = 0$. Continuing in this way we would get an infinite sequence of distinct hs in G, a contradiction. Obviously, $\mathbb{Z}$ is not a divisible group, since so no cyclic group (finite or infinite) can be divisible.

Example 1.14.10. The rational numbers $\mathbb{Q}$ form a divisible group under addition. More generally, the underlying additive group of any vector space over $\mathbb{Q}$ is divisible. In particular, $\mathbb{R}$ is divisible. Here we use the fact (dealt with in Chapter 11) that if K is a field and k a subfield then K is a vector space over k.

A subgroup of a divisible group need not be divisible. For example $\mathbb{Q}$ is divisible but $\mathbb{Z}$ is not. But, as we shall see, divisibility is a hereditary property under the quotients and the direct sums. This leads us to the following proposition.

Proposition 1.14.11. *If G is a divisible group and H a subgroup, then G/H is divisible.*

Proof. Take a coset $g + H$ in G/H, and let $n \neq 0$ be a positive integer. Since G is divisible, there is a $h \in G$ such that $g = nh$. Therefore $n(h + H) = nh + H = g + H$. $\square$

In particular, $\mathbb{Q}/\mathbb{Z}$ and $\mathbb{R}/\mathbb{Z}$ are divisible groups.

Corollary 1.14.12. *If the divisible group G is a direct sum, then each summand is also divisible.*

We now characterize the Abelian divisible groups by a universal property:

Definition 1.14.13. We say that the Abelian group G is *injective*, if for any Abelian groups A and B, and any homomorphism $\varphi : A \longrightarrow G$ and monomorphism $\iota : A \longrightarrow B$, there is a homomorphism $\psi : B \longrightarrow G$ such that the following diagram is commutative:

$$
\begin{array}{ccc}
A & \xrightarrow{\;i\;} & B \\
 & \varphi \searrow & \downarrow \psi \\
 & & G
\end{array}
$$

Theorem 1.14.14. *Let G be an Abelian group. The following two statements are equivalent:*

1. G is divisible.

2. G is injective.

Proof. $1 \implies 2$ Let G be divisible. Consider two Abelian groups A and B, $i : A \hookrightarrow B$ a 1:1 homomorphism, and let $\varphi : A \longrightarrow G$ be a homomorphism. Since i is a 1:1 homomorphism, there is no loss of generality if we replace the group A by its image, $\mathrm{Im}(i)$, in B, i.e. we consider A as a subgroup of B and as i we take the inclusion map. What we want is to extend the map φ to B.

Let Ξ be the set of all extensions $\xi : H \longrightarrow G$ of φ. By that we mean $A \subseteq H \subseteq B$ and ξ is a homomorphism such that $\xi(a) = \varphi(a)$ for all $a \in A$. Now we can order Ξ by setting

$$\xi \preceq \xi' \quad \text{if} \quad \xi' \text{ is an extension of } \xi,$$

where $\xi : H \longrightarrow G$, $\xi' : H' \longrightarrow G$ and $H \subseteq H'$. To apply Zorn's lemma to the Ξ, consider the chain

$$\{\xi_i \mid \xi_i : H_i \longrightarrow G, \ i \in I\}$$

and put

$$H = \bigcup_{i \in I} H_i.$$

Now, let $\sigma : H \longrightarrow G$ be defined by $\sigma(x) = \xi_i(x)$ for every $x \in H_i$. The map σ is well defined since ξ_i and ξ_j agree where both are defined. Then $\sigma \in \Xi$ and σ is an upper bound for the chain $(\xi_i)_{i \in I}$. By Zorn's lemma, let $\gamma : H \longrightarrow G$ be a maximal element of Ξ. If $H = B$ we would be done.

If $H \neq B$, there must exist an element $x \in B - H$. If we can prove that γ can be extended to $H+ < x >:= F$, this would contradict the fact that H is maximal, and therefore complete the proof.

There are two possibilities: either $H \cap < x >= 0$ and therefore $F = H \oplus < x >$. In this case we extend γ to $\gamma_1 : F \longrightarrow G$ by setting $\gamma_1(x) = 0$, or $H \cap < x > \neq 0$ in which case let n be the smallest positive integer such that $nx \in H$. Suppose that $\gamma(nx) := g$, $g \in G$. Since, by assumption, G is divisible, there must exist a $g_1 \in G$ with $g = ng_1$. Each element of F can be written in a unique way as $h + mx$, where $h \in H$ and $0 \leq m < n$. Because of the minimality of n we can define a map $\gamma_1 : F \longrightarrow G$ so that $\gamma_1(h + mx) = \gamma(h) + mg_1$. One checks easily that γ_1 is a homomorphism.

$2 \Longrightarrow 1$ To prove that G is divisible for any $g \in G$ and $n \in \mathbb{N}$, we must find an $h \in G$ with $n \cdot h = g$. Accordingly, choose a $g \in G$ and a $n \in \mathbb{Z}$ and consider the homomorphism $\varphi : \mathbb{Z} \longrightarrow G$, $m \mapsto \varphi(m) = mg$. Since G is injective, and $\mathbb{Z} \hookrightarrow \frac{1}{n}\mathbb{Z}$, ϕ must extend to a homomorphism $\tilde{\varphi} : \frac{1}{n}\mathbb{Z} \longrightarrow G$. But then $\varphi(1) = g = \tilde{\varphi}(1) = \tilde{\varphi}(\frac{n}{n}) = n\tilde{\varphi}(\frac{1}{n})$. Therefore, $h = \tilde{\varphi}(\frac{1}{n})$. $\qquad \square$

Corollary 1.14.15. *A divisible subgroup of an Abelian group G is always a direct summand of any Abelian group containing it.*

Proof. Let H be a subgroup of G. We shall show that

$$G = H \oplus F$$

for some subgroup F of G. Let $i : H \hookrightarrow G$ be the inclusion map. Consider the identity map $\mathrm{Id}_H : H \longrightarrow H$. Since H is divisible, according to 1.14.14 Id_H has an extension to G, i.e. there is a homomorphism $\psi : G \longrightarrow H$ such that $\psi \circ i = \mathrm{Id}_H$. Now, for any $g \in G$, $\psi(g) \in H$ and therefore $\psi^2(g) = \psi(g)$. This implies $\psi(g - \psi(g)) = 0$. In other words

$g - \psi(g)$ is in $\mathrm{Ker}(\psi) := F$. Hence $G = H + F$. Moreover, if $h \in H \cap F$, then $h = \psi(h) = 0$, hence $G = H \oplus F$.   □

We will now prove every Abelian group G is an extension of a divisible group by a group having no divisible subgroups. To do this, we need the following definition:

Definition 1.14.16. If G is an Abelian group we denote by $D(G)$ the subgroup of G generated by all the divisible subgroups of G.

Obviously, $D(G)$ contains each divisible subgroup of G.

Proposition 1.14.17. $D(G)$ *is itself a divisible subgroup of* G.

Proof. Let $n > 0$ and let $g \in D(G)$. Then, $g = g_1 + g_2 + \cdots + g_i$, where $g_k \in G_k$, and G_k is a divisible subgroup of G for each $k = 1, ..., i$. Hence, for each k, there exists an $h_k \in G_k$ with $nh_k = g_k$. Therefore $g = n(h_1 + h_2 + \cdots + h_i)$, where $h_1 + h_2 + \cdots + h_i \in D(G)$. Thus $D(G)$ is divisible.   □

Definition 1.14.18. If $D(G) = \{0\}$, we say that the group G is *reduced*.

Proposition 1.14.19. *Each Abelian group* $G = D(G) \oplus F$, *where* F *is a reduced group. In particular, any Abelian group has a divisible subgroup such that the quotient group by this subgroup has no divisible subgroups.*

Proof. Since $D(G)$ is a divisible group, according to 1.14.15, we must have $G = D(G) \oplus F$, for some subgroup F of G. Now, if F contains a divisible subgroup K, then $D(G) \oplus K$ is a divisible subgroup of G and since $D(G)$ contains all divisible subgroups of G, K must $\{0\}$. Hence F is reduced.   □

Exercise 1.14.20. Prove an Abelian group G is divisible if and only for each $a \in G$ and each prime p the equation $px = a$ has a solution in G.

1.14.3   Prüfer Groups

Here we apply Theorem 1.14.4 to the torsion group $\mathbb{Q}/\mathbb{Z}$.

$$(\mathbb{Q}/\mathbb{Z})_p = \{n + \mathbb{Z} \mid n + \mathbb{Z} \text{ of order some power of } p\}$$
$$= \{n + \mathbb{Z} \mid p^r n \in \mathbb{Z}\}$$
$$= \left\{\frac{m}{p^r} + \mathbb{Z} \mid m \in \mathbb{Z} : 0 \le m < p^r\right\}$$

and by the above theorem,

$$\mathbb{Q}/\mathbb{Z} = \bigoplus_p (\mathbb{Q}/\mathbb{Z})_p.$$

We recall that $(\mathbb{Q}/\mathbb{Z})_p = \mathbb{Z}(p^\infty)$.

Proposition 1.14.21. *Show that if G is a p-Prüfer group, then it is divisible.*

Proof. As $\mathbb{Q}$ is divisible, so is $\mathbb{Q}/\mathbb{Z}$. But since $\mathbb{Q}/\mathbb{Z} = \oplus(\mathbb{Q}/\mathbb{Z})_p$, each $(\mathbb{Q}/\mathbb{Z})_p$ is itself divisible as a direct summand of a divisible group. $\square$

The reader might find it interesting to consider the following alternative proof.

Proof. Let $g \in G = (\mathbb{Q}/\mathbb{Z})_p$. Then, $g = \frac{m}{p^r}$ for some $0 \le m \le p^r$. Let $n = p^s k$, where p and k are co-primes. Let $g_1 = \frac{m}{p^{r+s}} + \mathbb{Z}$. Then

$$p^s g_1 = g.$$

As p^{r+s} and k are co-prime, there are α and β such that

$$\alpha k + \beta p^{r+s} = 1$$

which implies that

$$g_1 = (\alpha k + \beta p^{r+s})g_1 = \alpha k g_1 + \beta p^{r+s} g_1$$
$$= \alpha k g_1 + \beta p^r g = \alpha k g_1 + \beta m + \mathbb{Z} = \alpha k g_1.$$

Set $h = \alpha g_1$. Then

$$nh = p^s kh = p^s k\alpha g_1 = p^s g_1 = g$$

i.e. G is divisible.   □

Here again we rely on Chapter 3 below. The vector spaces involved here and in the next section can be *infinite dimensional.*

Theorem 1.14.22. *If G is a divisible group, then for each prime p,*

$$\mathrm{Tor}_p(G) = \bigoplus_{m_p} \mathbb{Z}(p^\infty),$$

for some cardinal m_p.

Proof. Let

$$H = \{x \in \mathrm{Tor}_p(G) \mid px = 0\}.$$

Then $0 \in H$ and if $x \in H$, then also $-x \in H$. Moreover if $px = py = 0$, then $p(x + y) = 0$. Hence H is an Abelian subgroup of G. Since each element of H has order p, H is actually a vector space over the finite field $\mathbb{Z}_p$. Let m_p be its dimension.

Now, consider the direct sum

$$M = \bigoplus_{i \leq m_p} \mathbb{Z}(p^\infty).$$

This is a divisible group by Proposition 1.14.21 and just as before, the set

$$M_p = \{x \in M \mid px = 0\}$$

is a subgroup of M, as well as a vector space over $\mathbb{Z}_p$. By identifying

$$\pi_i(M) \cong \mathbb{Z}(p^\infty) = \langle x_{i,1}, \ x_{i,2}, \ ..., \ : \ px_{i,1} = 0, \ px_{i,2} = x_{i,1}, ...\rangle$$

for each $i \leq m$ we see that $\{x_{i,1}, \ 1 \leq i \leq m\}$ is a basis for M_p. Hence as vector spaces over $\mathbb{Z}_p$, H and M_p have the same dimension. They are therefore isomorphic. Let φ be this isomorphism. Then φ as an injective homomorphism $H \rightarrow M$ of the additive groups. Since M is divisible, Corollary 1.14.15, tells us that φ extends to a homomorphism

$$\widetilde{\varphi} : \mathrm{Tor}_p(G) \longrightarrow M.$$

We claim that $\widetilde{\varphi}$ is $1:1$. To see this, let $x \in \mathrm{Ker}\,\widetilde{\varphi}$, with $\mathrm{ord}(x) = p^k$. If $k > 0$, then the element $y = p^{k-1}x$ has order p and $\widetilde{\varphi}(y) = p^{k-1}\widetilde{\varphi}(x) = 0$. This means $y \in |(\,\mathrm{Ker}\,\widetilde{\varphi}) \cap H = \mathrm{Ker}\,\varphi = \{0\}$, which contradicts the fact that y has order p. Hence, $k = 0$ and thus $x = 0$, proving our claim.

Now, by Corollary 1.14.12, $\mathrm{Im}(\widetilde{\varphi})$ is a divisible subgroup of M, so

$$M \cong \mathrm{Im}(\widetilde{\varphi}) \times A, \ \text{ for some subgroup } A \text{ of } M, \text{ with } A \cap \mathrm{Im}(\widetilde{\varphi}) = \{0\}.$$

We claim that $A = \{0\}$. For this, let $x \in A$ with order p^k. If $k > 0$, then $y = p^{k-1}x$ has order p and so $y \in A \cap \mathrm{Im}(\widetilde{\varphi})$, a contradiction (since y has order p). Hence, $A = \{0\}$ and therefore $\mathrm{Im}(\widetilde{\varphi}) \cong M$. Hence,

$$\mathrm{Tor}_p(G) \cong \mathrm{Im}(\widetilde{\varphi}) \cong M = \bigoplus_{i \leq m_p} \mathbb{Z}(p^\infty).$$

$\square$

1.14.4   Structure Theorem for Divisible Groups

For this we shall require the following lemma. Here G is an arbitrary divisible group. The proof is similar to the above except here we will be dealing with vector spaces over $\mathbb{Q}$.

Lemma 1.14.23. *The torsion subgroup* $\mathrm{Tor}(G)$ *of a divisible group* G *is divisible. In addition,* $G/\mathrm{Tor}(G)$ *is a vector space over* $\mathbb{Q}$ *and* $G = \mathrm{Tor}(G) \oplus G/\mathrm{Tor}(G)$.

Proof. Let $x \in \mathrm{Tor}(G)$. Then, there is an $m \in \mathbb{N}$ such that $mx = 0$. Since G is divisible, there is an $n \in \mathbb{N}$ such that $x = ng$ for some $g \in G$. Thus $(mn)g = 0$ and so $g \in \mathrm{Tor}(G)$. Hence $\mathrm{Tor}(G)$ is divisible. By Proposition 1.14.2 $G/\mathrm{Tor}(G)$ is torsion-free, and by Proposition 1.14.11 it is divisible. Let $x \in G/\mathrm{Tor}(G)$ and y_1, y_2 be elements of $G/\mathrm{Tor}(G)$

such that $ny_1 = x = ny_2$. Since $G/\operatorname{Tor}(G)$ is torsion-free $y_1 - y_2 = 0$. Therefore, the (scalar multiplication) map

$$\mathbb{Q} \times \Big(G/\operatorname{Tor}(G) \Big) \longrightarrow G/\operatorname{Tor}(G), \quad \Big(\frac{m}{n}, x\Big) \mapsto \frac{m}{n}x := z$$

is well defined since z is the unique element of $G/\operatorname{Tor}(G)$ such that $nz = mx$. This shows $G/\operatorname{Tor}(G)$ is a vector space over $\mathbb{Q}$. ☐

As a torsion free divisible Abelian group the structure of $G/\operatorname{Tor}(G)$ is easy to understand as such a group is a vector space over $\mathbb{Q}$. Thus it is isomorphic to the (weak) direct sum of some cardinal number m of copies of $\mathbb{Q}$.

Theorem 1.14.24. (Classification of divisible groups.) *Divisible groups are direct sums of the groups* $\mathbb{Q}$ *and* $\mathbb{Z}(p^\infty)$, *in other words there exist cardinals m and m_p (one for each prime) so that,*

$$G = \bigoplus \mathbb{Q}^m \oplus \bigoplus_p \Big(\mathbb{Z}(p^\infty)\Big)^{m_p}.$$

Proof. From Lemma 1.14.23 we know $G = \operatorname{Tor}(G) \oplus G/\operatorname{Tor}(G)$. Since $G/\operatorname{Tor}(G)$ is a torsion free divisible group it is a vector space over $\mathbb{Q}$ and so it is isomorphic to

$$G/\operatorname{Tor}(G) \cong \bigoplus \mathbb{Q}^m.$$

In addition, $\operatorname{Tor}(G) = \bigoplus_{p \in P} \operatorname{Tor}_p(G)$, and by theorem 1.14.22, for each p there is some cardinal m_p. Therefore

$$\operatorname{Tor}(G) = \bigoplus_{m_p} \mathbb{Z}(p^\infty).$$

☐

An interesting application of divisibility is the following: As we shall see in Chapter 3, for a k-vector space V, the linear functionals, $\operatorname{Hom}(V, k)$, of V separates the points of V. One might compare this with the following fact about Abelian groups.

Proposition 1.14.25. *Let A be an arbitrary Abelian group, and let D be a divisible Abelian group containing $\mathbb{Q}/\mathbb{Z}$. Then the homomorphisms of $\operatorname{Hom}(A, D)$ separate the points of A.*

This applies in particular to $D = \mathbb{Q}/\mathbb{Z}$ and to $D = \mathbb{T} = \mathbb{R}/\mathbb{Z}$.

Proof. Assume that $0 \neq a \in A$. We must find a homomorphism $\chi : A \to D$ such that $\chi(a) \neq 0$. Let C be the cyclic subgroup $\mathbb{Z} \cdot a$ of A generated by a. If C is infinite, then C is free and for any non-zero element t in $\mathbb{Q}/\mathbb{Z}$ (for example $t = \frac{1}{2} + \mathbb{Z} \in \mathbb{Q}/\mathbb{Z}$) by the universal property of free groups, there is an $f : C \to D$ with $f(a) = t \neq 0$. If C has order n, then C is isomorphic to $\frac{1}{n}\mathbb{Z}/\mathbb{Z} \subset \mathbb{Q}/\mathbb{Z}$, and thus there is an injection $f : C \to D$. Let $\chi : A \to D$ be an extension of f which exists by the injectivity of $\mathbb{T}$. Then $\chi(a) = f(a) \neq 0$.   □

1.14.5   Maximal Subgroups

Definition 1.14.26. We say that the subgroup H of the group G is a *maximal* subgroup if for any other subgroup F of G we have $F \subseteq H$.

Proposition 1.14.27. *Let G be a group and H be a non-trivial normal subgroup. Then H is maximal (as a normal subgroup of G) if and only if G/H is simple.*

Proof. If H is a normal subgroup of a group G, then the subgroups of G/H are in 1:1 correspondence with the subgroups of G that contain H. But H is maximal, so G/H is simple. The converse also holds since all steps are reversible.   □

As we know $\mathbb{Z}_p$ is a simple Abelian group.

Proposition 1.14.28. *If G be a simple Abelian group, then G is isomorphic to $\mathbb{Z}_p$.*

Proof. Let $g \neq 1$ be an element of G. Then, since G is simple, the cyclic group $< g >$, as a nonempty subgroup of G, must coincide with G, so G is cyclic. Therefore, $G \cong \mathbb{Z}$, or $G \cong \mathbb{Z}_n$. The first case must be excluded, since $\mathbb{Z}$ has many non-trivial subgroups. In the second if p is a divisor of n then $\mathbb{Z}_n$ would have a subgroup isomorphic with $\mathbb{Z}_p$ and since G is simple, $G \cong \mathbb{Z}_p$. $\square$

Proposition 1.14.29. *Let G be a divisible Abelian group. Then G has no maximal subgroups.*

Proof. Let H be a subgroup of G. If H is a maximal subgroup, then according to 1.14.27, G/H must be simple. But, by 1.14.28, the only simple Abelian groups are of prime order p. Therefore $[G : H] = p$, and so $pH = G$. Now, take a $g \in (G - H)$. Since G is divisible, for this p there is a $k \in G$ such that $pk = g$, and so $pk \in pH$, i.e. $k \in H$, a contradiction. $\square$

1.14.5.1 Exercises

Exercise 1.14.30. Let H and K be subgroups of a group G. Show that the set

$$HK = \{hk \mid h \in H, \ k \in K\}$$

is a subgroup of G if and only if $HK = KH$.

Exercise 1.14.31. Show $\mathbb{Z}$ has infinite index in $\mathbb{Q}$ (that is, there are infinitely many (left or right) cosets of $\mathbb{Z}$ in $\mathbb{Q}$).

Exercise 1.14.32. Let G be the group of non-zero real numbers under multiplication and let $H = \{1, -1\}$. Prove G/H is isomorphic to the group of positive real numbers under multiplication. More generally, let $\mathrm{GL}(n, \mathbb{R})$ be the group of non-singular $n \times n$ real matrices and $\mathrm{GL}(n, \mathbb{R})_0$ be the (normal) subgroup of those of det > 0. Prove

$$\mathrm{GL}(n, \mathbb{R})/\mathrm{GL}(n, \mathbb{R})_0 \cong \{\pm 1\}.$$

Exercise 1.14.33. Prove that, if H is a subgroup of G, then

$$\bigcap_{g \in G} gHg^{-1}$$

is a normal subgroup of G.

Exercise 1.14.34. Find a subgroup of $\mathbb{Q}^+$ which is maximal with respect to not containing 1.

Exercise 1.14.35. A subgroup H of G of index 2 in G is normal.

In particular, the alternating group A_n is normal in S_n. Alternatively, show A_n is normal in S_n by showing it is the kernel of a homomorphism sgn $: S_n \to \{\pm 1\}$.

Exercise 1.14.36. If H is the unique subgroup of G of order m, then H is characteristic.

Exercise 1.14.37. Let G be a simple group. Show that any group homomorphism $\phi : G \to H$ either is the trivial homomorphism or $1 : 1$, and conversely if this is so for every homomorphism, then G is a simple group.

Exercise 1.14.38. Let G be a group.

1. Let $H_1, \ldots H_n$ be a finite number of subgroups each of finite index in G. Show $\cap_{i=1}^n H_i$ has finite index in G.

2. Let H be a subgroup of finite index. Show H contains a normal subgroup of finite index in G.

3. Show this index is a divisor of $[G : H]!$.

Exercise 1.14.39. Let G and H be simple groups. Show the only normal subgroups of $G \times H$ are 1, G, H and $G \times H$.

Exercise 1.14.40. The following exercises deal with ideas connected with direct products.

1. If $g = g_1 g_2$, the *fiber* over g is defined to be any pair in $G_1 \times G_2$ such that m maps it to g. Show the fiber is $\{(g_1 l, l^{-1} g_2) : l \in G_1 \cap G_2\}$.

2. Hence all fibers have the same cardinality, namely $|G_1 \cap G_2|$.

3. Show that $|G_1 G_2||G_1 \cap G_2| = |G_1||G_2|$.

4. Let G be a finite group and A and B non-void subsets. Show if $|A| + |B| > |G|$, then $AB = G$.

Exercise 1.14.41. Let $\{H_i : i \in I\}$ be a family of subgroups of G all containing a normal subgroup N of G. Show $\cap_{i \in I}(H_i/N) = \cap_{i \in I}(H_i)/N$.

Exercise 1.14.42. Let G be a finite group.

1. Show if G has even order then it contains an element of order 2.

2. Show if G has order $2r$ where r is odd, then G has a subgroup of index 2. (Let G act on itself by left translation.)

Exercise 1.14.43. Here we ask the reader to calculate a few automorphism groups.

1. Calculate the automorphism groups of $\mathbb{Z}_n$, S_3 and D_n.

2. Calculate the symmetry groups of the tetrahedron, cube and octahedron. Show that the last two are isomorphic.

Exercise 1.14.44. Let G be a group and α be an automorphism. For $g \in G$ we call $\alpha(g)g^{-1}$ the *displacement* of g which we abbreviate disp.

1. Show $\operatorname{disp}(g_1) = \operatorname{disp}(g_2)$ if and only if $g_2^{-1}g_1$ is α fixed.

2. Hence, if α fixes only the identity, disp is an injective map $G \to G$. In particular, if G is finite disp is bijective.

3. Show if $g_1 \in Z(G)$, then $\operatorname{disp}(g_1g_2) = \operatorname{disp}(g_1)\operatorname{disp}(g_2)$.

4. In particular, disp is a homomorphism $Z(G) \to Z(G)$ whose kernel is the points of $Z(G)$ fixed by α.

Exercise 1.14.45. Find the subgroup of all elements of finite order in $\mathbb{C}^\times$? Do the same for the groups $O(2, \mathbb{R})$ and $O(2, \mathbb{R}) \times \mathbb{R}^2$.

Exercise 1.14.46. Show $\mathbb{C}^\times$ is isomorphic with various proper quotient groups of itself, viz $\mathbb{C}^\times/R_n$, where R_n is the n-th roots of unity.

Exercise 1.14.47. This exercise requires a small bit of knowledge of Chapter 5 since here $\mathbb{Z}$ is regarded as a ring. Consider the function,

$$\pi_k : \mathrm{SL}(n, \mathbb{Z}) \to \mathrm{SL}(n, \mathbb{Z}_k),$$

where we map each of the i, j entries to its k-th residue class.

1. Show for each integer $k > 1$, π_k is a surjective group homomorphism. (Notice that the map to residue classes is a ring homomorphism.)

2. Show $\mathrm{SL}(n, \mathbb{Z}_k)$ is a finite group.

3. Show that $\mathrm{SL}(n, \mathbb{Z})$ contains a normal subgroup of finite index. This is called a *congruence subgroup*.[13]

4. Show there are an infinite number of normal subgroups of finite index in $\mathrm{SL}(n, \mathbb{Z})$ whose intersection is the identity and therefore we get an injection of $\mathrm{SL}(n, \mathbb{Z})$ into a direct product of finite groups.

---

[13]This is the easy part of a famous theorem of Meineke which states, that conversely, if Γ is any normal subgroup of $\mathrm{SL}(n, \mathbb{Z})$ of finite index it must contain a congruence subgroup.

Chapter 2

Further Topics in Group Theory

2.1 Composition Series

Let G be a group and H be a non-trivial normal subgroup. Since H and G/H are in some sense "smaller" than G, it is a standard technique in group theory to get information about G by using corresponding information about the smaller groups H and G/H. For this reason we make the following definitions. These will be particularly effective when G is finite (or when G is a Lie group).

As we know from Proposition 1.14.27, if G is a group and H is a non-trivial normal subgroup, then H is maximal (as a normal subgroup of G) if and only if G/H is simple.

Definition 2.1.1. Let G be a group and $G = G_0 \supseteq G_1 \supseteq \ldots \supseteq G_n = (1)$ be a *finite* sequence of subgroups each normal in its predecessor. We call this a *normal series*. If each G_i is a maximal normal subgroup of G_{i-1} we call it a *composition series*, or a Jordan-Hölder series. In this case we call n its *length* and the various G_{i-1}/G_i the *composition factors*. As we shall see these are invariants of G.

We remark that a group G always has at least one normal series. Namely, $\{1\} \lhd G$. But it may not always have a composition series. For

109

example, and we leave this to the reader as an exercise, the additive group $\mathbb{Z}$ does not have a composition series. However, evidently a finite group always has a composition series. Indeed, any normal series of a finite group can be *refined* to a composition series.

Example 2.1.2. A normal series for $\mathbb{Z}_k$ can be gotten by factoring k into its primes. Thus $k = p_1^{e_1} \cdots p_j^{e_j}$ where the p_i are distinct. Then as we know these are precisely the p primary components and,

$$\mathbb{Z}_k = \mathbb{Z}_{p_1^{e_1}} \cdots \oplus \cdots \mathbb{Z}_{p_j^{e_j}}.$$

Hence, a normal series is

$$\mathbb{Z}_k \supseteq \mathbb{Z}_{p_2^{e_2}} \cdots \oplus \cdots \mathbb{Z}_{p_j^{e_j}} \supseteq \mathbb{Z}_{p_3^{e_3}} \cdots \oplus \cdots \mathbb{Z}_{p_j^{e_j}} \cdots \supseteq \mathbb{Z}_{p_j^{e_j}}.$$

This normal series can then be refined to a composition series by considering each p primary component $G_p = \mathbb{Z}_{p^e}$. Because G_p is Abelian normality is not an issue and we have a series of subgroups, $\mathbb{Z}_{p^f} \supseteq \mathbb{Z}_{p^{f-1}}$, whose quotients are all the simple group $\mathbb{Z}_p$.

Example 2.1.3. A composition series for D_{2k}, the dihedral group with $2k$ elements, is given by, $D_{2k} \supsetneq \langle \sigma \rangle$ followed by a composition series for $\mathbb{Z}_k$ (as above, but written multiplicatively).

Exercise 2.1.4. Let G and H be two groups with composition series

$$G = G_r \supsetneq G_{r-1} \supsetneq \cdots \supsetneq G_0 = \{1\}$$

and

$$H = H_s \supsetneq H_{s-1} \supsetneq \cdots \supsetneq H_0 = \{1\}.$$

Then, we obtain the chain

$$G \times H = G_r \times H \supsetneq \cdots \supsetneq \{1\} \times H$$
$$= \{1\} \times H_s \supsetneq \cdots \{1\} \times H_{s-1} \supsetneq \cdots \supsetneq \{1\} \times \{1\}.$$

Show that this is a composition series of $G \times H$. In particular we conclude that the length of this composition series is the sum of the lengths of the composition series of G and H.

The Jordan-Hölder theorem is the following.

Theorem 2.1.5. *Let*

$$G = G_r \supsetneq G_{r-1} \supsetneq \cdots \supseteq G_0 = \{1\}$$

and

$$G = H_s \supsetneq H_{s-1} \supsetneq \cdots \supsetneq H_0 = \{1\}$$

be two composition series for the group G. Then $r = s$ and after an appropriate permutation of the indexes the composition factors G_{i-1}/G_i and H_{i-1}/H_i are isomorphic.

Notice that this theorem does not assert that there is a composition series. It merely states that, if there is one, it is essentially unique.

The proof of the Jordan-Hölder theorem itself will be deferred to Chapter 6. We do this because it will be convenient to have a more general version of this result of which the Jordan-Hölder theorem here is a corollary.

It is not surprising, in view of the example just above, that an application of the Jordan-Hölder theorem gives an alternative proof of the *uniqueness of prime factorization of integers*.

Theorem 2.1.6. *Any integer in $\mathbb{Z}^+$ can be factored uniquely into primes.*

Proof. Suppose the integer n has two decompositions into primes, i.e.

$$n = p_1 p_2 ... p_s \quad \text{and} \quad n = q_1 q_2 ... q_r.$$

Let $\mathbb{Z}_k$ be the cyclic group of order k. Then, we get the two composition series

$$\mathbb{Z}_n \supseteq \mathbb{Z}_{p_2...p_s} \supseteq \mathbb{Z}_{p_3...p_s} \supseteq \cdots \supseteq \mathbb{Z}_{p_s} \supseteq \{1\}$$

and

$$\mathbb{Z}_n \supseteq \mathbb{Z}_{q_2...q_r} \supseteq \mathbb{Z}_{q_3...q_r} \supseteq \cdots \supseteq \mathbb{Z}_{q_r} \supseteq \{1\}.$$

By the Jordan-Hölder theorem, these two must be equivalent. In other words, $s = r$ and, by a suitable rearrangement, $p_l = q_l$, with $1 \le l \le s$. $\square$

2.2   Solvability, Nilpotency

2.2.1   Solvable Groups

Let G be a group, and let $G_1 = [G, G]$ be its derived group. Consider the groups

$$G_2 = [G_1, G_1], \; G_3 = [G_2, G_2], ..., G_n = [G_{n-1}, G_{n-1}], ...$$

Thus we have the sequence

$$G \supseteq G_1 \supseteq G_2 \supseteq ... \supseteq G_n, ...$$

called the *derived series*.

Exercise 2.2.1. Show that the terms of the derived series are characteristic subgroups of G.

Definition 2.2.2. If, after a finite number of steps, we hit the identity, i.e. if $G_n = \{1\}$ for some $n \in \mathbb{Z}^+$, then we say that G is *solvable*, and we call n, the first place where this occurs, the *index of solvability*.

Of course, since $[x, y] = xyx^{-1}y^{-1}$, if $n = 1$, then G is Abelian. Thus Abelian groups are exactly the 1 step solvable groups.

Example 2.2.3. The affine group of the real line is 2-step solvable.

Proof. As we know, the affine group G is,

$$G = \left\{ \begin{pmatrix} a & b \\ 0 & 1 \end{pmatrix}, a \neq 0, \, a, b \in \mathbb{R} \right\}.$$

An easy calculation shows that the derived $[G, G]$ is:

$$[G, G] = \left\{ \begin{pmatrix} 1 & b \\ 0 & 1 \end{pmatrix}, b \in \mathbb{R} \right\}$$

which is Abelian. Hence $G_2 = [G_1, G_1] = \{1\}$. $\square$

Here are some groups which are not solvable.

Exercise 2.2.4. Let k be $\mathbb{Q}$, $\mathbb{R}$, $\mathbb{C}$, or even $\mathbb{Z}$. Then $\mathrm{SL}(2,k)$ is not solvable. Actually, the same holds for $\mathrm{SL}(n,k)$, where $n \geq 2$.

Proposition 2.2.5. *Each G_i is characteristic in G and in G_{i-1}.*

Proof. We shall prove G_i is characteristic in G. We leave the similar proof that G_i is characteristic in G_{i-1} to the reader as an exercise. Our proof is by induction on i.

To show $[G, G]$ is characteristic in G let

$$[x_1, y_1] \cdot [x_2, y_2]... \cdot [x_n, y_n] = g$$

be in $[G, G]$ and α be an automorphism of G. Then

$$\alpha([x_1, y_1]...[x_n, y_n]) = [\alpha(x_1), \alpha(y_1)]...[\alpha(x_n), \alpha(y_n)] = \alpha(g).$$

Thus $\alpha[G, G] \subseteq G$.

Now suppose G_{i-1} is characteristic in G. Exactly as above, G_i is generated by commutators $[x, y]$, where $x, y \in G_{i-1}$. Take an $\alpha \in \mathrm{Aut}(G)$ and apply it to $[x_1, y_1] \cdot [x_2, y_2]... \cdot [x_n, y_n]$. Because α is an automorphism,

$$\alpha([x_1, y_1]...[x_n, y_n]) = [\alpha(x_1), \alpha(y_1)]...[\alpha(x_n), \alpha(y_n)].$$

By inductive hypothesis this last term is in $[G_{i-1}, G_{i-1}] = G_i$. Thus G_i is characteristic in G. $\qquad\qquad\square$

The significance of $[G, G]$ is that it is the smallest normal subgroup of G with Abelian quotient.

Proposition 2.2.6. *$G/[G, G]$ is Abelian. In addition, if G/H is Abelian where H is a normal subgroup of G, then $[G, G] \subseteq H$.*

Proof. First, we will prove that $G/[G, G]$ is Abelian. That is, we show

$$g[G, G].g'[G, G] = g'[G, G].g[G, G].$$

Hence

$$g[G, G].g'[G, G] = gg'[G, G]$$

and
$$g'[G,G].g[G,G] = g'g[G,G].$$
Now, $gg'[G,G] = g'g[G,G]$ holds if and only if $(g'g)^{-1}gg' = [g^{-1}, g'^{-1}] \in [G,G]$, which it is.

For the second part, since G/H is Abelian arguing as above we see this means $[g^{-1}, g'^{-1}] \in H$ for all $g, g' \in G$. Since these generate $[G,G]$ we see $[G,G] \subseteq H$.   □

Since $[G,G]$ is a normal subgroup of G, we get,

Corollary 2.2.7. *A non-Abelian simple group cannot be solvable.*

The following result is often used to establish solvability.

Theorem 2.2.8. *Let G be a group and H a normal subgroup, then G is solvable if and only if H and G/H are solvable.*

That is, given a short exact sequence of groups

$$(1) \to A \to G \to C \to (1)$$

then, G is solvable if and only if A and C are solvable.

Proof. $\Longrightarrow$ Suppose that G is solvable, and $H \lhd G$. Then we have the following sequences of inclusions:

$$H_1 = [H,H] \subseteq [G,G] = G_1$$

$$\dotsb$$

$$H_n = [H_{n-1}, H_{n-1}] \subseteq [G_{n-1}, G_{n-1}] = G_n$$

Because G is solvable, $G_n = \{1\}$, and so, $H_n = \{1\}$, i.e. H is solvable. Now, consider the map

$$\pi : G \to G/H.$$

Then $(G/H)_1 = [G/H, G/H]$. Since $H \lhd G$,

$$[G/H, G/H] = \{gHg_1Hg^{-1}Hg_1^{-1}H, \ g, g_1 \in G\}$$
$$= \{gg_1g^{-1}g_1g_1^{-1}H, \ g, g_1 \in G\}$$
$$= ([G,G].H)/H.$$

Therefore, $\pi((G)_1) = (\pi(G))_1$, and, by induction,

$$\pi(G_n) = (\pi(G))_n).$$

Thus, since $G_n = \{1\})$, G/H is solvable.

$\Longrightarrow$ Suppose H and G/H are solvable. Since $(\pi(G))_n = \pi(G_n)$ it follows that $G_n \subseteq H$, for some sufficiently large n. Therefore, $G_{n+1} \subseteq H_1$, so $G_{n+k} \subseteq H_k$ for each k. Eventually, H_k must be trivial. Therefore, G_k must be trivial, i.e. G is solvable. $\qquad\square$

Corollary 2.2.9. *The group S_3 is solvable.*

Proof. We know $A_3 \subseteq S_3$. Since $A_3 \simeq \mathbb{Z}_3$ is Abelian, and $S_3/A_3 \simeq \mathbb{Z}_2$ is also Abelian, S_3 is solvable. $\qquad\square$

The following exercise is slightly more difficult, but the idea is the same. The interesting thing here is that everything changes with S_5!

Exercise 2.2.10. Show S_3 and S_4 are solvable.

Definition 2.2.11. When $r \leq n$ we call an r-cycle a permutation of n elements which leaves invariant the set of these r elements and permutes them cyclicly. For example, a transposition is a 2 cycle.

Our next proposition will be useful in proving the important fact that A_n is simple for $n \geq 5$. In contrast to the exercise just above,

Proposition 2.2.12. *The symmetric group S_n is not solvable for $n \geq 5$.*

To do this we need the following lemma:

Lemma 2.2.13. *Suppose a subgroup H of S_n $(n \geq 5)$ contains every 3-cycle. If F is a normal subgroup of H with H/F is Abelian, then F itself contains every 3-cycle.*

Proof. Let (k, j, s) be a 3-cycle in S_n and, by hypothesis in H. We want to prove that $(k, j, s) \in F$. Since $n \geq 3$ we can consider two more numbers, say i and r. Now, consider the quotient map $\pi : H \longrightarrow H/F$, and let $\sigma = (ijk)$ and $\tau = (krs)$. These are in H. Let $\tilde{\sigma} = \pi(\sigma)$, and $\tilde{\tau} = \pi(\tau)$. Since π is a homomorphism and H/F is Abelian,

$$\pi(\sigma^{-1}\tau^{-1}\sigma\tau) = \tilde{\sigma}^{-1}\tilde{\tau}^{-1}\tilde{\sigma}\tilde{\tau} = 1.$$

Therefore, $\sigma^{-1}\tau^{-1}\sigma\tau$ is in F. But

$$\sigma^{-1}\tau^{-1}\sigma\tau = (kji).(srk).(ijk).(krs) = (kjs).$$

Thus, each (kjs) is in F.   □

We now prove Proposition 2.2.12 above.

Proof. Let $n \geq 5$. If S_n were solvable then there would be a descending sequence of Abelian subgroups terminating at $\{1\}$. Since S_n contains every 3-cycle, the lemma above shows that so will each subgroup in this sequence. Therefore, the sequence can never end at $\{1\}$.   □

This brings us to a fact which has profound implications concerning the impossibility of solving a general polynomial equation of degree ≥ 5. (See Chapter 11 on Galois theory).

Corollary 2.2.14. *The groups A_n, $n \geq 5$, are all simple.*

We will give an alternative rather shorter proof of this result in Corollary 2.4.12. To prove 2.2.14 we need the following lemma:

Lemma 2.2.15. *The groups A_n, $n \geq 5$ are generated by its three cycles, and all three cycles are conjugate.*

Proof. Since S_n is generated by transpositions (ij), it follows that A_n is generated by the double transpositions $(ij)(kl)$. So for the first statement it suffices to show that every double transposition is a product of three cycles. One can verify that directly that $(ij)(kl) = (ilk)(ijk)$ where i, j, k, l are distinct and that $(ij)(ik) = (ikj)$, which proves the first claim.

Since $\sigma(ijk)\sigma^{-1} = (\sigma(i)\sigma(j)\sigma(k))$, for any permutation σ, it follows easily that every three cycle is conjugate to (123) in S_n. If σ is even, we are done. Otherwise, since $n \geq 5$ we can choose l, m different from $\sigma(i)$, $\sigma(j)$, $\sigma(k)$ and take $\tau = (lm)\sigma$. Then τ and $\sigma(ijk)\sigma^{-1}$ commute.   □

We can now complete the proof of Corollary 2.2.14.

Proof. We will use the previous lemma to show that every non-trivial normal subgroup of A_n with $n \geq 5$ contains a three cycle. To this end, we note that if $N \lhd A_n$ is non-trivial it contains a non-trivial permutation that fixes some letter (show that if a permutation σ fixes no letter there exists a permutation τ such that $[\sigma, \tau]$ fixes some letter).

We claim that if we pick such non-trivial permutation that is not a three cycle, we may conjugate it to obtain a permutation with more fixed points. It will follow that any permutation with a maximum number of fixed points is a three cycle. Indeed, let σ be non-trivial with fixed points, but not a three cycle. Using the cycle decomposition in S_n we can write σ as a permutation of the form

$$\sigma_1 = (123 \cdots) \cdots \quad \text{or} \quad \sigma_2 = (12)(34) \cdots .$$

Let $\tau = (345)$. We then have

$$\tau\sigma_1 = (124 \cdots) \cdots \quad \text{and} \quad \tau\sigma_2 = (12)(45) \cdots$$

and in either case $\tau\sigma \neq \sigma$ so that

$$\sigma' = [\tau, \sigma] = \tau\sigma\tau^{-1}\sigma^{-1} \neq 1.$$

Any letter > 5 is fixed by τ so if it is fixed by σ it is also fixed by σ'. The reader can verify that in the first case σ' fixes 2, and since σ_1 moves 1, 2, 3, 4 and 5 to obtain more fixed points. In the second case σ' fixed 1 and 2, and again we have more fixed points. This proves the Corollary. □

2.2.2 Nilpotent Groups

Here we consider the series:

$$G \supseteq [G, G] = G^1 \supseteq [G, G^1] = G^2 \supseteq \ldots \supseteq [G, G^{n-1}] = G^n \supseteq \ldots .$$

This is called the *descending central series* of G.

Definition 2.2.16. G is called *nilpotent* if the above series hits $\{1\}$ after a finite number of steps. The smallest of these numbers is called the *index of nilpotency*.

Since each term of the descending central series contains the corresponding term of the derived series we see that

Proposition 2.2.17. *A nilpotent group is solvable.*

Exercise 2.2.18. Show that the terms of the nilpotent series are characteristic subgroups of G.

We make the following simple, but important, observation.

Corollary 2.2.19. *A nilpotent group has a non-trivial center.*

Proof. Let G be nilpotent and consider the last non-trivial term of the descending central series. It is evidently contained in $\mathcal{Z}(G)$. □

Exercise 2.2.20. Show the solvable groups S_3 and S_4 are not nilpotent.

Example 2.2.21. Evidently, Abelian groups are nilpotent. A more typical example of a nilpotent group is:

$$N_k = \left\{ \begin{pmatrix} 1 & x_{12} & \cdots & & x_{1k} \\ 0 & 1 & x_{23} & & \cdots \\ \vdots & \vdots & \ddots & & x_{(k-1)(k-1)} \\ 0 & \cdots & \cdots & & 1 \end{pmatrix}, k \times k \text{ matrices} \right\}.$$

Exercise 2.2.22. Show index of $G = 1$ means G is Abelian, and index of $G = 2$ means $G/\mathcal{Z}(G)$ is Abelian. What is the index of N_k?

A somewhat similar theorem to 2.2.8 holds for nilpotent groups. We shall say H is a *central* subgroup of G if $H \subseteq \mathcal{Z}(G)$.

Theorem 2.2.23. *Let G be a group and $H \triangleleft G$. If G is nilpotent, then both H and G/H are nilpotent. The index of nilpotence of both H and G/H are $\leq$ index of G. If H is a central subgroup of G and G/H is nilpotent, then so is G. Moreover, index of $G \leq 1 + $ index of G/H.*

Proof. $\implies$ Let $H^1 = [H, H] \subseteq [G, G] = G^1$. Then

$$H^2 = [H, H^1] \subseteq [H, G^1] \subseteq [G, G^1] = G^2.$$

Thus $H^2 \subseteq G^2$. Continuing by induction it follows that

$$H^n \subseteq G^n, \text{ for all } n.$$

Since eventually G^n is trivial so is H^n.

Let L be a group and $\pi : G \to L$ be a surjective homomorphism. Then $\pi([G, G]) = [L, L]$. Hence $\pi(G^1) = L^1$. Continuing by induction we find that $\pi(G^n) = L^n$ for all n. Therefore index of nilpotency of $H \leq$ index of G, and index of $G/H \leq$ index of G.

$\impliedby$ Suppose that $L = G/H$ is nilpotent. Then as we have seen $\pi(G^n) = L^n$. So, since L is supposed to be nilpotent, $\pi(G^n) = (1)$, for some n. Thus $G^n \subseteq \operatorname{Ker} \pi = H$. Since H is a central subgroup, $G^{n+1} = [G, G^n] \subseteq [G, H] = (1)$. So G is nilpotent. Moreover we also see that,

$$\text{index of } G \leq 1 + \text{index of } G/H.$$

$\square$

We now give an alternative (but equivalent) definition of nilpotency.

Definition 2.2.24. G is a *nilpotent* group if there is an integer $n \geq 0$, and normal subgroups $G_i \subseteq G$, $i = 0, ... n$, forming the *lower central series*

$$(1) = G_0 \supseteq G_1 \supseteq G_2 \supseteq ... \supseteq G_n = G,$$

where for $i = 0, ..., n - 1$, G_{i+1}/G_i lies in $\mathcal{Z}(G/G_i)$.

Since the conditions defining the lower central series are clearly equivalent to $[G_{i+1}, G] \subseteq G_i$ for each i, The existence of such a series is equivalent to G being nilpotent.

Now, let G be a group and $\mathcal{Z}(G)$ be its center. Let $\mathcal{Z}^2(G) \supseteq \mathcal{Z}(G)$ be the subgroup of G corresponding to $\mathcal{Z}(G/\mathcal{Z}(G))$, i.e.

$$g \in \mathcal{Z}^2(G) \iff [g, x] \in \mathcal{Z}(G),$$

for each $x \in G$.

Continuing in this fashion we get a sequence of subgroups, which is called the *ascending central series*,

$$\{1\} \subseteq \mathcal{Z}(G) \subseteq \mathcal{Z}^2(G) \subseteq \ldots$$

where,

$$g \in \mathcal{Z}^i(G) \Longleftrightarrow [g,x] \in \mathcal{Z}^{i-1} \ \forall x \in G.$$

Again, if $\mathcal{Z}^n(G) = G$, for some n, then G is a nilpotent group.

Proposition 2.2.25. *Let G be a non-Abelian nilpotent group. Let $g \in G$ be any element and H be the subgroup generated by g and G_1. Then H is a normal subgroup and $n(H) < n(G)$, where $n(H)$ and $n(G)$ denote the respective indices of nilpotence.*

Proof. We prove this by induction on n. If $n = 1$, then $G_1 = \mathcal{Z}(G)$ and G/G_1 is Abelian. If $x \in G$, then $xgx^{-1} = g \bmod G_1$ which implies that H is normal. Obviously, H is Abelian and therefore $n(H) = 0 < n(G)$.

Now, for the inductive step. Assume $n \geq 2$ and let $H^* \subseteq G/G_n$ be the subgroup generated by G_1/G_n and gG_n. The inductive hypothesis is:

$$H^* \text{ is normal and } n|H^*| < n|G/G_n| - 1.$$

But H is the inverse image of H^* under the projection map $G \longrightarrow G/G_n$. Hence, H is normal. In addition, $n|G/G_n| = n(G) - 1$ and so $n(H) \leq n|H^*| + 1$. Therefore $n(H) \leq n(G) - 1$. $\qquad\square$

2.2.2.1   Exercises

Exercise 2.2.26. Prove this last definition of nilpotence is equivalent to the others and gives the same index of nilpotency.

Exercise 2.2.27.

1. Suppose A and B are normal subgroups of G and both G/A and G/B are solvable. Prove that $G/(A \cap B)$ is solvable.

2. Let G be a finite group. Show that G has a subgroup that is the unique smallest subgroup with the properties of being normal with solvable quotient. This subgroup is denoted by G^∞.

3. If G has a subgroup S isomorphic to A_5 (not necessarily normal), show that S is a subgroup of G^∞.

2.3   Group Actions

In this chapter we will discuss a notion which is extremely useful in many areas of mathematics, namely, that of a *group action*. Let G be a group, and let X be a set.

Definition 2.3.1. We say that the group G *acts* on the set X, or equivalently, that we have an *action* of G on X, if there is a map called the *action*:

$$\varphi : G \times X \longrightarrow X$$

such that:

1. For all $x \in X$, $\varphi(1, x) = x$.

2. For all $g, h \in G$ and $x \in X$, $\varphi(g, \varphi(h, x)) = \varphi(gh, x)$.

As the reader can easily imagine, these axioms will enable us to treat group actions as similar to groups themselves. For convenience we shall write $g \cdot x$ (or sometimes even just gx) instead of $\varphi(g, x)$.

Definition 2.3.2. We shall call X a *G-space* or equivalently, say that G acts as a *transformation group* on X.

Sometimes G, or X, or both, will carry additional structure.

Definition 2.3.3. We say that an action $\varphi : G \times X \longrightarrow X$ is *faithful* or *effective*, if and only if $g \cdot x = x$ for all $x \in X$ implies $g = 1$.

In applications X could, for example, be some geometric space and G a group of transformations of X which preserve some geometric property such as length, or angle, or area, etc. As we shall see, such things always form a group and it is precisely the properties of this group which is the key to understanding length, or angle, or area, respectively. This is the essential idea of Klein's Erlanger Program. Here however, our applications will be to algebra. In fact we will turn this on itself and use it as a tool to study group theory. To this end we will give a few useful example of actions which arise from groups.

Example 2.3.4. Let G be a group and take $X = G$. Consider as action φ the group multiplication, i.e. $\varphi(g, x) = gx$. We can check that this is an action. In this case we say that G acts on itself by left multiplication.

Example 2.3.5. Let G be a group, H a subgroup, *not necessarily normal*, and $X = G/H$. Then G acts on G/H by: $\varphi(g, g_1 H) = g g_1 H$. In this case we say G acts on G/H by left translation. Sometimes this is called *the left regular action of G on G/H*.

Example 2.3.6. Let G be a group, take $X = G$ and as action consider the map $\varphi(g, x) = gxg^{-1}$. In this case we say that G acts on itself by conjugation.

Example 2.3.7. Let G be a group, and consider the group $\text{Aut}(G)$ of automorphisms of G. Then the map

$$\varphi : \text{Aut}(G) \times G \to G \ : \ \varphi(\alpha, g) = \alpha(g)$$

is an action of $\text{Aut}(G)$ on G.

Example 2.3.8. Consider a group G, and let ρ be a representation of G on the vector space V (see Chapter 12), i.e. a homomorphism

$$\rho : G \to GL(V).$$

Taking $X = V$, the map $\varphi(g, v) = \rho(g)(v)$ is an action. This is called a *linear action* or a *linear representation*.

We leave the proof of the following two statements as an exercise.

Proposition 2.3.9. *If $\varphi : G \times X \to X$ is an action of G on X and H is a subgroup of G, then the restriction $\varphi|_{H \times X}$ of φ on H, is an action of H on X.*

Definition 2.3.10. Let G act on X. A subset Y of X is called a *G-invariant* subspace, if and only if $G \cdot Y \subseteq Y$.

Proposition 2.3.11. *If $\varphi : G \times X \to X$ is an action and Y is a G-invariant subspace of X, then $\varphi|_{G \times Y} : G \times Y \to Y$ is also an action.*

Now let G act on X. We can associate to this a number of other important actions.

- The action $G \times \mathcal{P}(X) \to \mathcal{P}(X)$ where $\mathcal{P}(X)$ is the power set of X. This is defined by $g \cdot Y = gY$, where $gY = \{gy : y \in Y\}$. It is called the *induced action on the power set*.

- We could also consider the action $G \times S(X) \to S(X)$, where $S(X)$ is the set of permutations of X. This is defined by $g \cdot \sigma = \sigma_g$, where $\sigma_g(x) = \sigma(gx)$. This is called the *induced action on permutations*.

- A final example is provided as follows: Let $\mathcal{F}(X)$ be some vector space of functions defined on a space X with values in a field, or even in some vector space over this field. The action $G \times \mathcal{F}(X) \to \mathcal{F}(X)$ is defined by $g \cdot f = f_g$, the *left translate of the function* f, where $f_g(x) = f(g^{-1}x)$. This is called the *induced action on functions*.

We leave it to the reader to check that each of these is an action; the last one, called the *left regular representation*, being a *linear* action. We shall not study this last representation, but merely list it for the record. Rather, we shall concern ourselves later with the other two regular representations.

Suppose the group G is acting on each of two spaces X and Y:

$$\varphi : G \times X \to X$$

and,

$$\psi : G \times Y \to Y.$$

When are these actions essentially the same? A natural answer to this question is when there is a *bijection* $f : X \to Y$ such that the following diagram is commutative:

$$X \xrightarrow{\ f\ } Y$$

$$g \downarrow \qquad\qquad \downarrow g$$

$$X \xrightarrow{\ f\ } Y$$

That is, $g \cdot f(x) = f(g \cdot x)$ for all $g \in G$ and $x \in X$.

We then say these actions are *equivariantly equivalent* and we call such a f a *G-equivariant isomorphism* or an equivariant equivalence.

Definition 2.3.12. When f is merely a map, (but not injective or bijective), then we call it a *G-equivariant map*.

2.3.1   Stabilizer and Orbit

Consider an action of the group G on a space X. Let x be a point of X. Then,

Definition 2.3.13. The *stabilizer*, or equivalently the *isotropy group*, of x for an action of G on X is defined as

$$\text{Stab}_G(x) = \{g \in G \ : \ g \cdot x = x\}.$$

If there is no confusion about the group G, we will write simply $\text{Stab}(x)$, or G_x.

Proposition 2.3.14. *The stabilizer of a point x of X is a subgroup of G.*

Proof. First, $1 \in \text{Stab}_G(x)$, since $1 \cdot x = x$. Also, if g_1, g_2 are in $\text{Stab}_G(x)$ then, by the second property of the action, $(g_1 g_2)(x) = g_1 \cdot (g_2 x) = x$, so $g_1 g_2 \in \text{Stab}_G(x)$. Finally, if $g \in \text{Stab}_G(x)$, then $g^{-1} \cdot x = g^{-1}(g \cdot x) = 1 \cdot x = x$, so $g^{-1} \in Stab_G(x)$. $\qquad\qquad\qquad\square$

In general the stabilizer of a point x is not a normal subgroup of G. Indeed we have,

Proposition 2.3.15. *For any point x in X*

$$\text{Stab}_{gx} = g \cdot \text{Stab}_x \cdot g^{-1}.$$

Proof. Here

$$\mathrm{Stab}(gx) = \{h \in G \mid h \cdot (gx) = g \cdot x\} = \{h \in G \mid (hg)x = gx\}$$
$$= \{h \in G \mid g^{-1}hg \cdot x = x\} = g \cdot \mathrm{Stab}(x) \cdot g^{-1}.$$

$\square$

When two subgroups H and H' of a group G are related by $H' = gHg^{-1}$ (as in Proposition 2.3.15 above) we say these subgroups are *conjugate*. Of course conjugate subgroups of a group are isomorphic. In particular if one is finite, they have the same order. They also have the same index, $[G : gHg^{-1}] = [G : H]$.

Now, let $G \times X \to X$ be an action, and x and y be two points of X. Define $x \sim y$, if and only if, there is a g in G such that $g \cdot x = y$.

Proposition 2.3.16. *The relation $\sim$, defined as above, is an equivalence relation.*

Proof. Suppose $x \sim y$. Then $gx = y$ for some g in G. This implies that $1 \cdot x = g^{-1}(g \cdot x) = g^{-1} \cdot y = x$, so $x \sim x$.
Also, if $x = gy$ and $y = hz$, then $x = g(hz) = (gh)z$, for any x, y and z in X. $\square$

Definition 2.3.17. We call the equivalence class of the point x of X, the *orbit* of x, and we denote it by $\mathcal{O}_G(x) = \{g \cdot x, \ g \in G\}$.

Proposition 2.3.18. *If G acts on a space X, then, the orbit of any point $x \in X$ is a G-invariant subspace of X. (Hence we get an action of G on each orbit.) The orbit is the smallest G-invariant subspace containing x.*

Proof. Take any point $g.x$ of the orbit $\mathcal{O}_G(x)$. Then $g'(g.x) = (gg').x \in \mathcal{O}_G(x)$. $\square$

Example 2.3.19. For the standard action of $GL(n, k)$ on k^n there are only two orbits: $\{0\}$ and $k^n - \{0\}$.

Proposition 2.3.20. *Given an action of G in X, then $X = \bigcup_{disjoint} \mathcal{O}_G(x)$. In particular, when X is finite, $|X| = \sum_{x \in X} |\mathcal{O}_G(x)|$.*

In the case of the action by conjugation, combining propositions 2.3.15 and 2.3.20, we get the important *Class Equation.*

Corollary 2.3.21. (The Class Equation.) *For a finite group G*

$$|G| = |Z(G)| + \sum [G : Z_G(x)],$$

where this sum is over representatives of the non-trivial conjugacy classes.

Corollary 2.3.22. *Any finite group G of prime power order is nilpotent.*

Proof. Let $|G| = p^k$. If the center were trivial, the class equation would say $p^n - 1$ is divisible by p, an obvious contradiction. Therefore $\mathcal{Z}(G)$ is non-trivial. Hence,

$$|G/\mathcal{Z}(G)| = p^r \text{ with } r \leq k - 1.$$

Since $G/\mathcal{Z}(G)$ is a group of lower prime power order, by induction it is nilpotent and hence so is G. $\square$

2.3.2   Transitive Actions

We now study one of the most important types of actions.

Definition 2.3.23. A group G is said to act *transitively* on X, or that the action is *transitive*, if for any two points x and y of the space X, there is a g in G such that $g \cdot x = y$. That is, there is only a single orbit. Alternatively, the set X is said to be a *homogeneous space*.

We now turn to what is essentially the only transitive action. That is, example 2.3.5. Let G be a group and H be a subgroup (not necessarily normal). Consider the action $G \times G/H \to G/H$ given by $(g_1, g_2 H) \mapsto (g_1 g_2) H$. This action is transitive. Indeed, take two points $g_1 H$, and $g_2 H$ in G/H. To find a g' in G such that $g' \cdot g_1 H = g_2 H$, take the point

$g' = g_2 g_1^{-1}$. Then, $g' \cdot g_1 H = g_2 g_1^{-1} g_1 H = g_2 H$. As we shall see, this example is prototypical of a transitive action.

If the action of G on X is transitive, then the set X is a quotient (as set, not as group) of G modulo the stabilizer Stab_x of any point $x \in X$. Precisely,

Theorem 2.3.24. *If the action of G on X is transitive, then for some subgroup H of G it is G-equivariantly equivalent to the action of G on G/H by left translation.*

Proof. Let $x_0 \in X$. Since the action is transitive $\mathcal{O}_G(x_0) = X$. Take for H the stabilizer of x_0 which is a subgroup of G (see 2.3.14) and consider the map

$$f : \mathcal{O}_G(x_0) \longrightarrow G/\text{Stab}_G(x_0) \quad \text{such that} \quad g \cdot x_0 \mapsto g \cdot \text{Stab}_G(x_0).$$

This map is well defined, since $g \cdot x_0 = g' \cdot x_0$, i.e $(g')^{-1} g \cdot x_0 = x_0$. Therefore, $(g')^{-1} g \in \text{Stab}_G(x_0)$, and so $g \cdot \text{Stab}_G(x_0) = g' \cdot \text{Stab}_G(x_0)$. One checks easily that f is bijective. Finally, f is G-equivariant because the following diagram

$$
\begin{array}{ccc}
g \cdot x_0 & \xrightarrow{\ f\ } & g \cdot \text{Stab}_G(x_0) \\
\Big\downarrow{\scriptstyle g_1} & & \Big\downarrow{\scriptstyle g_1} \\
g_1(g \cdot x_0) & \xrightarrow{\ f\ } & g_1 g \cdot \text{Stab}_G(x_0)
\end{array}
$$

is commutative for each $g_1 \in G$.   □

This theorem states that essentially there is only one type of transitive action, namely the left regular action of G on G/H. Here we have dealt with X only as a set. However, if for example G were a Lie group acting transitively on a smooth compact manifold X, and $H = \text{Stab}_G(x)$ is the stabilizer of a point x, then we could conclude G/H is compact and the action is equivariantly equivalent to the smooth action of G on G/H by left translation. Or if G were a topological group and X a compact topological space, then this action would be equivariantly equivalent by a homeomorphism to the continuous action of G on G/H by left translation.

2.3.3   Some Examples of Transitive Actions

Our first example is:

Let $f : G_1 \to G_2$ be a *group homomorphism* and let G_1 act on $f(G_1)$ by left multiplication through f. Then this is a transitive action whose stability group is Ker f. Moreover the G_1-equivariant equivalence given in the theorem above of this action with the action of G_1 on the orbit $f(G_1)$ gives the first isomorphism theorem. Here the isotropy group is normal. In the succeeding examples this will not be so.

2.3.3.1   The Real Sphere S^{n-1}

The group $\mathrm{SO}(n, \mathbb{R})$ acts *by rotations* on the unit sphere S^{n-1}:

$$\mathrm{SO}(n, \mathbb{R}) \times S^{n-1} \longrightarrow S^{n-1} \quad : \quad (g, x) \mapsto g \cdot x.$$

This action is transitive, for any $n \geq 2$. The reason is that for any two points x and y on the sphere, there is a rotation whose axis is perpendicular to the plane defined by x, y and the origin (the center of the sphere) mapping in that way x to y (this plane is not unique if the two points are antipodal.) Since the action is transitive we can take any base point, but in order to determine the stabilizer in a convenient way, we take as our base point in S^{n-1}, the point $e_1 = (1, 0, ..., 0)$. Now, in order for a rotation $g \in \mathrm{SO}(n, \mathbb{R})$ to leave this point fixed, the first column of g must be e_1, so g is an orthogonal matrix of the form

$$g = \begin{pmatrix} 1 & a \\ 0 & B \end{pmatrix}$$

with $\det(B) = 1$. Since g is orthogonal, every invariant subspace has an invariant orthocomplement so $a = 0$ and $B \in \mathrm{SO}(n - 1, \mathbb{R})$. Therefore, the stabilizer of the point e_1 is isomorphic to $\mathrm{SO}(n - 1, \mathbb{R})$, and so by abuse of notation we say $\mathrm{SO}(n - 1, \mathbb{R})$ this is the stabilizer.

When $n = 2$ since $\mathrm{SO}(1, \mathbb{R}) = \{1\}$, we get $\mathrm{SO}(2, \mathbb{R}) = S^1$. When $n = 3$, $\mathrm{SO}(3, \mathbb{R})/\mathrm{SO}(2, \mathbb{R}) \cong S^2$. This shows one can obtain the group $\mathrm{SO}(3, \mathbb{R})$ just by gluing circles to the surface of the sphere in $\mathbb{R}^3$, in such a way that these circles do not intersect!

2.3.3.2   The Complex Sphere S^{2n-1}

The complex sphere[1] is Σ^{n-1} where,

$$\Sigma^{n-1} = \{(z_1, ..., z_n) \in \mathbb{C}^n \ : \ |z_1|^2 + \cdots + |z_n|^2 = 1\}$$

Equivalently, setting $z_k = x_k + iy_k$,

$$\Sigma^{n-1} = \{(x_1, ..., x_n, y_1, ..., y_n) \in \mathbb{R}^{2n} \ : \ x_1^2 + \cdots + x_n^2 + y_1^2 + \cdots + y_n^2 = 1\}.$$

Thus, the complex sphere in $\mathbb{C}^n$ is the real sphere S^{2n-1} in $\mathbb{R}^{2n}$. As in the real case, we define the action

$$\mathrm{SU}(n, \mathbb{C}) \times S^{2n-1} \longrightarrow S^{2n-1}$$

which we prove is transitive just below. The stabilizer of a point is $\mathrm{SU}(n - 1, \mathbb{C})$. Therefore,

$$\mathrm{SU}(n, \mathbb{C})/SU(n - 1, \mathbb{C}) \cong S^{2n-1}.$$

Here again as in the real case, we identify $\mathrm{SU}(n - 1, \mathbb{C})$ with a subgroup of $\mathrm{SU}(n, \mathbb{C})$.

Proposition 2.3.25. *The special unitary group* $\mathrm{SU}(n)$ *acts transitively on the sphere* S^{2n-1} *in* $\mathbb{C}^n$, $n \geq 2$.

Proof. It suffices to show that $\mathrm{SU}(n)$ maps $e_1 = (1, 0, \ldots, 0)$ to any other vector v_1 of length 1 in $\mathbb{C}^n$. We first show that, given $x \in S^{2n-1}$ there is $g \in \mathrm{U}(n)$ such that $ge_1 = x$, where $e_1, \ldots, e_n$ is the standard basis for $\mathbb{C}^n$. That is, we construct $g \in \mathrm{U}(n)$ such that the left column of g is x. Indeed, complete x to a $\mathbb{C}^n$-basis $x, x_2, x_3, ..., x_n$ for $\mathbb{C}^n$. Then apply the Gram-Schmidt process to obtain an orthonormal with respect to the standard Hermitian inner product basis $x, u_2, ..., u_n$.

---

[1]Similarly to the real case, we could try to define the complex unit sphere as the complex hyper-surface

$$S_{\mathbb{C}}^{n-1} = \{(z_1, ..., z_n) \in \mathbb{C}^n \ : \ z_1^2 + \cdots + z_n^2 = 1\}.$$

However, this is a non-compact set in $\mathbb{C}^n$. Therefore, a compact group such as $\mathrm{SU}(n, \mathbb{C})$ cannot act transitively on it.

The condition $g^*g = 1$, is the assertion that the columns of g form an orthonormal basis. Thus, taking x, $u_2,...,$ u_n as the columns of g gives $g \in \mathrm{U}(n)$ such that $ge_1 = x$. To make $\det(g) = 1$, replace u_n by $(\det g)^{-1}u_n$. Since $|\det(g)| = 1$, this change does not harm the orthonormality.

The isotropy group Stab_{e_n} of the last standard basis vector $e_n = (0, \ldots, 0, 1)$ is:

$$\left\{ g = \begin{pmatrix} A & 0 \\ 0 & 1 \end{pmatrix}, \text{ with } A \in \mathrm{SU}(n-1) \right\}.$$

$\square$

Therefore
$$\mathrm{SU}(n)/\mathrm{SU}(n-1) \cong S^{2n-1}.$$
When $n = 2$, since $\mathrm{SU}(1, \mathbb{C}) = \{1\}$, $\mathrm{SU}(2, \mathbb{C})) \cong S^3$.

2.3.3.3   The Quaternionic Sphere S^{4n-1}

Let $\mathbb{H}$ be the skew field of quaternions (which will be discussed in Chapter 5) and let

$$\mathbb{H}^n = \{(q_1, q_2, \ldots, q_n), \ q_i \in \mathbb{H}, \ i = 1, \ldots, n\}.$$

We shall equip $\mathbb{H}^n$ with the following Hermitian inner product

$$\langle (q_1, q_2, \ldots, q_n), (w_1, w_2, \ldots, w_n) \rangle = \sum_{i=1}^{n} \bar{q}_i w_i$$

where $\bar{q}_i$ is the conjugate of q_i for all $i = 1, \ldots, n$. Now, the group of isometries of $\mathbb{H}^n$ is the called the compact symplectic group of rank n denoted by $\mathrm{Sp}(n)$.

As an example, $\mathrm{Sp}(1)$ is the multiplicative group of unit quaternions acting on $\mathbb{H}$ by left multiplication, in other words

$$\mathrm{Sp}(1) = \{q \in \mathbb{H}, : q\bar{q} = 1\} \cong S^3.$$

The action is given by

$$S^3 \times \mathbb{H} \longrightarrow \mathbb{H} : (q, q_1) \mapsto qq_1.$$

Proposition 2.3.26. *The group* $\mathrm{Sp}(n)$ *acts transitively on the sphere* S^{4n-1} *in* $\mathbb{H}$.

Proof. Let $e_1 = (1, 0, \ldots, 0, 0), \ldots, e_n = (0, 0, \ldots, 0, 1)$ in $\mathbb{H}^n$ and let $g \in \mathrm{Sp}(n)$. Then, $\{q_i = g(e_i),\ 1 \le i \le n\}$ is an orthonormal basis of $\mathbb{H}^n$, i.e.

$$\langle q_i, q_j \rangle = \langle g(e_i), g(e_j) \rangle = \langle e_i, e_j \rangle = \delta_{ij}.$$

Conversely, if $(q_i)_{i=1,\ldots,n}$ is an orthonormal basis in $\mathbb{H}^n$, then it exists a unique $g \in \mathrm{Sp}(n)$ such that $g(e_i) = q_i$ for all $i = 1, \ldots, n$. In addition, using the Gram-Schmidt orthogonalization process we can extend any unit vector $q_1 \in \mathbb{H}^n$ to an orthonormal basis $(q_i)_{i=1,\ldots,n}$ in $\mathbb{H}^n$. This implies that the compact symplectic group $\mathrm{Sp}(n)$ acts transitively on the unit sphere

$$S^{4n-1} = \left\{ v = (q_1, \ldots, q_n),\ :\ |v|^2 = \sum_{i=1}^{n} \bar{q}_i q_i = 1 \right\}.$$

$\square$

Let Stab_{e_n} be the stabilizer of this action. Clearly

$$\mathrm{Stab}_{e_n} \cong \mathrm{Sp}(n-1).$$

Here again as in the complex case, we identify $\mathrm{Sp}(n-1, \mathbb{H})$ with a subgroup of $\mathrm{Sp}(n, \mathbb{H})$. Therefore,

$$S^{4n-1} \cong \mathrm{Sp}(n)/\mathrm{Sp}(n-1).$$

2.3.3.4 The Poincaré Upper Half-Plane

The *Poincaré upper half-plane*, $\mathcal{H}$, is defined as the open subset of $\mathbb{R}^2$

$$\mathcal{H} := \{z \in \mathbb{C} \mid \mathrm{Im}(z) > 0\}.$$

The group $\mathrm{SL}(2, \mathbb{R})$ acts on this space by

$$\mathrm{SL}(2, \mathbb{R}) \times \mathcal{H} \longrightarrow \mathcal{H}, \quad (A, z) \mapsto A \cdot z := \frac{az + b}{cz + d}$$

where $A = \begin{pmatrix} a & b \\ c & d \end{pmatrix}$, with $ad - bc = 1$. We can see easily that this action is transitive. Now, to find the stabilizer of a point, we first observe that since the action is transitive it suffices to find the stabilizer of the point i. So, we are looking to find a matrix

$$A = \begin{pmatrix} a & b \\ c & d \end{pmatrix} \quad \text{such that} \quad A \cdot i := \frac{ai + b}{ci + d} = i.$$

From this we see $a = d$ and $-b = c$, and so A is

$$A = \begin{pmatrix} a & b \\ -b & a \end{pmatrix}.$$

Since $A \in \mathrm{SL}(2, \mathbb{R})$, $a^2 + b^2 = 1$. In other words A is special orthogonal. Hence for each $z \in \mathcal{H}$, $\mathrm{Stab}_z \cong \mathrm{SO}(2, \mathbb{R})$. Therefore, $\mathrm{SL}(2, \mathbb{R}) / \mathrm{SO}(2, \mathbb{R}) \cong \mathcal{H}$.

Notice here that the group acting is non-compact, as is the space being acted upon.

2.3.3.5   Real Projective Space $\mathbb{RP}^n$

The *real projective space* $\mathbb{RP}^n$, is defined as the set of all lines through the origin in $\mathbb{R}^{n+1}$ (i.e. the set of all 1-dimensional subspaces). Since a 1-dimensional subspace of $\mathbb{R}^{n+1}$ as spanned by a non-zero vector v, we could also view projective space as the quotient space

$$\mathbb{R}^{n+1} - \{0\} / \sim, \text{ where } v \sim u \iff u = \lambda v, \text{ for some } \lambda \in \mathbb{R} - \{0\}.$$

We denote the equivalence class of v by $[v]$.

A third equivalent definition of $\mathbb{RP}^n$ is the following: Since every line l in $\mathbb{R}^{n+1}$, intersects the sphere S^n in two antipodal points, we can consider real projective space to be the quotient of S^n by identifying antipodal points. This shows S^n is a double cover of the space $\mathbb{RP}^n$.

Now, we let the group $\mathrm{SO}(n + 1, \mathbb{R})$ act on $\mathbb{RR}^n$ as follows: any element R of $\mathrm{SO}(n + 1, \mathbb{R})$ (i.e. any rotation R) takes any line $l = [v]$ and sends it to $R \cdot [v] = [R \cdot v]$. This action clearly is well defined (it does not depend on the representative v of the equivalence class $[v]$). It is also transitive (check!).

The stabilizer of the line $l = [e_1]$ where $e_1 = (1, 0, ..., 0) \in \mathbb{R}^{n+1}$ is the orthogonal group $O(n, \mathbb{R})$ and therefore,

$$\mathbb{RP}^n \cong SO(n+1, \mathbb{R})/O(n, \mathbb{R}).$$

This gives a picture of real projective space for $n = 1$, and 2. If $n = 1$. The equivalence relation identifies opposite points on S^1. Any line through the center of the circle will intersect at least one point in the upper half of the circle. Furthermore, the only line that intersects it in two points is the horizontal line (the x-axis). So this means that a set of representatives can be taken to be those (x, y) on the circle with $y > 0$, along with one of the points $(1, 0)$ or $(-1, 0)$. This shows that

$$\mathbb{RP}^1 \cong S^1.$$

The reader should think about what $\mathbb{RP}^2$ *looks like*, using an analogous construction where S^1 is replaced by S^2. Here, what happens at the boundary (the equator of the sphere) gets more interesting. One finds

$$\mathbb{RP}^2 \cong S^2.$$

2.3.3.6   Complex Projective Space $\mathbb{CP}^n$

In a similar way, we define the *complex projective space* $\mathbb{CP}^n$ as the space of all 1-dimensional subspaces of $\mathbb{C}^n$ i.e. the set of all complex lines passing through the origin. We can view $\mathbb{CP}^n$ as the quotient space $\mathbb{C}^{n+1} - \{0\}/ \sim$, where the equivalence relation $\sim$ is defined by

$$v \sim u \Leftrightarrow v = \lambda u, \text{ for some } \lambda \in \mathbb{C} - \{0\}.$$

As in the real case we denote by $[v]$ the equivalence class of v, i.e, the line defined by the non-zero vector v. Now, let l be the line $l = [z]$, with $z = (z_1, ..., z_{n+1}) \in \mathbb{C}^{n+1} - \{0\}$. Then, a point $v \in \mathbb{C}^{n+1}$ lies on the line l if and only if $z \sim v$, i.e.

$$v = \lambda(z_1, ..., z_{n+1}), \quad \lambda \in \mathbb{C} - \{0\}.$$

This implies

$$l \cap \Sigma^n = \{\lambda(z_1, ..., z_{n+1}) \in \mathbb{C}^{n+1} \; : \; \lambda\bar{\lambda}(z_1\bar{z}_1, ..., z_{n+1}\bar{z}_{n+1}) = 1\}.$$

Therefore,

$$l \cap \Sigma^n = \left\{\lambda(z_1, ..., z_{n+1}) \in \mathbb{C}^{n+1} \; : \; |\lambda| = \frac{1}{(|z_1|^2 + \cdots + |z_{n+1}|^2)^{1/2}}\right\}.$$

In other words, there is a 1 : 1 correspondence between $l \cap \Sigma^n$ and the circle, S^1, i.e., geometrically, $l \cap \Sigma^n \cong S^1$. But, as we observed in example 2.3.3.2, Σ^n can be seen as the real sphere S^{2n+1} in $\mathbb{R}^{2n+2}$. This implies that complex projective space is the quotient space of the sphere S^{2n+1} by the equivalence relation $\sim$, defined by

$$z \sim v \;\Leftrightarrow\; z, v \in l \cap \Sigma^n,$$

for some line l passing through the origin. Hence,

$$\mathbb{CP}^n \cong S^{2n+1}/S^1.$$

In the case $n = 1$, using the fact that $\mathbb{RP}^2 \cong S^2 \cong \mathbb{CP}^1$, we get the isomorphism

$$S^3/S^1 \cong S^2$$

known as the *Hopf fibration*.[2] The geometric meaning of this is that the fiber (the inverse image) over any point p of the sphere S^2 is a circle in S^3, so S^3 can be viewed as the union of a family of disjoint circles.

---

[2]Heinz Hopf (1894-1971) German-Jewish mathematician was a major figure in topology. His dissertation classified simply connected complete Riemannian 3-manifolds of constant sectional curvature. He proved the famous Poincaré-Hopf theorem on the indices of zeros of vector fields on compact manifolds connecting their sum to the Euler characteristic. Hopf discovered the Hopf invariant of maps $S^3 \to S^2$ and proved that the Hopf fibration has invariant 1. He also invented what became known as Hopf algebras. After he left Germany during the Nazi period and was fortunate to find a position at the ETH in Zurich where he received a letter from the German foreign ministry stating that if he didn't return to Germany he would lose his German citizenship! His lifelong friendship with Paul Alexandrov resulted in their well known topology text. Three volumes were planned, but only one was finished. It was published in 1935.

2.3.3.7 The Grassmann and Flag Varieties

We now turn to the Grassmann and flag varieties. These turn out to be compact, connected, real (or complex) manifolds spaces. Indeed, they are projective varieties. The complex and algebraic structures are closely related. Compactness ensures the finite dimensionality of cohomology spaces and allows for integration on the manifold. They are homogeneous spaces and as such they can be represented as coset spaces of Lie groups.[3]

We make one remark concerning dimension. Since $GL(n, \mathbb{R})$ is *open* in $M(n, \mathbb{R})$ and $GL(n, \mathbb{C})$ is *open* in $M(n, \mathbb{C})$ (both because det is continuous), if $S \subseteq GL(V)$ its real or complex dimension regarded as a subset of $End(V)$, is its dimension as a manifold.

Let V be a real or complex vector space of dimension n. For each integer $1 \le r \le n$. The *Grassmann space* $\mathcal{G}(r, n)$ is the set of all subspaces of V of dimension r. Thus when $r = 1$ this is just the real or complex projective space. We give it a topology and manifold structure as follows. Evidently $GL(V)$ acts transitively and continuously on $\mathcal{G}(r, n)$ under its natural action. Hence $\mathcal{G}(r, n)$ is a $GL(V)$ space. What is the isotropy group? If $g \in GL(V)$ stabilizes an r-dimensional subspace W, *i.e. a point in the Grassmann space*, then

$$g = \begin{pmatrix} A & B \\ 0 & C \end{pmatrix}$$

and obviously any such g will do. This makes $\mathcal{G}(r, n)$ a homogeneous space, i.e. a quotient space of $G = GL(V)$ modulo the isotropy group above. In the complex case, since both $GL(V)$ and the isotropy group are complex Lie groups, $\mathcal{G}(r, n)$ is actually a complex manifold. Also, (both over $\mathbb{R}$ or $\mathbb{C}$) we see that $\dim(\mathcal{G}(r, n)) = r(n - r)$. Moreover, in both the real and complex cases $GL(V)/\operatorname{Stab}_{GL(V)}(W) = \mathcal{G}(r, n)$ is compact. To see this, we need only show that a compact subgroup of $GL(V)$ acts transitively and as we shall see immediately below, the compact groups $SO(n, \mathbb{R})$ (respectively $U(n, \mathbb{C})$) act transitively.

---

[3]We encounter these manifolds in many different branches of mathematics: differential geometry, algebraic geometry, Lie groups, representation theory, Coxeter groups, twistor theory, and mathematical physics.

The group $O(n)$ acts naturally on $\mathcal{G}(r,n)$ by matrix multiplication. This action is transitive. Indeed, let W be a $GL(V)$ invariant subspace and let $w_1, \ldots w_r$ be an orthonormal basis of W. Extend this to a complete orthonormal basis of V. If we have a different invariant subspace of the same dimension and apply the same procedure. Then there is an *orthogonal* map of V taking the first basis to the second in the same order (for all this see Chapter 4). Thus $O(n, \mathbb{R}))$ acts transitively. If g is an orthogonal matrix stabilizing W, then it has the form,

$$\begin{pmatrix} A & 0 \\ 0 & C \end{pmatrix}$$

where $A \in O(r)$ and $C \in O(n-r)$. Hence,

$$\mathcal{G}(r,n) \cong O(n)/O(r) \times O(n-r)$$

(a completely similar calculation works in the complex case using the unitary group).

Actually, even $SO(n)$ acts transitively on $\mathcal{G}(r,n)$, and we get

$$\mathcal{G}(r,n) \cong SO(n+1)/S\big(O(n) \times O(k)\big).$$

Here, by $S\big(O(n) \times O(1)\big)$ we denote the subgroup of $SO(r+k)$ containing all matrices of the form,

$$\begin{pmatrix} A & 0 \\ 0 & C \end{pmatrix}$$

with $\det = 1$.

Finally as we have just seen, in both the real and complex cases, there is a connected group acting transitively. It follows that $\mathcal{G}(r,n)$ is connected.

We now turn to Flag Varieties. Here, for brevity, we shall restrict ourselves to the complex case.

Let V be a complex vector space of dimension n. A *flag* is a sequence of nested subspaces:

$$\{0\} \subset V_1 \subset V_2 \subset \cdots \subset V_n = \mathbb{C}^n,$$

where each dim $V_i = i$. The set of all such flags is a manifold called the *flag manifold*.

The simplest flag manifold is complex projective space $\mathbb{CP}^1$ (the set of lines).

Given a flag we can associate a basis to it by choosing any non-zero vector $v_1 \in V_1$, then a vector $v_2 \in V_2$ so that v_1 and v_2 are linearly independent, and so on. Conversely, given a basis $v_1,...,v_n$ of $V = \mathbb{C}^n$ we define a flag by taking

$$(0) \subset V_1 = \text{lin.sp}(\{v_1\}) \subset \cdots \subset V_{n-1} = \text{lin.sp}(\{v_1, ..., v_{n-1}\}) \subset V.$$

This relationship of flags to bases of V shows the action of $\text{GL}(V) = \text{GL}(n, \mathbb{C})$ on V gives rise to a transitive action of $\text{GL}(n, \mathbb{C})$ on $\mathcal{F}_n$. Given a fixed flag, $g \in \text{GL}(n, \mathbb{C})$ preserves this flag exactly when g $in T_n$, the full group of upper triangular matrices. This complex group is called a Borel subgroup B of G.[4] Hence,

$$\mathcal{F}_n = \text{GL}(n, \mathbb{C})/B,$$

making $\mathcal{F}_n$ into a complex, connected, homogeneous manifold. From this we see $\mathcal{F}_n$ has complex dimension $\frac{n(n-1)}{2}$.

Proposition 2.3.27. *Actually,* $\text{U}(n)$ *already acts transitively on* $\mathcal{F}_n$, *with isotropy group* T *(a maximal torus of* $\text{U}(n)$*), the diagonal matrices with entries of modulus* 1. *Thus* $\text{U}(n)/T$ *is also the space of all flags. Since* $\text{U}(n)$ *is a compact connected group,* $\mathcal{F}_n$ *is compact and connected.*

Proof. Given any flag, one can choose an orthonormal basis which generates the flag. Since the group $\text{U}(n)$ is transitive on orthonormal bases, it is also transitive on flag manifolds. This results in a diffeomorphism:

$$\mathcal{F}_n \cong \text{U}(n)/T,$$

where T is the stabilizer of a point. Evidently $T = U(n) \cap T_n$ which is a maximal torus in $U(n)$. $\qquad\qquad \square$

---

[4]Armand Borel (1923-2003) Swiss mathematician, worked in algebraic topology, Lie groups, and was one of the creators of the theory of linear algebraic groups. He proved a fixed point theorem concerning the action of a Zariski connected solvable group on a complete variety which generalized Lie's theorem. His 1962 paper with Harish-Chandra on algebraic groups over $\mathbb{Q}$ and their lattices is very important.

As an exercise, we invite the reader to work out the situation for the real flag manifold.

2.4 PSL(n, k), A_n & Iwasawa's Double Transitivity Theorem

Lemma 2.4.1.

1. $\mathcal{Z}(\text{GL}(n, \mathbb{R}))$ *consists of the non-zero real scalar matrices and* $\mathcal{Z}(\text{GL}(n, \mathbb{C}))$ *of the non-zero complex scalar matrices.*

2. *Here:*

$$\text{SL}(n, \mathbb{R}) \cap \{\lambda I : \lambda \in \mathbb{R}\} = \begin{cases} \{I\}, & \textit{if } n \textit{ is odd} \\ \{\pm I\} & \textit{if } n \textit{ is even.} \end{cases}$$

3. *Also,*

$$\text{SL}(n, \mathbb{C}) \cap \{\lambda I : \lambda \in \mathbb{C}\}$$
$$= \{\omega^j I : \omega \textit{ primitive } n\textit{-th root of unity, } j = 1, \ldots, n\}.$$

4. *For $k = \mathbb{R}$ or $\mathbb{C}$,*

$$\text{GL}(n, k) = \text{SL}(n, k) \cdot \{\lambda I : \lambda \neq 0, \lambda \in k\}.$$

Proof. To compute $\mathcal{Z}(\text{GL}(n, \mathbb{R}))$ and $\mathcal{Z}(\text{GL}(n, \mathbb{C}))$, we first note that as we will see in Chapter 5 if A commutes with $M(n, \mathbb{R})$ (respectively $M(n, \mathbb{C})$), then A is a scalar matrix. However, since $\text{GL}(n, \mathbb{R}))$ is dense in $M(n, \mathbb{R})$ and $\text{GL}(n, \mathbb{C})$ is dense in $M(n, \mathbb{C}))$ (in the Euclidean topology) and matrix multiplication is continuous. The same is true for the groups. If $\lambda \in \mathbb{R}$ and $\lambda^n = 1$, then if n is odd $\lambda = 1$ while if n is even $\lambda = \pm 1$. If $\lambda \in \mathbb{C}$ and $\lambda^n = 1$, then λ can be any of the n-th roots of unity. Finally, $\text{SL}(n, k) \cdot \{\lambda I : \lambda \neq 0, \lambda \in k\}$ does fill out all of $\text{GL}(n, k)$ by a dimension count since the intersection of the subgroups is finite (discrete). $\square$

Exercise 2.4.2. Let $k = \mathbb{R}$, $\mathbb{C}$ and consider the homomorphism

$$\det : \mathrm{GL}(n,k) \longrightarrow k^\times = \{\lambda \in k \ : \ \lambda \neq 0\}.$$

This gives the following exact sequence:

$$(1) \longrightarrow \mathrm{SL}(n,\ k) \hookrightarrow \mathrm{GL}(n,k) \longrightarrow k^\times.$$

Does this sequence split?

1. Show $\mathrm{GL}(n,\ k)$ always equals $\mathrm{SL}(n,\ k) \cdot \lambda I$.

2. Show this is a direct product if and only if $k = \mathbb{R}$ and n is odd.

Let $E_{ij}(\lambda)$, $i,\ j = 1, ..., n$, $i \neq j$ be the matrices that have 1's on the diagonal, $\lambda \in k$ in the (i,j)- entry, and 0 everywhere else. We call these matrices the *elementary matrices*. Clearly, $\det E_{ij}(\lambda) = 1$, so $E_{ij}(\lambda) \in \mathrm{SL}(n,k)$. For every $n \times n$ matrix A the product matrix $E_{ij}(\lambda)A$ is obtained from A by adding λ times the j-th row of A to the i-th row and leaving the remaining entries unchanged. Similarly $AE_{ij}(\lambda)$ is obtained by adding λ times the i-th column of A to the j-th column and leaving the remaining entries unchanged.

The *projective linear group* $\mathrm{PGL}(n,k)$ and the *projective special linear group* $PSL(n,k)$ are defined as the quotients of $\mathrm{GL}(n,k)$ and $\mathrm{SL}(n,k)$ by their centers:

$$\mathrm{PGL}(n,k) := \mathrm{GL}(n,k)/\mathcal{Z}(\mathrm{GL}(n,k)) = \mathrm{GL}(n,k)/k^\times,$$

$$\mathrm{PSL}(n,k) = \mathrm{SL}(n,k)/\mathcal{Z}(\mathrm{SL}(n,k)) = \mathrm{SL}(n,k)/\{z \in k^\times \mid z^n = 1\}.$$

When k is the finite field $\mathbb{Z}_p$, we write $\mathrm{PGL}(n,p)$ and $\mathrm{PSL}(n,p)$ respectively.[5]

Now, the group $\mathrm{GL}(n,k)$ acts on the projective space $k\mathbb{P}^{n-1}$, since an invertible linear transformation maps a subspace to another subspace of the same dimension.

---

[5]The groups $\mathrm{PSL}(2,p)$ were constructed by Evariste Galois in the 1830s. Galois viewed them as fractional linear transformations, and in his last letter to Chevalier, he observed that they were simple except if $p = 2$, 3. In the same letter Galois constructed the general linear group over a prime field. The groups $\mathrm{PSL}(n,p)$ were constructed in the classic text (1870) by Camille Jordan, Traité des substitutions et des équations algébriques.

Proposition 2.4.3. *The kernel of this action of* $\mathrm{GL}(n,k)$ *is*

$$\{\lambda I, \quad \lambda \neq 0 \in k\}.$$

Proof. Let $A \in \mathrm{GL}(n,k)$ be a matrix which stabilizes any 1-dimensional subspace of k^n. Hence, if e_1, ..., e_n are the standard basis vectors of k^n, then, $Ae_i = \lambda_i e_i$ for some $\lambda_i \in k^\times$ which means that $A = \mathrm{diag}(\lambda_1, \ldots, \lambda_n)$. We claim that $\lambda_i = \lambda_j$ for all i, j since, otherwise, A will not stabilize $k(e_i + e_j)$:

$$A(e_i + e_j) = \lambda_i e_i + \lambda_j e_j \in k(e_i + e_j) \quad \Longleftrightarrow \quad \lambda_i = \lambda_j.$$

Thus $A = \lambda I_n$. □

Definition 2.4.4. The action of a group G on a set X is called *doubly transitive*, or *twofold transitively* if G acts transitively on the set of all ordered pairs of distinct elements of X. In other words, given any x_1, x_2, y_1, $y_2 \in X$ so that $x_1 \neq x_2$ and $y_1 \neq y_2$, there is a $g \in G$ so that $gx_i = y_i$ for $i = 1$, 2. (Any pair of distinct points in X can be carried to any other pair of distinct points by some element of G.)[6]

Example 2.4.5. If k is any field, the affine group

$$\mathrm{Aff}(G) = \left\{ \begin{pmatrix} a & b \\ 0 & 1 \end{pmatrix}, \quad a, \ b \in k \right\}$$

acts on k by $\begin{pmatrix} a & b \\ 0 & 1 \end{pmatrix} x = ax + b$ and this action is doubly transitive.

Exercise 2.4.6. Show that the affine group of k^n acts twofold transitively on k^n.

Proposition 2.4.7. *If G acts doubly transitive on a set X, then the stabilizer,* Stab_x, *of any $x \in X$ is a maximal subgroup of G.*

---

[6]When G has more structure, for example if X has a metric and G acts by isometries, then for twofold transitivity we would, of course, also require $d(x_1, x_2) = d(y_1, y_2)$.

Proof. Suppose that the stabilizer Stab_x is not maximal. Then, there is a subgroup H of G so that $\text{Stab}_x < H < G$. Thus, there is a $g \in G$, $g \notin H$ and some $h \in H$, $h \notin \text{Stab}_x$. As g and h are not in Stab_x, neither gx nor hx is equal to x. Since G acts doubly transitive on X, there is an $l \in G$ so that $lx = x$ and $l(gx) = hx$. But then, l and $h^{-1}lg$ are both in Stab_x. Therefore, $g \in l^{-1}h\,\text{Stab}_x \subseteq H$, a contradiction. $\square$

We remark that the converse of the above proposition is false. Indeed, let H be a maximal subgroup of a finite group G and let G act by left multiplication on G/H. Then this action has as stabilizer H, but it is not doubly transitive if the order of G is odd. This is because a finite group with a doubly transitive action must have even order.

Proposition 2.4.8. *Let k be a field and $n \geq 2$. Then the action of $\text{SL}(n,k)$ and thus of $\text{PSL}(n,k)$, on projective space $P^{n-1}(k)$ is doubly transitive.*

Proof. Suppose that l_1, l_2 are distinct elements of $k\mathbb{P}^{n-1}$ ($l_i \subseteq k^n$, $i = 1,2$). Then any non-zero vectors $v_1 \in l_1$ and $v_2 \in l_2$ will be linearly independent. Therefore, one can extend this set to obtain a basis for k^n (see Chapter 3). This implies that there is a matrix $A = \text{diag}(a_1, a_2, 1, \ldots, 1) \in \text{GL}(n,k)$ so that $Ae_i = v_i$ with $i = 1, 2$. Now, setting

$$B = A \cdot \text{diag}(a_1^{-1}, a_2^{-1}, 1, \ldots, 1) \in \text{SL}(n,k),$$

B sends e_1 to $a_1^{-1}v_1$ and e_2 to $a_2^{-1}v_2$. Therefore, B takes the line through v_i to l_i, for $i = 1, 2$. $\square$

Proposition 2.4.9. *If the group G acts doubly transitively on a set X, then any normal subgroup N of G acts on X either transitively or trivially.*

Proof. Suppose that N does not act trivially on X. Then, there is some $x \in X$ such that $nx \neq x$ for some $1 \neq n \in N$. Now, pick any two points y and y_1 in X, with $y \neq y_1$. Since G acts doubly transitively, there is a

$g \in G$ such that $gx = y$ and $g(nx) = y_1$. Hence,

$$y_1 = (gng^{-1})(gx) = (gng^{-1})y,$$

and $gng^{-1} \in N$, which shows that N acts transitively on X.   □

**Theorem 2.4.10. (*Iwasawa Double Transitivity Theorem*[7]) *Sup-
pose the group G acts doubly transitively on X and that*

  1. *For some $x \in X$ the stabilizer Stab_x has an Abelian normal sub-
     group whose conjugate subgroups generate G.*

  2. $[G, G] = G$.

Then, G/Stab_x is a simple group.

Proof. First, we will show that there is no normal subgroup of G lying
between Stab_x and G. To see this, assume that there is one, say N such
that $\mathrm{Stab}_x \subset N \subset G$. Now, consider $x \in X$. By Proposition 2.4.7,
this is a maximal subgroup of G. Since N is normal, $N\,\mathrm{Stab}_x = \{nh :
n \in N,\ h \in \mathrm{Stab}_x\}$ is a subgroup of G, and it contains Stab_x, so by
maximality either $N\,\mathrm{Stab}_x = \mathrm{Stab}_x$, or $N\,\mathrm{Stab}_x = G$. By Proposition
2.4.9, N acts trivially or transitively on X.

If $N\,\mathrm{Stab}_x = \mathrm{Stab}_x$ then $N \subset \mathrm{Stab}_x$, so N fixes x. Therefore, N
does not act transitively on X, which implies that N must act trivially
on X so that $N \subset \mathrm{Stab}_x$. Since $\mathrm{Stab}_x \subset N$ by hypothesis we get
$N = \mathrm{Stab}_x$. Now suppose $N\,\mathrm{Stab}_x = G$. Let A be the Abelian normal
subgroup of Stab_x as in hypothesis (1). Then its conjugates generate
G. Since $A \subset \mathrm{Stab}_x$, $NA \subset N\,\mathrm{Stab}_x = G$. Then, for $g \in G$,

$$gAg^{-1} \subset g(NA)g^{-1} = NA.$$

---

[7]A variant of Iwasawa's theorem is the following: Suppose that the group G acts
primitively (see definition 2.5.2) on a set X, that $G = [G, G]$ and that for some
$x \in X$ the stabilizer G_x contains a normal Abelian subgroup A whose conjugates
generate G. Let K be the kernel of the action. Then the quotient group G/K is
simple. Both versions were published in the early 1940s.

Thus NA contains all the conjugates of A and since by hypothesis, $NA = G$, we see

$$G/N = (NA)/N \cong A/(N \cap A).$$

Since A is Abelian, this isomorphism shows G/N is also Abelian. Therefore $[G, G] \subset N$, and since by hypothesis, $[G, G] = G$, we conclude $N = G$.   □

Corollary 2.4.11. *The group* PSL(n, k) *is simple*[8] *for any field k whose order is at least 3 and for any integer $n \geq 3$.*

Proof. This is a consequence of Iwasawa's double transitivity theorem. Since we know that PSL(n, k) acts doubly transitively (see 2.4.8), it remains to show:

1. The stabilizer of some point in $k\mathbb{P}^{n-1}(k)$ has an Abelian normal subgroup A.

2. The subgroups of SL(n, k) that are conjugate to A generate SL(n, k),

3. $[\text{SL}(n, k) , \text{SL}(n, k)] = \text{SL}(n, k)$.

Proof of 1.

To show that the stabilizer of some point in $k\mathbb{P}^{n-1}(k)$ has an Abelian normal subgroup, using just transitivity we can look at the stabilizer Stab_p of the point $p = (1, 0, ..., 0) \in k\mathbb{P}^{n-1}(k)$. This is a group of $n \times n$-matrices of the form

$$\begin{pmatrix} \lambda & \star \\ 0 & B \end{pmatrix}$$

with $\lambda \in k - (0)$ and $B \in \text{GL}(n-1, k)$.

A matrix of that form is in SL(n, k). Taking $\lambda = \frac{1}{\det B}$ we see this matrix has det $= 1$.

---

[8]In the case of an abstract group the notion of simplicity is stronger than in the case of a Lie group. That is, there are no non-trivial normal subgroups at all, as opposed to possibly having discrete central subgroups. Thus, in particular, this theorem proves that PSL(n, k) and SL(n, k) are simple Lie groups when $k = \mathbb{R}$ or $\mathbb{C}$.

Now, the homomorphism

$$\pi : \text{Stab}_p \rightarrow \text{GL}(n-1,k) \quad : \quad \begin{pmatrix} \lambda & \star \\ 0 & B \end{pmatrix} \mapsto B$$

has as kernel the subgroup

$$A = \left\{ \begin{pmatrix} \lambda & \star \\ 0 & I_{n-1} \end{pmatrix} \right\}.$$

This is the Abelian normal subgroup we want.

Proof of 2.

The elementary matrices $E_{ij}(\lambda)$ generate $\text{SL}(n,k)$.

Since $n \geq 2$, let $A \in \text{SL}(n,k)$. It suffices to reduce A to the identity matrix by elementary row and column operations. So, suppose first that $a_{21} \neq 0$. Then the $(1,1)$- entry of $E_{12}(\lambda)A$ will be equal to 1 for some (unique) $\lambda \in k$. If $a_{21} = 0$, then we can make it non-zero by a row operation $E_{2j}(1)$ for some j. Thus, we may assume that $a_{11} = 1$. Now multiplying on the left by the elementary matrices $E_{i1}(-a_{i1})$ and on the right by $E_{1j}(-a_{1j})$ will clear the first row and column of A producing a matrix $\begin{pmatrix} 1 & 0 \\ 0 & B \end{pmatrix}$, where $B \in \text{SL}(n-1,k)$. We now proceed by induction on n.

Proof of 3.

We will prove that the elementary matrices $E_{ij}(\lambda)$ are commutators in $\text{SL}(n,k)$ except when $n = 2$ and $|k| \geq 3$.

If $n \geq 3$, there is a third index l and

$$[E_{il}(\lambda) , E_{lj}(1)] = E_{ij}(\lambda).$$

If $n = 2$, we use the commutator relation:

$$\left[\begin{pmatrix} a & 0 \\ 0 & a^{-1} \end{pmatrix} , \begin{pmatrix} 1 & b \\ 0 & 1 \end{pmatrix} \right] = \begin{pmatrix} 1 & (a^2 - 1)b \\ 0 & 1 \end{pmatrix}.$$

Now, if $\lambda \in k$ the equation

$$\lambda = (a^2 - 1)b$$

can be solved for b if there is a unit $a \in k^\times$ so that $a \neq \pm 1$. This can happen only if k has at least three elements. $\square$

Corollary 2.4.12. *The alternating group A_n is simple for $n \geq 5$.*

Proof. The group A_n acts on the set $\{1, 2, 3, 4, 5\}$ doubly transitively. For this we can assume $n = 5$. Let $x = 5$, so $\text{Stab}_x = A_4$ which has the Abelian normal subgroup

$$\{(1), (12)(34), (13)(24), (14)(23)\}.$$

As we learned in Lemma 2.2.15), the group A_n, $n \geq 5$ is generated by its three cycles, and all three cycles are conjugate. Hence $[A_n , A_n] = A_n$, and by Iwasawa' theorem A_n is simple. $\square$

2.5   Imprimitive Actions

This notion has much to do with infinite dimensional group representations and ergodic theory (as well as physics). It is mostly due to G.W. Mackey, see e.g. [74] p.253 for the continuous imprimitivity theorem and p.285 for the converse. This section is also related to the section of Chapter 12 concerning Clifford's theorem.

Definition 2.5.1. A family $\{X_i : i \in I\}$ of subsets of the space X is called a *partition* of X if $X = \cup_{i \in I} X_i$ and $X_i \cap X_j = \varnothing$, whenever $i \neq j$, where i and $j \in I$.

A partition is called *G*-invariant if for each $g \in G$ and $i \in I$, gX_i is in the partition. That is $gX_i = X_j$ for some $j \in I$. Of course the j depends on both $g \in G$ and $i \in I$. Now there are three partitions of X which are clearly invariant: The partition of X into G-orbits, the partition, $X = \varnothing \cup X$, and the partition of X into its individual points. These last are to be regarded as trivial so the first one tells us we must consider only transitive actions and then we rule out the other two.

Definition 2.5.2. A transitive action which has only the two trivial invariant partitions is called *primitive*; otherwise it is called *imprimitive*.

Theorem 2.5.3. *Let $G \times X \to X$ be a transitive action of G on a set X with at least two elements. This action is primitive if for each $x \in X$ the stabilizer, $\text{Stab}_G(x)$, is a maximal subgroup of G.*

Proof. We first show that G acts imprimitively on X if and only if there exists a proper subset Y of X, with at least two elements, such that for any $g \in G$ either $gY = Y$, or $gY \cap Y = \varnothing$.

Suppose there is such a set Y and $g_1, g_2 \in G$. Then either

$$g_1(Y) = g_2(Y), \quad \text{or} \quad g_1(Y) \cap g_2(Y) = \varnothing.$$

Now clearly, $W = \cup \{gY : g \in G\}$ is a G-invariant set in X. Its compliment, Z, is also G-invariant. For suppose $z \in Z$ and $g \in G$. Then $gz \in Z$. If not, then $gz = g'y$. So $z \in \cup \{gY : g \in G\}$, a contradiction. Thus $gZ \subseteq Z$ and hence $gZ = Z$. Then W and Z give an invariant partition of X. W, in fact each of the gY, clearly consist of more than a single point. If Z is non-empty, this is a non-trivial partition. If $Z = \varnothing$ then W exhausts X and we have a disjoint union $X = \cup \{gY : g \in G\}$. By the above X consists of a disjoint union $\{gY : g \in G\}$ which itself is a non-trivial invariant partition of X. Thus, in any case, we have a system of imprimitivity. Conversely, suppose we have a non-trivial G-invariant partition of $X = \cup X_i$ of X. One of the X_k has at least two elements. Then for $g \in G$, either $gX_k = X_k$ or $gX_k \cap X_k = \varnothing$.

Turning to the proof of the theorem, suppose for some $x_0 \in X$, $\text{Stab}_G(x_0)$ is not a maximal subgroup of G. Then there is a subgroup H, with $\text{Stab}_G(x_0) < H < G$. By transitivity, the given action is equivariantly equivalent to the action of G on $G/\text{Stab}_G(x_0)$ by left translation. Thus it suffices to show this action is imprimitive. Consider the subset $H/\text{Stab}_G(x_0)$ of $G/\text{Stab}_G(x_0)$. This set has at least two points and is proper. Let $g \in G$. If g lies in H then $g \cdot H/\text{Stab}_G(x_0) = H/\text{Stab}_G(x_0)$. But if g lies outside H, and $h \text{Stab}_G(x_0) \in H/\text{Stab}_G(x_0)$, then $gh \text{Stab}_G(x_0)$ cannot lie in $H/\text{Stab}_G(x_0)$. For if $gh \text{Stab}_G(x_0) = h' \text{Stab}_G(x_0)$ then $h'^{-1}gh \in \text{Stab}_G(x_0) \subseteq H$. But then, $g \in H$, a contradiction. Thus $g \cdot H/\text{Stab}_G(x_0) = H/\text{Stab}_G(x_0)$ or $g \cdot H/\text{Stab}_G(x_0) \cap H/\text{Stab}_G(x_0) = \varnothing$ according to whether $g \in H$ or not.

Conversely, suppose G acts imprimitively. Then there is a non-trivial subset Y of X with the property that for $g \in G$ either $gY = Y$, or $gY \cap Y = \varnothing$. Let $y \in Y$ be fixed and $H = \{g \in G : gY = Y\}$. Then H is a subgroup of G. If $g(y) = y$ then $gY \cap Y$ cannot be empty so $gY = Y$. Thus $\mathrm{Stab}_G(y) \subseteq H$. We show $\mathrm{Stab}_G(y)$ is not maximal by showing $\mathrm{Stab}_G(y) < H < G$. Since G acts transitively on X and Y is strictly smaller than X some g must take y outside of Y. For this g, $gY \cap Y = \varnothing$ so $H < G$. Finally, choose a $z \neq y$, but in Y. Then some g takes z to y, by transitivity. Then, clearly $g \in H$. But g does not stabilize y. $\qquad\qquad\qquad\qquad\qquad\qquad\qquad\qquad\qquad\qquad\qquad\quad\square$

Remark: We might call a transitive action of G on X *ergodic* if X has no *non-trivial* G-invariant set X_0. The condition that there is such a set clearly implies the action is imprimitive. For, we know that G acts imprimitively if and only if there exists a non-trivial subset Y of X, such that for any $g \in G$ either $gY = Y$, or $gY \cap Y = \varnothing$ and clearly X_0 is such a set. Hence if a transitive action is primitive, then it is ergodic.

We also get the following corollary of the imprimitivity theorem.

Corollary 2.5.4. *If G acts transitively on a set X with p elements, where p is a prime, then this action is primitive.*

Proof. For any $x \in X$ we know $G/\mathrm{Stab}_G(x)$ has the cardinality of X. Thus $\mathrm{Stab}_G(x)$ has index p in G. This means $\mathrm{Stab}_G(x)$ is a maximal subgroup for each $x \in X$. $\qquad\qquad\qquad\qquad\qquad\qquad\qquad\qquad\qquad\square$

2.6   Cauchy's Theorem

We now generalize the Abelian version of Cauchy's theorem to arbitrary finite groups. The very short proof given here, is due to McKay [72].

Theorem 2.6.1. *Cauchy's theorem*[9] *If the prime p divides the order of a group G, then G has an element of order p.*

---

[9]Augustin Cauchy (1789-1857) French mathematician was a professor at the Ecole Polytechnique and the College de France. His mathematical accomplishments would take several pages to detail. Suffice it to say, he was one of the inventors of complex

Proof. Let X be the set consisting of sequences $(g_1, g_2, ..., g_p)$ of elements of G such that $g_1 g_2 \cdots g_p = 1$. Note that here g_1 through g_{p-1} can be chosen arbitrarily, and $g_p = (g_1 g_2 \cdots g_{p-1})^{-1}$. Thus the cardinality of X is $|G|^{p-1}$. Recall that if for two elements x and y of G we have $ab = 1$ then we have also $ba = 1$. Hence if $(g_1, g_2, ..., g_p) \in X$, then $(g_p, g_1, g_2, ..., g_{p-1}) \in X$ as well. Therefore, the cyclic group $\mathbb{Z}_p$ acts on X by cyclic permutations of the sequences. Because p is prime, each element of X is either fixed under the action of $\mathbb{Z}_p$, or it belongs to an orbit of size p. Thus $|X| = n + kp$, where n is the number of fixed points and k is the number of orbits of size p. Note that $n > 1$, since $(1_G, 1_G, ..., 1_G)$ is a fixed point of X. But p divides $|X| - kp = n$, so X has a fixed point $(x, x, ..., x)$ with $x \neq 1$. But then x has order p. $\square$

2.7 The Three Sylow Theorems

Here we prove the very important Sylow[10] theorems for finite groups. We do this using transformation groups.

---

analysis, the Cauchy criterion for convergence and its relation to completeness and the modern understanding of limits and continuity. He made contributions to optics, elasticity, astronomy, hydrodynamics, and partial differential equations and the theorem just above. However, the 15-year delay in the publication of Abel's major paper (from 1826 to 1841) was largely due to Cauchy's cavalier treatment of it. Abel died in 1829, the same year in which Cauchy contributed to the suppression of the young Galois's epoch discoveries. Galois died in 1832. It was this contemptuous attitude toward younger mathematicians, together with his religious and political bigotry, that made Cauchy's reputation for being bigoted, selfish and narrow minded. In the case of Galois, this may have been a consequence of the latter having been a radical republican, whereas in the case of Abel, it might simply have been snobism, or the fact that he was a foreigner. Baron Cauchy was an aristocrat, a royalist and quite wealthy. When his prolific mathematical output couldn't all be published, in 1826 he founded a private journal, Exercices de Mathématiques. Its twelve issues a year were filled by Cauchy himself.

[10]Peter Ludwig Sylow (1832-1918) was a Norwegian mathematician who proved fundamental results in finite group theory. He was a high school teacher from 1858 to 1898, and a substitute lecturer at Christiania University in 1862, teaching Galois theory when he proved what we now call the Sylow theorems. Together with Sophus Lie, he then turned to editing the mathematical works of Niels Henrik Abel. He was finally appointed professor of Christiania University (now Oslo) in 1898!

Let G be a finite group and $p_1^{e_1} p_2^{e_2} ... p_k^{e_k}$ be the prime factorization of $|G|$.

Definition 2.7.1. For each p_i, a *Sylow p_i-subgroup* is a subgroup of G of order $p_i^{e_i}$ if it exists.

Here we ask the following questions:

- Do Sylow subgroups always exist?

- If so, how unique are they?

- Are the Sylow subgroups nilpotent?

We prove the first Sylow theorem by induction on the order of G.

Theorem 2.7.2. *Let G be a finite group. For each such $i = 1 \ldots k$ there is subgroup P_i of G whose order is $p_i^{e_i}$.*

Proof. Our proof is by induction. Let p be a prime divisor on the order of G and p^n, its highest prime power divisor, as above. Consider the class equation. If p did not divide the order of $\mathcal{Z}(G)$, then it would not divide the index of the centralizer $Z_G(g_0)$ of some non-central element $g_0 \in G$. Since $|G| = |Z_G(g_0)| \cdot |[G : Z_G(g_0)]|$, p^n must divide $|Z_G(g_0)|$. Clearly no higher power of p can do so. By inductive hypothesis $Z_G(g_0)$ has a subgroup of order p^n which is also a subgroup of G. Thus we may assume p divides the order of $\mathcal{Z}(G)$. By the lemma just above $\mathcal{Z}(G)$ contains an element of order p. Let N be the (normal) subgroup generated by this element. Clearly, G/N has lower order than G and the highest power of p dividing G/N is p^{n-1}. Again, by inductive hypothesis, G/N has a subgroup of order p^{n-1} whose inverse image has order p^n. $\qquad\square$

As we observed earlier, for any automorphism α and any subgroup H of G we have

$$[G : H] = [G : \alpha(H)] \text{ and } |H| = |\alpha(H)|.$$

In particular, the conjugate of a Sylow p group P is again a Sylow p group and if there were only one Sylow p subgroup of G it would be normal, and in fact characteristic.

Corollary 2.7.3. *Let G be a finite group and p a prime divisor of $|G|$, then G contains an element of order p.*

Proof. Let P be a Sylow p group of G. Since p divides the order of P we may clearly replace G by P. Thus we may assume G is a p group. Then G has a center whose order is divisible by p. To get an element in G of order p we may replace G by its center and since this last group is Abelian we are done.   □

Theorem 2.7.4. *In a finite group any two Sylow p groups associated with the same prime p are conjugate. Any p subgroup of G is contained in a Sylow p subgroup.*

We prove this as a corollary of our previous results on actions as follows.

Proof. Suppose H is the p subgroup and P is a Sylow p subgroup of G. Let H act on G/P by left translation. Then, as above, we see that p divides $|G/P| - |(G/P)^H|$. But p clearly does not divide $|G/P|$. Therefore, p also does not divide $|(G/P)^H|$. In particular, the order of $(G/P)^H$ is non-zero, since zero is definitely divisible by p. Now the fixed point set is clearly $\{gP : g^{-1}Hg \subseteq P\}$. Thus there is some $g \in G$ such that $g^{-1}Hg \subseteq P$. That is, $H \subseteq gPg^{-1}$, proving the second assertion. In particular, taking H itself to be a Sylow p group shows any two Sylow p subgroups are conjugate.   □

For the third Sylow theorem we study the action of G by conjugation.

Theorem 2.7.5. *In a finite group the number of Sylow p subgroups is a divisor of the index of any one of them and is congruent to $1 \bmod p$.*

Proof. Let G act by conjugation on the finite set X of all Sylow p groups. By Theorem 2.7.4 this action is transitive. As above, the stabilizer of a P is $N_G(P)$, its normalizer. Hence by Theorem 2.7.2 the number of Sylow p subgroups is $[G : N_G(P)]$, which is a divisor of $[G : P]$.

In order to prove the last statement of the theorem we need the following lemma.

Lemma 2.7.6. *Let P be a Sylow p group of G and H a p subgroup of G contained in $N(P)$. Then $H \subseteq P$.*

Proof. Since H normalizes P, HP is a subgroup of G and HP/P is isomorphic with $H/H \cap P$. In particular, it has order p^j where $j \leq i$. Since P also has order a power of p, it follows that the same is true of HP. But the latter contains P and this subgroup is maximal with this property. Thus $HP = P$. This means $H \subseteq P$.   □

Continuing the proof of Theorem 2.7.5, let P be a fixed Sylow p subgroup and let it act on X by conjugation. By our previous results on actions, p divides $\mid X \mid - \mid X^P \mid$. Keeping in mind that X consists of conjugates gPg^{-1} we see that X^P consists of all gPg^{-1} such that $pgPg^{-1}p^{-1} = gPg^{-1}$ for all $p \in P$. That is, $g^{-1}Pg \subseteq N(P)$. Since $g^{-1}Pg$ is a p group, the lemma above tells us $g^{-1}Pg \subseteq P$. This means that $g \in N(P)$. Hence, $gPg^{-1} = P$. Thus X^P consists of the single point $\{P\}$, and so p divides $\mid X \mid -1$.   □

Corollary 2.7.7. *If P is any Sylow p group, then $N(N(P)) = N(P)$.*

Proof. Since P is a Sylow subgroup of G it is clearly also one of $N(P)$. In fact, since P is normal in $N(P)$ it is the unique Sylow subgroup of $N(P)$. Let $g \in N(N(P))$. Then by this uniqueness $\alpha_g(P) = P$. But this means $g \in N(P)$. Thus $N(N(P)) \subseteq N(P)$. The opposite inclusion is always true.   □

Our final result of this section is a prelude to the important $p^a q^b$ theorem of Burnside in Chapter 12. It also enables us to amplify our list of all finite groups up to and including order 11. The stronger hypothesis enables us to conclude the group of order 10 is Abelian instead of merely solvable.

Corollary 2.7.8. *(Burnside's Theorem) Let G be a finite group of order pq the product of two distinct primes, where p is not congruent to 1 mod q and q is not congruent to 1 mod p. Then G is Abelian and is the direct product of the cyclic groups $\mathbb{Z}_p$ and $\mathbb{Z}_q$.*

Proof. Let P be a Sylow p subgroup of G. The number n of such P is a divisor of pq and is congruent to 1 mod p. But the divisors of pq are 1, p, q and pq. The last three being impossible, we see that $n = 1$ and so P is normal. Similarly, if Q is a Sylow q subgroup of G it is also normal. Now $P \cap Q$ is a subgroup of P and therefore has order a divisor of p and similarly the order is a divisor of q. This means $P \cap Q = (1)$. PQ is a subgroup of G and $PQ/Q = P/P \cap Q = P$. Thus PQ has order pq. This means $G = PQ$. Now $P = \mathbb{Z}_p$ and $Q = \mathbb{Z}_q$ so they are Abelian. To see that G itself is Abelian we show that $[P, Q] = (1)$. Indeed, let $[g, h] = ghg^{-1}h^{-1}$ be a generating commutator. Then $[g, h] \in P \cap Q = (1)$. Since P is isomorphic with $\mathbb{Z}_p$ and Q is isomorphic with $\mathbb{Z}_q$ the result follows. □

2.7.0.1   Exercises

Exercise 2.7.9. Prove that a group of order 99 has a non-trivial normal subgroup.

Exercise 2.7.10. Let G be a group of order $10,989$ (note that $10989 = 3^3 \times 11 \times 37$).

1. Compute the number, n_p, of Sylow p-subgroups permitted by Sylow's Theorem for each of $p = 3$, 11, and 37. For each of these n_p give the order of the normalizer of a Sylow p-subgroup.

2. Show that G contains either a normal Sylow 37-subgroup or a normal Sylow 3-subgroup.

3. Explain why (in all cases) G has a normal Sylow 11-subgroup.

4. Deduce that the center of G is nontrivial.

Exercise 2.7.11. Let G be a group of odd order and let σ be an automorphism of G of order 2.

1. Prove that for every prime p dividing the order of G there is some Sylow p-subgroup P of G such that $\sigma(P) = P$ (i.e., σ stabilizes the subgroup P. (Note that σ need not fix P elementwise).

2. Suppose G is a cyclic group. Prove that $G = A \times B$ where

$$A = C_G(\sigma) = \{g \in Gj \mid \sigma(g) = g\} \text{ and } B = \{x \in G \mid \sigma(x) = x^{-1}\}.$$

(Remark: This decomposition is true more generally when G is Abelian.)

Chapter 3

Vector Spaces

In this chapter we shall study vector spaces over a (commutative) field k together with their linear operators. For those readers who are not yet familiar with the theory of fields it will be sufficient to just think of vector spaces over say $\mathbb{Q}$, $\mathbb{R}$, $\mathbb{C}$, or $\mathbb{Z}_l$ (but, not over the ring $\mathbb{Z}$ which is not a field). On occasion we may also look at vector spaces over a division algebra (a non-commutative field such as the quaternions).

Definition 3.0.1. We define V as a *vector space V* over a field k, if it has an operation $+$ called *addition*, making $(V, +)$ an Abelian group, and if there is a function
$$k \times V \to V$$
called *scalar multiplication*, written $c \cdot v$ (or even just cv), satisfying the following axioms:

1. $(c + d)v = cv + dv$

2. $c(v + w) = cv + cw$

3. $c(dv) = (cd)v$

4. $1v = v$

For example, let n be a positive integer and $V = k^n$. For $v = (v_1, \ldots v_n)$ and $w = (w_1, \ldots w_n) \in V$ we take $v + w = (v_1 + w_1, \ldots v_n + w_n)$

and for $c \in k$, $cv = (cv_1, \ldots cv_n)$. The reader should verify that this makes V into a vector space over k. As we shall see, in some sense these are all the (finite dimensional) vector spaces over k. Notice that when $n = 1$ we actually have just the field, k when we identify c with (c), $c \in k$.

Another example of a vector space is given by taking a set X and considering all functions $f : X \to k$. We denote this space of functions by $\mathcal{F}(X, k)$. It forms a vector space over k with pointwise operations which is infinite dimensional if X is infinite.

Definition 3.0.2. Let V and W be vector spaces over the same field k and $T : V \to W$ be a map which preserves addition, i.e.

$$T(x + y) = T(x) + T(y)$$

and scalar multiplication, i.e.

$$T(cx) = cT(x),$$

for all $x, y \in V$ and $c \in k$. We call T a *linear transformation*. If T is a bijection we call it an *isomorphism* of vector spaces.

Just as with groups, in effect an isomorphism of two vector spaces means that they share all the same properties as vector spaces.

We now construct some additional examples of vector spaces which will be crucial in the sequel. Let V and W be vector spaces over the same field, k. Then $\mathrm{Hom}(V, W)$, the set of all k-linear mappings from $V \to W$, with pointwise operations, forms a vector space over k. In particular, if $V = W$ we write this vector space as $\mathrm{End}_k(V)$. At the other extreme we also have the k-vector space $\mathrm{Hom}(V, k) = V^*$, where here we regard the field k as a k-space in its own right. Its elements are called *linear functionals* and V^* is called the *dual space*.

Exercise 3.0.3. We ask the reader to verify that all of the above are indeed vector spaces over k.

Definition 3.0.4. Let V be a vector spaces over k and $W \subseteq V$. We say W is a *subspace* of V if W is closed under addition and scalar multiplication.

Evidently V is a subspace of itself, as is (0).

If $V = k^n$ and W is the set of points in V where at certain indices the coordinate is required to be zero, W is also a subspace of V.

Exercise 3.0.5. Prove the following:

1. Let $\mathcal{F}(X, k)$ be the function space above, where the set X is infinite. Consider the functions, $\mathcal{F}_0(X, k)$, in it which are 0 except at perhaps finitely many points of X. Show $\mathcal{F}_0(X, k)$ is a subspace.

2. Let $k = \mathbb{R}$, or $\mathbb{C}$ and X be a topological space, for example an interval. Let $\mathcal{C}(X, k)$ denote the continuous functions $X \to k$. Show $\mathcal{C}(X, k)$ is a subspace of $\mathcal{F}(X, k)$.

3.1   Generation, Basis and Dimension

Let V be a vector space over k. Given a finite set $S = \{v_1, \ldots, v_m\}$ of V, we call a *linear combination* any expression of the form

$$c_1 v_1 + \ldots + c_m v_m = \sum_{i=1}^{m} c_i v_i,$$

where the $c_i \in k$. Sometimes we shall write $\text{lin.sp}_k S$. (Of course, when all the $c_i = 0$ this is just the zero vector.) More generally, one can consider any subset $S \subseteq V$ (not necessarily finite). Then, $\text{lin.sp}_k S$ means *all possible finite* linear combinations with coefficients from k of elements of S.

Definition 3.1.1. Let V be a vector space over k and $S \subseteq V$. We say S is a *generating set* for V, or S *generates* (or *spans*) V if every element of V is some finite linear combination of elements of S. That is, every $v \in V$ can be written $v = \sum_{i \in F} c_i s_i$, where F is a finite set.

Now given $S \subseteq V$, it might not generate V. However, it certainly generates some subspace. That is, $\text{lin.sp}_k S$ is a subspace of V called the *subspace generated by S*. For example, if $V = k^n$ there is a standard generating set for V consisting of $\{e_1, \ldots, e_n\}$, where e_i is the n-tuple with zeros everywhere except in the i-th place and 1 in that place. The reader should verify that this representation is *unique*. That is,

Exercise 3.1.2. Any vector $v \in V$ can be written *uniquely* as $v = \sum_{i=1}^{n} c_i e_i$. Thus, $\{e_1, \ldots, e_n\}$ spans V.

We now turn to the concepts of linear independence and basis.

Definition 3.1.3. Let V be a vector space and S be a subset of V (not necessarily finite). We say S is *linearly independent*, if for any finite subset $\{s_1, \ldots, s_m\}$ of S, the only way $c_1 s_1 + \ldots + c_m s_m = 0$ is if all the $c_i = 0$.

For example, $\{e_1, \ldots, e_n\}$ is a linearly independent subset of k^n, by virtue of the uniqueness part of the previous exercise.

Definition 3.1.4. We shall call $B \subset V$ a *basis* if it is linearly independent and spans V. The *dimension* of V is defined as the number of elements in this basis. (We shall show that the number of elements of *any* basis is the same, making the definition of dimension an invariant of V; actually the only invariant.)

For example $\{e_1, \ldots, e_n\}$ is linearly independent and also spans, hence it is a basis for k^n.

The reader should notice that in the following proposition it is essential that we are over a field.

Proposition 3.1.5. *Let V be a vector space over k and S a subset of independent vectors that is not properly contained in any larger such subset (such a set is called maximal). Then S is a basis for V.*

Proof. We must show that S spans V. Let $v \in V$. If $v \in S$ we simply write $v = 1v$. Assume v is not in S. Then S is a proper subset of $S \cup \{v\}$. Hence, by hypothesis $S \cup \{v\}$ is not linearly independent. So there must be a (finite) non-trivial linear relation among its elements. This relation must involve v since S is linearly independent. Thus

$$cv + c_1 s_1 + \ldots + c_m s_m = 0,$$

or $cv = -c_1 s_1 + \ldots - c_m s_m$.

Now $c \neq 0$, for if it were zero there would again be a non-trivial linear relation among the elements of S. Since k is a field

$$v = \frac{-c_1}{c} s_1 + \ldots + \frac{-c_m}{c} s_m.$$

Thus S spans V.   □

Corollary 3.1.6. *Given a spanning set of vectors in V we can extract a linearly independent subset of them which continues to span V. Such a subset is a basis of V.*

Proof. A finite set $v_1, \ldots, v_m$ of vectors in V satisfies a non-trivial dependence relation $c_1 s_1 + \ldots + c_m s_m = 0$ if and only if one of them is a linear combination of the others. In that case throwing that one away will not change the span. We can continue throwing away any of these superfluous vectors until there are only linearly independent ones.   □

Corollary 3.1.7. *If the vector space V has a finite basis $x_1, \ldots, x_n$ and another one $y_1, \ldots, y_m$, then $n = m$.*

Proof. Since the y_i's span V, and the x_j's are linearly independent, by Corollary 3.1.6, $m \geq n$. Reversing the roles of x_j's and y_i's, we see that $n \geq m$. Therefore, $m = n$.   □

In the above corollary we restricted our attention to finite dimensional vector spaces. Actually, in the infinite dimensional case, the same result holds.

Our next result concerns all vector spaces, finite or infinite dimensional.

Theorem 3.1.8. *Every vector space has a basis.*

Proof. By proposition 3.1.5, to prove this we need only find a maximally independent set of vectors (that is, one not properly contained in any larger such subset). To do so consider linearly independent subsets $S \subseteq V$ partially ordered by inclusion. We denote this family by $\mathcal{F}$ which as we noted earlier this is a partially ordered set. $\mathcal{F}$ is non-empty since it clearly contains the singletons of all non-zero vectors. Let $\{\mathcal{S}_i : i \in I\}$, where I is some index set, be a chain in $\mathcal{F}$. Then $\cup_I \{S_i : i \in I\}$ is in $\mathcal{F}$. For if $s_1, \ldots, s_m$ are chosen from this union, each s_i comes from some S_i and because there are only finitely many of them they form a chain. Hence there is a single $S \in \mathcal{F}$ which contains all the others and therefore $s_1, \ldots, s_m \in S$. But because $S \in \mathcal{F}$ it consists of linearly independent vectors. Thus every chain has a cap. By Zorn's lemma (0.4.4), $\mathcal{F}$ has a maximal element B. Proposition 3.1.5 tells us B is a basis. $\square$

When V has a finite basis we say it is *finite dimensional*.

Proposition 3.1.9. *In a vector space V every linearly independent set $\{y_1, \ldots, y_m\}$ can be extended to a basis of V.*

Proof. Since V is a finite dimensional vector space it has a basis $\{x_1, \ldots, x_n\}$. Consider the set $S = \{y_1, \ldots, y_m, x_1, \ldots, x_n\}$. S spans V since the x_i already span and S is linearly dependent since the x_i form a basis. Therefore some vector $s \in S$ is a linear combination of the *preceding ones*. Since the y_i are linearly independent, $s \neq y_i$ for all i. Therefore, $s = x_j$ for some j. So we can throw away that x_j and what's left will still span V. Continuing in this way we can throw away as many of the x's as necessary until we end up with a linearly independent set $S' = \{y_1, \ldots, y_m, x_{j_1}, \ldots, x_{j_k}\}$ which still spans V. Therefore S' is a basis. Since it contains all the y_i this completes the proof. $\square$

Proposition 3.1.10. *In a finite dimensional vector space V let $\{y_1, \ldots, y_m\}$ be linearly independent and $\{x_1, \ldots, x_n\}$ span. Then $n \geq m$.*

Proof. Applying the previous proposition to the set $\{y_m, x_1, \ldots, x_n\}$ which spans and is linearly dependent we see we can throw away some x_j and still span. Now add y_{m-1} in front of y_m. This set also spans. Continuing in this way we can keep eliminating x's and replacing them by y *in front*. The x's will not be exhausted before the y's because otherwise the remaining y's would be linear combinations of the ones we have already inserted and this cannot be since the y's are linearly independent. $\square$

From this we see that every basis of a finite dimensional vector space has the same number of elements.

Theorem 3.1.11. *In a finite dimensional vector space V every basis has the same number of elements.*

Proof. Let $\{x_1, \ldots, x_n\}$ and $\{y_1, \ldots, y_m\}$ be two bases of V. Since $\{y_1, \ldots, y_m\}$ is linearly independent and $\{x_1, \ldots, x_n\}$ spans $n \geq m$ by Proposition 3.1.10. Reversing the roles of the x's and the y's we also have $m \geq n$, therefore $n = m$.   □

Since the e_i form a basis of k^n (which has dimension n) we get the following corollary.

Corollary 3.1.12. *A vector space V over k of dimension n is isomorphic to k^n.*

This is because if $\{v_1, \ldots, v_n\}$ is a basis of V and therefore each $v \in V$ can be uniquely written as $v = \sum_{i=1}^{n} c_i v_i$, then $v \mapsto \sum_{i=1}^{n} c_i e_i$, is an isomorphism (check this). Thus, not only is $\dim V$ an invariant, it is the only invariant of the finite dimensional vector space V.

Corollary 3.1.13. *Let V be a vector space and W be a subspace, then W is a direct summand of V. That is, there is some subspace U of V so that $V = W \oplus U$.*

Proof. Let w_i be a basis of W. Then w_i is linearly independent. Hence by the corollary above w_i can be extended to a basis w_i, u_j of V. Let $U = \text{lin.sp.} u_j$. Then U is a subspace of V and every $v \in V$ can be written as a unique linear combination of vectors in w_i, u_j, that is unique vectors in W and U.   □

Exercise 3.1.14. Show a set S is a basis for V if and only if each $x \in V$ can be written *uniquely* as a finite linear combination of elements of S.

Now this concept of a (possibly infinite) basis is too unwieldy and also not very helpful in an algebraic setting (although it is in an analytic one!). For this reason we will now restrict ourselves to vector spaces V

which have a finite generating set. Thus to *finite dimensional* vector spaces.

Let S be such a set. That is, $V = \text{lin.sp}_k S$. We now apply the reasoning of Theorem 3.1.8 to conclude that there is a maximal linearly independent subset $B \subseteq S$. Then as above, B is a basis. Since S is finite so is B. Conversely if V has a finite basis this is a generating set so V is finite dimensional. Thus we get

Corollary 3.1.15. *A vector space is finite dimensional if and only if it has a finite basis and so every basis is finite and has the same number of elements.*

Since an isomorphism of vector spaces carries a basis of one onto a basis of the other, we get the following proposition.

Proposition 3.1.16. *Isomorphic vector spaces have the same dimension.*

3.1.0.1   Exercises

Exercise 3.1.17. Let V be the vector space of all continuous functions $\mathbb{R} \to \mathbb{R}$. For each of following subspaces find its dimension and give an explicit basis.

1. The subspace consisting of all functions f satisfying $f(a + b) = f(a) + f(b)$ for all $a, b \in \mathbb{R}$.

2. The subspace spanned by all functions of the form $f(x) = \sin(x + a)\sin(x + b)$ where $a, b \in \mathbb{R}$.

Exercise 3.1.18. Let V be a 4-dimensional vector space over a field k, and let $\mathcal{B}_1$ and $\mathcal{B}_2$ be two bases for V. Show that there exists a basis for V consisting of two members of $\mathcal{B}_1$ and two members of $\mathcal{B}_2$.

3.2   The First Isomorphism Theorem

We now come to the notion of quotient space leaving the easy proof of the following proposition to the reader.

Proposition 3.2.1. *Let W be a subspace of V and form the coset space $V/W = \{v + W : v \in V\}$. This coset space also carries the structure of a k vector space under the operations $(v + W) + (v' + W) = v + v' + W$ and $c(v + W) = cv + W$. These functions are well defined (that is, they are indeed functions). Moreover, the natural map $\pi : V \to V/W$ given by $\pi(v) = v + W$, $v \in V$ is a k-linear map.*

Exercise 3.2.2. Show if V has finite dimension, then $\dim(V/W) = \dim(V) - \dim(W)$. Suggestion: Choose a basis for W and extend this to a basis of V.

The *first isomorphism* theorem for vector spaces is the following:

Theorem 3.2.3. *Let $T : V \to W$ be a linear transformation of vector spaces over k. Then $\operatorname{Ker} T = \{x \in V : T(x) = 0\}$ is a subspace of V and $T(V)$ is a subspace of W. Moreover, $V/\operatorname{Ker} T$ is isomorphic with $T(V)$, the isomorphism $\widetilde{T}$ is induced from T making the following diagram commutative.*

$$
\begin{array}{ccc}
V & \xrightarrow{\ T\ } & W \\
\pi \downarrow & & \uparrow i \\
V/\operatorname{Ker}(T) & \xrightarrow{\ \widetilde{T}\ } & T(V)
\end{array}
$$

Here π is the projection map and i the inclusion map.

Proof. We leave the easy check that $\operatorname{Ker} T = \{x \in V : T(x) = 0\}$ is a subspace of V and $T(V)$ is a subspace of W to the reader. For $x + \operatorname{Ker} T \in V/\operatorname{Ker} T$, define $\widetilde{T}(x + \operatorname{Ker} T) = T(x)$. Then $\widetilde{T}$ is well defined since if $x + \operatorname{Ker} T = y + \operatorname{Ker} T$, then $x - y \in \operatorname{Ker} T$. Hence $T(x) = T(y)$. Because $\pi \cdot \widetilde{T} = T$ and π and T are linear, so is $\widetilde{T}$. Finally, $\widetilde{T}$ is surjective because T is, and its Ker is (0) because if $\widetilde{T}(x + \operatorname{Ker} T) = 0 + \operatorname{Ker} T$, then by definition $T(x) = \pi(0)$. But $\pi(0) = 0$ so $x \in \operatorname{Ker} T$. Therefore $x + \operatorname{Ker} T = 0$. $\qquad\square$

Corollary 3.2.4. *Let $T : V \to W$ be a linear transformation between the vector spaces,*

$$\dim(V) = \dim(\operatorname{Ker} T) + \dim T(V).$$

Proof. Since $V/\operatorname{Ker}(T)$ is isomorphic to $T(V)$, by Proposition 3.1.16

$$\dim(V/\operatorname{Ker}(T)) = \dim T(V).$$

But $\dim(V/\operatorname{Ker}(T)) = \dim V - \dim(\operatorname{Ker}(T)).$ □

An immediate consequence of this is:

Corollary 3.2.5. *Let* $T : V \to W$ *be a linear map.*

1. *If* T *is injective, then* $\dim(V) \leq \dim(W)$.

2. *If* T *is surjective, then* $\dim(V) \geq \dim(W)$.

Corollary 3.2.6.   1. *Let* $T : V \to W$ *be a linear map. If* T *is bijective, then* T *takes a basis of* V *to a basis of* W *and* $\dim V = \dim W$.

2. *If* $\dim V = \dim W$, *then there is a bijective linear map* $V \to W$ *which takes a basis of* V *to a basis of* W.

3. *If* $\{v_1, \ldots, v_n\}$ *is a basis of* V *and* $\{w_1, \ldots, w_n\}$ *is a basis of* W, *then the map* $v_i \mapsto w_i$ *extends to a linear isomorphism* $P : V \to W$. *In particular such a* P *is invertible. We denote its inverse by* P^{-1}.

3.2.1   Systems of Linear Equations

We now turn to an important fact concerning systems of m homogeneous linear equations over a field in n unknowns. Namely, if $n > m$ there is always a non-trivial solution.

Consider the system of m homogeneous linear equations in n unknowns,

$$a_{11}x_1 + a_{12}x_2 + \cdots + a_{1n}x_n = 0$$
$$a_{21}x_1 + a_{22}x_2 + \cdots + a_{2n}x_n = 0$$
$$\ldots\ldots\ldots\ldots\ldots\ldots\ldots\ldots\ldots\ldots\ldots\ldots\ldots$$
$$a_{m1}x_1 + a_{m2}x_2 + \cdots + a_{mn}x_n = 0,$$

whose coefficients, a_{ij}, come from the field k.

Definition 3.2.7. A *solution* of the system above, is a set $x_1, \ldots x_n$ which satisfies all the equations. Of course, taking each $x_i = 0$ will do this. A solution is called *non-trivial* if at least one $x_i \neq 0$.

Exercise 3.2.8. The reader is asked to check that the solution set of a system of homogeneous equations forms a vector space.

Theorem 3.2.9. *If $m < n$ there is a non-trivial solution to this system.*

Proof. Let $V = k^n$, $W = k^m$ and $T : V \to W$ be defined by taking $T(e_i)$ to be the i-th row of the matrix above, which is in W. We extend T to all of V by linearity. Since $\dim(V) > \dim W$, by Corollary 3.2.5, T cannot be injective. $\qquad\square$

We make two remarks concerning Theorem 3.2.9. First, homogeneity is essential here. The system $x + y = 0$, $x + y = 1$ has no solutions and therefore no non-trivial solutions. Also, $n > m$ is also essential since $x + y = 0$, $x - y = 0$ has only the trivial solution.

We remark that Theorem 3.2.9 also works over a division algebra.

Theorem 3.2.9 is non-constructive. That is, it just says that a non-trivial solution exists without telling how one is to find it, or how many (linearly independent) solutions there are. However, there is a method for dealing with this that even works in the case of non-homogeneous equations called *Gaussian elimination* which we briefly explain now.

Consider the same system of equations as above, except they need not be homogeneous. Any equation with all coefficients 0 can be thrown away. Take the first equation and by renumbering the unknowns, if necessary, the a_{11} term is non-zero. Dividing the equation by this coefficient, leaving the solutions unchanged, we may assume $a_{11} = 1$. By multiplying and subtracting we can arrange for all the first column of coefficients, other than a_{11} to be 0. Continuing in this way we get a matrix of the form

$$A = \left(\begin{array}{c|c} I_r & \star \\ \hline 0 & \star\star. \end{array} \right)$$

Here $r \leq \min n, m$. This procedure gives an algorithm for solving such a system which will, if carried out fully, give all possible solutions (there may be more than one or also possibly none).

3.2.2   Cramer's Rule

Here we both categorize invertible matrices and give a method to solve systems of non-homogeneous linear equations when the matrix of the system is non-singular. This last is called Cramer's rule. Let A be a $n \times n$ matrix with $\det(A) \neq 0$, then we show its inverse A^{-1} is given by

$$A^{-1} = \frac{1}{\det A} C,$$

where C is the transpose of the matrix of co-factors, (co-factors of row i of A become the i-th column of C).

To do so we prove $AC = \det(A)I$ (actually this is true even if $\det(A) = 0$).

$$AC = \begin{pmatrix} a_{11} & \cdots & a_{1n} \\ \vdots & \ddots & \vdots \\ a_{n1} & \cdots & a_{nn} \end{pmatrix} \begin{pmatrix} c_{11} & \cdots & c_{n1} \\ \vdots & \ddots & \vdots \\ c_{1n} & \cdots & c_{nn} \end{pmatrix}$$

An easy calculation shows that the entry in the first row and first column of the product matrix is

$$\sum_{i=1}^{n} a_{1i} c_{i1} = \det(A).$$

(This is just the co-factor formula for the determinant). Similarly, this occurs for every entry on the diagonal of AC. To complete the proof, we check that all the off-diagonal entries of AC are zero. In general the product of the first row of A and the last column of C equals the determinant of a matrix whose first and last rows are identical. This happens with all the off diagonal entries. Hence

$$A^{-1} = \frac{1}{\det A} C.$$

Corollary 3.2.10. *A square matrix A is invertible if and only if $\det(A) \neq 0$.*

Proof. Suppose A is invertible. Then $AA^{-1} = I$. Taking determinants we see $\det(A)\det(A^{-1}) = \det I = 1$. Hence $\det(A) \neq 0$. Conversely, if $\det(A) \neq 0$ the formula above gives A^{-1}. $\square$

Another approach to this question is by the Jordan Canonical Form over the algebraic closure of k. A can be put in triangular form (see Chapter 6). Therefore $\det(A)$ is the product of all the eigenvalues. Hence $\det(A) \neq 0$ if and only A has only non-zero eigenvalues. But this is equivalent to A being invertible.

Corollary 3.2.11. *Suppose we want to solve the system*

$$Ax = b,$$

where A is a non-singular $n \times n$ matrix. Then

$$x = A^{-1}b.$$

The solution is unique.

Proof. Applying 3.2.10 we get

$$\mathbf{x} = \frac{1}{\det(A)}C\mathbf{b}.$$

Thus there is a unique solution

$$x_1 = \tfrac{1}{\det(A)} \begin{pmatrix} \mathbf{b} & \begin{matrix} \text{last } n-1 \\ \text{columns} \\ \text{of } A \end{matrix} \end{pmatrix}$$

$$\cdots\cdots\cdots\cdots\cdots\cdots\cdots\cdots\cdots$$

$$\cdots\cdots\cdots\cdots\cdots\cdots\cdots\cdots\cdots$$

$$x_n = \tfrac{1}{\det(A)} \begin{pmatrix} \begin{matrix} \text{first } n-1 \\ \text{columns} \\ \text{of } A \end{matrix} & \mathbf{b} \end{pmatrix}.$$

$\square$

3.3   Second and Third Isomorphism Theorems

The other two isomorphism theorems will follow from Theorem 3.2.3 just as they do for groups (with the attendant diagram isomorphisms). We will only state them and leave the proofs to the reader.

We do the third isomorphism theorem and then the second.

Corollary 3.3.1. *Let $V \supseteq U \supseteq W$ be vector spaces. Then U/W is a subspace of V/W and V/U is isomorphic to $(V/W)/(U/W)$.*

Corollary 3.3.2. *Let V be a vector space and U and W be subspaces. Then $U + W$ and $U \cap W$ are subspaces and $U + W/U$ is isomorphic to $W/U \cap W$.*

Corollary 3.3.3. *Let V be a vector space and U and W be subspaces. Then*
$$\dim U + \dim W = \dim(U + W) + \dim(U \cap W).$$
In particular, if $V = U \oplus W$ is a direct sum then $\dim U + \dim W = \dim V$.

Proof. This follows immediately from the fact that $U + W/U$ is isomorphic to $W/U \cap W$. □

Corollary 3.3.4. *Let V be a finite dimensional vector space and U and W be subspaces. Then the following conditions are equivalent*

1. *$V = U \oplus W$,*

2. *$\dim(U + W) = \dim U + \dim W$,*

3. *$U \cap W = (0)$.*

Exercise 3.3.5. Prove:

1. If $V = U \oplus W$, each $v \in V$ can be written uniquely as $v = u + w$.

2. Let $V = U \oplus W$. For $v \in V$ write $v = u + w$ and define $P_W : V \to W$ be defined by $P_W(v) = w$. Show P_W is a linear operator and that $P_W^2 = P_W$.

3. Show P_W is surjective and $\operatorname{Ker} P_W = U$. P_W is called the *projection* of V onto W.

4. Show $P_W P_U = 0 = P_U P_W$.

Exercise 3.3.6. Conversely, let W_1 and W_2 be subspaces of V and $P_i : V \to W_i$ be linear transformations from V onto W_i for $i = 1, 2$. Formulate conditions on these maps which force $V = W_1 \oplus W_2$.

Corollary 3.3.7. *Let V be a finite dimensional vector space and W be a subspace. Suppose there is a linear operator $P : V \to W$ satisfying $P^2 = P$. Then $P(V) = W$ and $V = \operatorname{Ker} P \oplus W$.*

Proof. Let $v \in V$ and write $v = v - P(v) + P(v)$. Now $v - P(v) \in \operatorname{Ker} P$ because $P(v - P(v)) = P(v) - P^2(v) = 0$ and $P(v) \in W$. Hence $V = \operatorname{Ker} P + W$. Let $w \in W$ and apply P. Then $P(w) = P^2(w)$. Hence P is the identity on $P(W)$. Now let $v \in \operatorname{Ker} P \cap W$. Then $P(v) = 0$ and since $v \in W$, $P(v) = v$. Hence $v = 0$ and so $\operatorname{Ker} P \cap W = (0)$ and $V = \operatorname{Ker} P \oplus W$. The same argument shows $V = \operatorname{Ker} P \oplus P(V)$. Hence $\dim \operatorname{Ker} P + \dim W = \dim V = \dim \operatorname{Ker} P + \dim P(V)$ and so $\dim W = \dim P(V)$. Since $P(V) \subseteq W$, $P(V) = W$. $\qquad\square$

Exercise 3.3.8. Formulate and then prove the extension of this situation of the direct sum of a finite number of subspaces, $V = V_1 \oplus \cdots \oplus V_j$.

We conclude this section with an interesting decomposition of a vector space where we suppose the field k has characteristic $\neq 2$ (this just means one can divide by 2, see Chapter 5). All the examples of fields given at the beginning of Chapter 3, except for $\mathbb{Z}_2$ have characteristic $\neq 2$.

Consider $\mathcal{F}(X, k)$ as above, where X is an Abelian group. We denote the symmetric functions by,

$$\mathcal{S}(X, k) = \{ f \in \mathcal{F}(X, k) : f(-x) = f(x), x \in X \}$$

and the anti-isymmetric functions by,

$$\mathcal{A}(X, k) = \{ f \in \mathcal{F}(X, k) : f(-x) = -f(x), x \in X \}.$$

Then, $\mathcal{S}(X, k)$ and $\mathcal{A}(X, k)$ are subspaces of $\mathcal{F}(X, k)$. Also, $\mathcal{F}(X, k) = \mathcal{S}(X, k) + \mathcal{A}(X, k)$ and $\mathcal{S}(X, k) \cap \mathcal{A}(X, k) = (0)$. Hence, $\mathcal{F}(X, k) = \mathcal{S}(X, k) \oplus \mathcal{A}(X, k)$.

We leave the verification that these are subspaces to the reader. To see $\mathcal{F}(X, k) = \mathcal{S}(X, k) + \mathcal{A}(X, k)$ and $\mathcal{S}(X, k) \cap \mathcal{A}(X, k) = (0)$, observe $f(x) = \frac{f(x)+f(-x)}{2} + \frac{f(x)-f(-x)}{2}$ and that if $f(x) = f(-x)$ and $f(x) = -f(-x)$, then $f(x) = -f(x)$ so $f = 0$.

3.4   Linear Transformations and Matrices

Let V and W be finite dimensional vector spaces and $T : V \rightarrow W$ be a linear transformation. Then T is determined by its values on a basis $v_1, \ldots v_n$ of V. For if $v \in V$, then $v = \sum_{i=1}^n c_i v_i$, where the c_i are unique and since T is k-linear, $T(v) = \sum_{i=1}^n c_i T(v_i)$.

Let $w_1, \ldots w_m$ be a basis of W. Then for each $i = 1 \ldots n$, $T(v_i) = \sum t_{i,j} w_j$ for unique $t_{i,j} \in k$. This $n \times m$ matrix $(t_{i,j})$ is called *the matrix of the linear transformation* T with respect to the bases $v_1, \ldots v_n$ of V and $w_1, \ldots w_m$ of W. Notice that given any matrix $(t_{i,j})$, the above formulas also define a k-linear transformation whose matrix is $(t_{i,j})$. The reader should verify this last statement. Thus we have a bijection between the k vector space, $\mathrm{Hom}(V, W)$, and the $m \times n$ matrices, $\mathrm{M}(m, n)$ which also form a k vector space under addition of matrices adding their components and multiplication of a scalar by a matrix. We ask the reader to check this. The vector space $\mathrm{M}(m, n)$ has dimension nm and so can be realized as k^{nm} (just not arranged all in a column, but that does not matter). We now study the relationship of $\mathrm{M}(m, n)$ with $\mathrm{Hom}(V, W)$.

Theorem 3.4.1.

1. Let T and $S \in \mathrm{Hom}(V, W)$. Then the matrix of $T + S$ is the matrix of T plus the matrix of S

2. Let $T \in \mathrm{Hom}(V, W)$ and $c \in K$. Then the matrix of cT is c times the matrix of T.

3. *Let $T \in \mathrm{Hom}(V, W)$ and $S \in \mathrm{Hom}(W, U)$. Then $ST \in \mathrm{Hom}(V, U)$, and the matrix of ST is the matrix of S times the matrix of T.*

Proof. The first two of these are easy and are left to the reader as an exercise. For the (3) we have $V \longrightarrow W \longrightarrow U$. Let $v_1, \ldots v_n$ be a basis of V, $w_1, \ldots w_m$ a basis of W and $u_1, \ldots u_p$ a basis of U. Now, T corresponds to $(t_{i,j})$, where $i = 1, \ldots n$, $j = 1 \ldots m$, S corresponds to $(s_{j,k})$, where $j = 1, \ldots m$, $k = 1, \ldots p$ and ST is a linear map $V \to U$. The question is, what is its matrix with respect to the bases $v_1, \ldots v_n$ of V and $u_1, \ldots u_p$ of U? (Notice that $(t_{i,j})$ and $(s_{j,k})$ can only be multiplied because the number of columns of T is the same as the number of rows of S both being equal to m.)

Now

$$S(T(v_i)) = S\left(\sum_{j=1}^{m} t_{i,j} w_j\right) = \sum_{j=1}^{m} t_{i,j} S(w_j) = \sum_{j=1}^{m} t_{i,j}\left(\sum_{k=1}^{p} s_{j,k} u_k\right)$$

$$= \sum_{k=1}^{p}\left(\sum_{j=1}^{m} t_{i,j} s_{j,k}\right) u_k.$$

Thus, $ST(v_i) = \sum_{k=1}^{p}(ST)_{i,k} u_k$. This means the matrix of ST with respect to $v_1, \ldots v_n$ and $u_1, \ldots u_p$ is the product of the matrix of T and that of S. $\qquad\square$

Corollary 3.4.2. *If V and W are finite dimensional vector spaces, then $\mathrm{M}(m, n) \cong \mathrm{Hom}(V, W)$, $\dim \mathrm{Hom}(V, W) = \dim V \dim W$. In particular, $\dim \mathrm{End}_k(V) = \dim^2 V$ and $\dim V^* = \dim V$.*

Proof. Since the dimension of a vector space is independent of the basis we may take convenient bases, $v_1, \ldots v_n$ of V and $w_1, \ldots w_m$ of W. Then each T is uniquely determined by its matrix $(t_{i,j})$. As we noted, $\mathrm{M}(m.n)$ has dimension $mn = \dim V \dim W$. On the other hand, by 1 and 2 of Proposition 3.4.1 the map, $T \mapsto (t_{i,j})$ is k-linear and bijective. Since these spaces have the same dimension, $\mathrm{M}(m, n) \cong \mathrm{Hom}(V, W)$. The rest of the corollary follows from this. $\qquad\square$

Exercise 3.4.3. Let $P_W : V \to W$ be a projection of a vector space V onto a subspace W of V. Choose a basis of W and extend this to a basis of V. What is the matrix of P_W with respect to this basis?

Exercise 3.4.4. What are the linear maps $T : V \to W \oplus U$? Let P_W be the projection of $W \oplus U$ onto W and P_U be the projection of $W \oplus U$ onto U. Then $(P_W T, P_U T) \in \mathrm{Hom}(V, W) \oplus \mathrm{Hom}(V, U)$. Thus we have a k-linear map $T \mapsto (P_W T, P_U T)$ from $\mathrm{Hom}(V, W \oplus U) \to \mathrm{Hom}(V, W) \oplus \mathrm{Hom}(V, U)$. Show this map is an isomorphism.

The most important special case of $Hom(V, W)$ is when $V = W$. Then we write it as $\mathrm{End}_k(V)$, where k is the underlying field. As we just saw $\mathrm{End}_k(V)$ is isomorphic as a vector space to $M_k(n)$ (sometimes written $\mathrm{M}(n, k)$), the isomorphism being given by $T \mapsto (T_{ij})_{i,j}$, where $(T_{ij})_{i,j}$ is the matrix of T with respect to a fixed basis of V and as we also saw, this isomorphism also respects composition $\mapsto$ matrix multiplication.

Corollary 3.4.5. *Matrix multiplication is associative and has a unit.*

Proof. Evidently composing of maps is associative so in particular composing of linear maps is associative. Hence by 3 of Theorem 3.4.1, multiplying matrices is also associative. Obviously, the matrix I corresponding to the identity map of V is the unit. □

Let T be a linear transformation $V \to V$. Let T_1 be the matrix of T with respect to a basis $v_1, \ldots v_n$ of V. What happens when we change the basis $v_1, \ldots v_n$ of V to a different basis $w_1, \ldots w_m$?

We leave the verification of the following proposition to the reader. We note that P is invertible by Corollary 3.2.6.

Proposition 3.4.6. *If T_2 is the matrix of T relative to the basis $w_1, \ldots w_m$ of V, then*

$$T_1 = P^{-1} T_2 P,$$

where P is the invertible matrix taking the v_i's to the w_i's.

Definition 3.4.7. Two matrices (respectively, linear transformations) S and T are called *similar* if

$$S = PTP^{-1}$$

for some invertible matrix P (respectively, linear transformation).

In the following exercise we define the symmetric and anti-symmetric parts of a linear operator.

Exercise 3.4.8. In a similar vein to decomposing a function into its symmetric and anti-symmetric parts, consider $x \in \text{End}_k(V)$, regarded as a matrix, and let x^t denote its transpose. For char $k \neq 2$ write $x = \frac{x+x^t}{2} + \frac{x-x^t}{2}$. Prove

1. $(x+y)^t = x^t + y^t$ and $(cx)^t = c(x)^t$. Thus ^{t} is a linear transformation from $\text{End}_k(V)$ to itself.

2. $(x^t)^t = x$.

3. Let $S = \{x \in \text{End}_k(V) : x = x^t\}$ and $A = \{x \in \text{End}_k(V) : x = -x^t\}$. Show S and A are subspaces of $\text{End}_k(V)$.

4. Show $\text{End}_k(V) = S \oplus A$.

3.4.1   Eigenvalues, Eigenvectors and Diagonalizability

Let V be a finite dimensional vector space over the field k and $T : V \to V$ be a k linear transformation. Here $n = \dim_k(V)$. We begin by noting that if T is either injective or surjective it must be bijective. Therefore T is injective if and only if T is invertible. By Volume II, Corollary 7.3.12, T is singular (not invertible) if and only if $\det(T) = 0$.

Definition 3.4.9. We shall say $v \neq 0 \in V$ is an eigenvector of V if $T(v) = \lambda v$ for some $\lambda \in k$. The scalar λ is called the corresponding eigenvalue.

The *characteristic polynomial* of T is $\chi_T(x) = \det(T - xI)$. We leave as an exercise to the reader to check that χ_T is a monic polynomial (leading coefficient 1) in x with coefficients in k, of degree n. The reader should also investigate the interesting coefficients of χ_T, for example when $n = 2$. In general, the coefficient of x^{n-1} is $-\operatorname{tr}(T)$, where $\operatorname{tr}(T) = \sum_{i=1}^{n} t_{i,i}$.

We now show the roots of χ_T are the eigenvalues of T. What does it mean for λ to be a root of χ_T? That is, $\det(T - \lambda I) = 0$. Equivalently, the linear transformation $T - \lambda I$ is singular. Equivalently, there is a non-trivial kernel, i.e there is a $v \neq 0 \in V$ so that $T(v) = \lambda v$. Hence v is an eigenvector of T corresponding to the eigenvalue λ. For each eigenvalue λ we let V_λ be the set of all eigenvectors associated with λ, together with the zero vector. V_λ is called the eigenspace associated with λ. It is clearly a T-invariant subspace of V. There are finitely many of them. The number $\leq n$. The dimension of V_λ is called the *geometric multiplicity* of λ. For completeness sometimes we write $V_\lambda = (0)$, when λ is not an eigenvalue.

Theorem 3.4.10. *Let $V_{\lambda_1}, \ldots, V_{\lambda_r}$ be the distinct eigenspaces of T. Then these are linearly independent, that is*

$$V_{\lambda_1} \cap \sum_{i=2}^{r} V_{\lambda_i} = (0). \tag{3.1}$$

Proof. First we deal with two eigenspaces V_λ and V_μ. If $v \in V_\lambda \cap V_\mu$, then $T(v) = \lambda(v) = \mu(v)$. So $(\lambda - \mu)(v) = 0$ and since $\lambda \neq \mu$ it follows $v = 0$.

Moving to the general case, we prove equation 3.1 by induction. Let $v_i \in V_{\lambda_i}$ and suppose $v_1 = \sum_{i=2}^{r} v_i$. Applying T we see that $\lambda_1 v_1 = \sum_{i=2}^{r} \lambda_i v_i$. But since $v_1 = \sum_{i=2}^{r} v_i$, we get $\lambda_1 v_1 = \sum_{i=2}^{r} \lambda_1 v_i$. Subtracting yields $\sum_{i=2}^{r} (\lambda_1 - \lambda_i) v_i = 0$ and since the λ_i are all distinct, solving for v_2, shows v_2 is in the linear span of the eigenspaces for $i \geq 3$. Therefore the induction proof works. $\square$

Definition 3.4.11. An operator T is said to be *diagonalizable* if V has a basis consisting of eigenvectors of T. Or, equivalently, if there are n distinct eigenspaces.

Evidently the hypothesis of our next corollary is equivalent to this last condition, thus giving an important sufficient condition for diagonalizability.

Corollary 3.4.12. *If χ_T has n distinct roots, that is, if T has n distinct eigenvalues, then T is diagonalizable.*

The reader should exercise caution here since the roots of χ_T may not lie in k, but rather in some extension field of k. However, in any case these roots must lie in the algebraic closure of k. In Chapter 4 we will formulate conditions having to do with positive definite symmetric and Hermitian operators which guarantee certain operators have all their eigenvalues in the appropriate ground field to give satisfactory results and at the end of this chapter, when the ground field is $\mathbb{R}$, using complexification, we provide a remedy in case they do not.

3.4.1.1 The Fibonacci Sequence

We now apply our result on diagonalizability to study the Fibonacci numbers.

Definition 3.4.13. The *Fibonacci numbers*[1] are the numbers of the following integer sequence:

$$x_0 = 0, \quad x_1 = 1, \quad x_{n+1} = x_n + x_{n-1}, \quad n \in \mathbb{N}.$$

In other words, the sequence of integers

$$0, \ 1, \ 1, \ 2, \ 3, \ 5, \ 8, \ 13, \ 21, \ 34, \ 55, \ 89, \ 144, \ ...$$

We want to approximate the n-th term of this sequence. To do this let us write

$$X_k = \begin{pmatrix} x_{k+1} \\ x_k \end{pmatrix}, \quad \text{and so} \quad X_0 = \begin{pmatrix} x_1 \\ x_0 \end{pmatrix} = \begin{pmatrix} 1 \\ 1 \end{pmatrix}.$$

---

[1]Leonardo Pisano Fibonacci (1170-1250) introduced Europe to Hindu-Arabic notation for numbers. In his book Liber Abaci (1202), he posed a problem about the growth of a rabbit's population, whose solution (now called the Fibonacci sequence, but known in Indian mathematics centuries before).

The point is that

$$X_{k+1} = \begin{pmatrix} x_{k+2} \\ x_{k+1} \end{pmatrix} = \begin{pmatrix} 1 & 1 \\ 1 & 0 \end{pmatrix} \begin{pmatrix} x_{k+1} \\ x_k \end{pmatrix}.$$

Therefore, if we put $A = \begin{pmatrix} 1 & 1 \\ 1 & 0 \end{pmatrix}$ we get $X_{k+1} = AX_k$. This means that

$$X_1 = AX_0, \quad X_2 = AX_1 = A^2 X_0, \dots X_n = A^n X_0, \dots \quad \text{for all } n \in \mathbb{N}.$$

Now the matrix A is diagonalizable. Indeed

$$\det \begin{pmatrix} 1-x & 1 \\ 1 & -x \end{pmatrix} = x^2 - x - 1 = \chi_A.$$

The roots of the characteristic polynomial χ_A are

$$\lambda_+ = \frac{1 + \sqrt{5}}{2} \quad \text{and} \quad \lambda_- = \frac{1 - \sqrt{5}}{2}.$$

These two eigenvalues are in $\mathbb{Q}(\sqrt{5})$ and they are distinct. Hence A is diagonalizable. Thus, there is a non-singular 2×2 real matrix P such that

$$P^{-1}AP = \begin{pmatrix} \lambda_+ & 0 \\ 0 & \lambda_- \end{pmatrix}.$$

Hence, $(P^{-1}AP)^n = P^{-1}A^n P$, and so

$$\begin{pmatrix} \lambda_+ & 0 \\ 0 & \lambda_- \end{pmatrix}^n = \begin{pmatrix} \lambda_+^n & 0 \\ 0 & \lambda_-^n \end{pmatrix}.$$

From this we get

$$A^n = P \begin{pmatrix} \lambda_+^n & 0 \\ 0 & \lambda_-^n \end{pmatrix} P^{-1},$$

and hence,

$$X_n = A^n X_0 = P \begin{pmatrix} \lambda_+^n & 0 \\ 0 & \lambda_-^n \end{pmatrix} P^{-1} X_0. \tag{3.2}$$

Now, setting $P = \begin{pmatrix} \lambda_+ & \lambda_- \\ 1 & 1 \end{pmatrix}$, we get $AP = P \begin{pmatrix} \lambda_+ & 0 \\ 0 & \lambda_- \end{pmatrix}$. Indeed,

$$\begin{pmatrix} 1 & 1 \\ 1 & 0 \end{pmatrix} \begin{pmatrix} \lambda_+ & \lambda_- \\ 1 & 1 \end{pmatrix} = \begin{pmatrix} \lambda_+ & \lambda_- \\ 1 & 1 \end{pmatrix} \begin{pmatrix} \lambda_+ & 0 \\ 0 & \lambda_- \end{pmatrix}$$

or equivalently

$$\begin{pmatrix} \lambda_+ + 1 & \lambda_- + 1 \\ \lambda_+ & \lambda_- \end{pmatrix} = \begin{pmatrix} \lambda_+^2 & \lambda_-^2 \\ \lambda_+ & \lambda_-, \end{pmatrix}$$

which is true since $\lambda_+^2 - \lambda_+ - 1 = 0$ and $\lambda_-^2 - \lambda_-^2 - \lambda_- - 1 = 0$. In addition, $\det(P) = \lambda_+ - \lambda_- = \sqrt{5} \neq 0$, and so the matrix P is invertible. An easy calculation yields,

$$P^{-1} = \frac{1}{\det P} \begin{pmatrix} 1 & -\lambda_- \\ -1 & \lambda_+ \end{pmatrix}.$$

Now, substituting P and P^{-1} in equation 3.2 we get:

$$
\begin{aligned}
X_n = \begin{pmatrix} x_{n+1} \\ x_n \end{pmatrix} &= \frac{1}{\sqrt{5}} \begin{pmatrix} \lambda_+ & \lambda_- \\ 1 & 1 \end{pmatrix} \begin{pmatrix} \lambda_+^n & 0 \\ 0 & \lambda_-^n \end{pmatrix} \begin{pmatrix} 1 & -\lambda_- \\ -1 & \lambda_+ \end{pmatrix} \begin{pmatrix} 1 \\ 1 \end{pmatrix} \\
&= \frac{1}{\sqrt{5}} \begin{pmatrix} \lambda_+ & \lambda_- \\ 1 & 1 \end{pmatrix} \begin{pmatrix} \lambda_+^n & -\lambda_+^n \lambda_- \\ -\lambda_-^n & \lambda_+ \lambda_-^n \end{pmatrix} \begin{pmatrix} 1 \\ 1 \end{pmatrix} \\
&= \frac{1}{\sqrt{5}} \begin{pmatrix} \lambda_+^{n+1} - \lambda_-^{n+1} & -\lambda_+^{n+1} \lambda_- + \lambda_+ \lambda_-^{n+1} \\ \lambda_+^n - \lambda_-^n & -\lambda_+^n \lambda_- + \lambda_+ \lambda_-^n \end{pmatrix} \begin{pmatrix} 1 \\ 1 \end{pmatrix} \\
&= \frac{1}{\sqrt{5}} \begin{pmatrix} \lambda_+^{n+1} - \lambda_-^{n+1} - \lambda_+^{n+1} \lambda_- + \lambda_+ \lambda_-^{n+1} \\ \lambda_+^n - \lambda_-^n - \lambda_+^n \lambda_- + \lambda_+ \lambda_-^n \end{pmatrix}.
\end{aligned}
$$

We have, $\lambda_+ - 1 = \frac{1+\sqrt{5}}{2} - 1 = -\frac{1-\sqrt{5}}{2} = -\lambda_-$ and $1 - \lambda_- = 1 - \frac{1-\sqrt{5}}{2} = \frac{1+\sqrt{5}}{2} = \lambda_+$, and so we get

$$\lambda_+^{n+1} - \lambda_-^{n+1} - \lambda_+^{n+1} \lambda_- + \lambda_+ \lambda_-^{n+1} = (1-\lambda_-)\lambda_+^{n+1} + (\lambda_+ - 1)\lambda_-^{n+1} = \lambda_+^{n+2} - \lambda_-^{n+2}$$

and

$$\lambda_+^n - \lambda_-^n - \lambda_+^n \lambda_- + \lambda_+ \lambda_-^n = (1 - \lambda_-)\lambda_+^n + (\lambda_+ - 1)\lambda_-^n = \lambda_+^{n+1} - \lambda_-^{n+1}.$$

Replacing, eventually we get

$$\begin{pmatrix} x_{n+1} \\ x_n \end{pmatrix} = \begin{pmatrix} \lambda_+^{n+2} - \lambda_-^{n+2} \\ \lambda_+^{n+1} - \lambda_-^{n+1} \end{pmatrix}.$$

Therefore,

$$x_n = \frac{1}{\sqrt{5}}\left[\left(\frac{1+\sqrt{5}}{2}\right)^{n+1} - \left(\frac{1-\sqrt{5}}{2}\right)^{n+1}\right].$$

The point is that $|\lambda_-| < 1$ and so if we take limits, as $n \to \infty$, $\lambda_-^n \longrightarrow 0$, and so

$$x_n \simeq \frac{1}{\sqrt{5}}\left(\frac{1+\sqrt{5}}{2}\right)^{n+1}.$$

3.4.2   Application to Matrix Differential Equations

Here we present a proof of Liouville's theorem for Matrix Ordinary Differential Equations.[2]

Before doing so we remark that if $f(x) = \sum_{i=0}^{\infty} a_i x^i$ is a convergent power series over $\mathbb{R}$ or $\mathbb{C}$ (an entire function) and A is a linear operator on $\mathbb{R}^n$ or $\mathbb{C}^n$, then we can form $f(A)$, which will also be a linear operator on $\mathbb{R}^n$ or $\mathbb{C}^n$, respectively. This process is called the functional calculus. Thus we have the important linear operator $\mathrm{Exp}(A)$.

Consider the *matrix differential equation*

$$\frac{dX(t)}{dt} = A(t) \cdot X(t), \tag{3.3}$$

where $X(t)$ is the unknown matrix valued function of t and $A(t) = (a_{ij}(t))$ is the matrix of coefficients. Here t lies in some connected interval I containing 0 in $\mathbb{R}$ which could be a finite, or half-infinite interval,

---

[2]J. Liouville (1809-1882), French mathematician founded the *Journal de Mathématiques Pures et Appliquées*. He was the first to read, and recognize the importance of the unpublished work of Évariste Galois which appeared there in 1846. Liouville worked in complex analysis and proved Liouville's theorem. In number theory he was the first to prove the existence of transcendental numbers. He also worked in differential geometry and topology, mathematical physics as well as in ODE, integral equations and dynamical systems where he made several fundamental contributions one of which is the Sturm-Liouville theory.

or $\mathbb{R}$ itself. As initial condition we take $X(0) = I$. We consider the linear homogeneous ordinary differential equation (3.3), where $X(t)$ is the unknown path in the space of $n \times n$ real (or complex) matrices and $A(t)$ is a given smooth $n \times n$ matrix valued function of $t \in I$ (the coefficients). In coordinates this is a system of n^2 differential equations in n^2 unknowns whose coefficients are the $a_{ij}(t)$ and so always has local solutions.

Liouville's theorem states and we will prove that if $A(t)$ is real analytic and $X(t)$ satisfies the differential equation above, then $\det(X(t))$ satisfies its own numerical differential equation.

Theorem 3.4.14. *Let $A(t)$ be a real analytic function of t and suppose $X(t)$ satisfies the matrix ode with initial condition as above. Then on the interval I,*

$$\frac{d}{dt}\det(X(t)) = \operatorname{tr}(A(t))\det(X(t)),$$

with initial condition $\det(X(0)) = 1$.

We shall do this using two lemmas.

Lemma 3.4.15. *Let A be a fixed $n \times n$ real matrix. Then*

$$\det(I + tA) = 1 + t\operatorname{tr}(A) + O(t^2), t \in \mathbb{R}.$$

In particular, $\frac{d}{dt}|_{t=0}\det(I + tA + O(t^2)) = \operatorname{tr}(A)$.

Proof. Let $\lambda_1, \ldots, \lambda_n$ be the eigenvalues of A counted with multiplicity. By the third Jordan form (see Chapter 6) A can be put in triangular form over $\mathbb{C}$ with the λ_i on the diagonal. Therefore, $I + tA$ is also in triangular form with the $1 + t\lambda_i$ on the diagonal. Hence, $\det(I + tA) = \prod_{i=1}^{n}(1 + t\lambda_i)$. This last term is $1 + t\operatorname{tr}(A) + O(t^2)$. $\quad\square$

Lemma 3.4.16. *If the coefficient function $A(t)$ is analytic, then the places where $\det(\frac{dX(t)}{dt}) \neq 0$ are an open dense set in I.*

Proof. Using the local theory one knows that since $A(t)$ is analytic as a solution $X(t)$ is locally analytic, and therefore (real) analytic. Hence so is $\frac{dX(t)}{dt}$ and since det is a polynomial, the same is true of $\det(\frac{dX(t)}{dt})$. The conclusion then follows for example from Corollary (6.4.4) of [83]. $\quad\square$

Proof of the Theorem.

Proof. Choose $t_0 \in I$ in the open dense set I_0 of Lemma 3.4.16 as well. Then the linear approximation for $X(t)$ at t_0 is:

$$X(t) = X(t_0) + (t - t_0)X'(t_0) + O((t - t_0)^2).$$

Since $X(t)$ is a solution to the ode, the right hand side is $X(t_0) + (t - t_0)A(t_0)X(t_0) + O((t - t_0)^2)$. Because of our choice of t_0, $X'(t_0)$ is invertible. Hence this last term is $[I + (t-t_0)A(t_0) + O((t-t_0)^2)]X(t_0)$ so that, $X(t) = [(I + (t - t_0)A(t_0) + O(t - t_0)^2]X(t_0)$. Taking determinants of both sides yields,

$$\det(X(t)) = \det[I + (t - t_0)A(t_0) + O(t - t_0)^2)] \det(X(t_0)).$$

By Lemma 3.4.15 the derivative at t_0 of $\det[I + (t-t_0)A(t_0) + O(t-t_0)^2)]$ is $A(t_0)$. Thus,

$$\frac{d}{dt}(\det(X(t))) = \operatorname{tr}(A(t)) \det(X(t)),$$

for all $t \in I_0$. By continuity this then holds for all $t \in I$. $\square$

We now consider the Liouville equation itself. This numerical ode can be solved by separation of variables,

$$\int \frac{d \det(X(s))}{\det(X(s))} = \int \operatorname{tr}(A(s))dt.$$

Using the initial condition tells us $\log(\det(X(t))) = \int_0^t \operatorname{tr}(A(s))ds$ and so $\det(X(t)) = \exp(\int_0^t \operatorname{tr}(A(s))ds)$.

Of particular interest is the case when the coefficients are constant $A(t) \equiv A$. Then the solution is global; it is $X(t) = X(0) \operatorname{Exp} tA$, $t \in I$. In this case, $I = \mathbb{R}$ and our ode is $\frac{dX(t)}{dt} = A \cdot X(t)$ with initial condition $X(0) = I$. Then (see [1] p. 34 bottom) $X(t) = \operatorname{Exp} tA$. By the Liouville equation $\det(\operatorname{Exp} tA) \equiv \exp(t \operatorname{tr}(A))$ for all t and taking $t = 1$ yields the important *Liouville's formula*

$$\det(\operatorname{Exp} A) = \exp(\operatorname{tr}(A)).$$

Here we proved Liouville's formula by differential equations. In Vol. II we shall give a different proof using multilinear algebra and algebraic geometry in the form of the Zariski topology.

3.4.2.1 Exercises

Exercise 3.4.17. Let G be a subgroup of the multiplicative group of real invertible 2×2 matrices. If $A^2 = I$ for every $A \in G$, prove that the order of G divides 4.

Exercise 3.4.18. Let V be the vector space of all real 3×3 matrices. Show that every 4-dimensional subspace U of V contains a nonzero diagonalizable matrix.

3.5   The Dual Space

Let V be a finite dimensional vector space over k of dimension n, and $v_1, \ldots v_n$ be a basis of V. The dual space V^* as a set is all the linear maps $V \to k$. We make V^* a vector space over k by adding linear functionals and multiplying by scalars both pointwise. We leave it to the reader to check that this makes V^* into a vector space over k. (Actually we have already done this in general for $\mathrm{Hom}(V, W)$ as well as having calculated the dimension.) It is also easy to see $\dim_k V^* = n$. However, it would be helpful if we could get an explicit basis for V^* directly linked to $v_1, \ldots, v_n$. Here is how one can do this. Let $v_1^*, \ldots, v_n^*$ in V^* be defined by $v_i^*(v_j) = \delta_{ij}$, where δ_{ij} is 1 if $i = j$ and 0 otherwise. This defines each v_i^* on a basis of V and since this functional is to be linear, this completely determines it.

Proposition 3.5.1. $v_1^*, \ldots v_n^*$ *is a basis of* V^*.

It is called the *dual basis* to $v_1, \ldots v_n$.

Proof. Suppose $\sum_{i=1}^n c_i v_i^* = 0$. Then applying this functional to each v_j we get $c_j = 0$. Thus $v_1^*, \ldots v_n^*$ is linearly independent. Since there are n of them, they constitute a basis by 3.1.9. $\qquad \square$

Exercise 3.5.2. Show directly that $v_1^*, \ldots, v_n^*$ is a basis for V^*.

Lemma 3.5.3. V^* *separates the points of* V.

Proof. Suppose $v_0 \neq 0$. Write $V = <v_0> \oplus W$, where $<v_0>$ is the k-line through v_0. Define λ_0 by $\lambda_0(cv_0 + w) = c \in k$. One checks easily that $\lambda_0 \in V^*$. Since $\lambda_0(v_0) = 1$, this contradiction proves the lemma. $\square$

We now consider the second dual, V^{**}, of V. Of course this space, being the dual of a space V^* of dimension n itself also has dimension n and so is isomorphic with V. But as we shall see an isomorphism can be given without choosing a basis of anything (as opposed to the case of V^*). For this reason such an isomorphism is called *intrinsic*.

Define $\phi : V \to V^{**}$ by $\phi_v(\lambda) = \lambda(v)$, $v \in V$, $\lambda \in V^*$. A moment's reflection tells us that $\phi_v \in V^{**}$ for each $v \in V$ and that $v \mapsto \phi_v$ is k-linear. Now observe that if $\phi_{v_0} = 0$, then $\phi_{v_0}(\lambda) = 0$ for all $\lambda \in V^*$. That is, $\lambda(v_0) = 0$ for all $\lambda \in V^*$. By Lemma 3.5.3 V^* separates the points of V. That is, if $\lambda(v_0) = 0$ for all $\lambda \in V^*$, then $v_0 = 0$. Thus $\operatorname{Ker} \phi = (0)$. Since $\dim V^{**} = n$ it follows that ϕ is also surjective. This leads us to the following theorem.

Theorem 3.5.4. $\phi : V \to V^{**}$ *is an isomorphism.*

Thus any finite dimensional vector space V can be identified with its double dual, V^{**}. In particular, any V is (naturally isomorphic to) the dual of another such vector space, namely V^*.

Definition 3.5.5. We shall say a bilinear form $\beta : V \times W \to k$ is *non-degenerate* if whenever $\beta(v, w) = 0$ for all $w \in W$, then $v = 0$ and whenever $\beta(v, w) = 0$ for all $v \in V$, then $w = 0$.

Now there is a natural bilinear form, or pairing $V \times V^* \to k$ putting V and V^* into duality. It is given by $(v, \lambda) \mapsto \lambda(v)$. We leave to the reader the easy verification that $(v, \lambda) \mapsto \lambda(v)$ is such a form.

Exercise 3.5.6.

1. Show this form is bilinear and non-degenerate.

2. Let $\beta : V \times W \to k$ be a non-degenerate bilinear form. Prove β induces an injective linear map $f : V \to W^*$, given by $f(v)(w) = \beta(v, w)$, $w \in W$. Similarly there is an injective linear map $g ::$ $W \to V^*$. Hence if $\dim V = \dim W = \dim W^*$, then f (and g) is also surjective and therefore are isomorphisms.

3.5.1   Annihilators

Let V be a vector space of dimension n, S be a subset of V and W be the subspace generated by S.

Definition 3.5.7. $S^{\perp} = \{\lambda \in V^{\star} : \lambda(S) = 0\}$. $S^{\perp}$ is called the *annihilator* of S.

Evidently, $S^{\perp}$ is a subspace of $V^{\star}$ and $S^{\perp} = W^{\perp}$. Also, if $S_1 \subseteq S_2$, then $S_1^{\perp} \supseteq S_2^{\perp}$.

Theorem 3.5.8. *Let W be a subspace of V and $W^{\perp}$ be its annihilator.*

1. $W^* \cong V^*/W^{\perp}$.

2. $\dim W^{\perp} = \dim V - \dim W$.

3. $W^{\perp\perp} = W$ (after identifying V with V^{**}).

4. If U is another subspace and $U^{\perp} = W^{\perp}$, then $U = W$.

5. $W_1 \subseteq W_2$ if and only if $W_1^{\perp} \supseteq W_2^{\perp}$.

Proof. Consider the k-linear map $V^* \to W^*$ given by $\lambda \mapsto \lambda|W$. This linear map is surjective. To see this let $w_1, \ldots, w_k$ be a basis of W and extend this to a basis $w_1, \ldots, w_k, x_1, \ldots, x_m$ of V. Let $w_1^*, \ldots, w_k^*, x_1^*, \ldots, x_m^*$ be the dual basis. Then $V^* = $ lin.sp.$\{w_1^*, \ldots, w_k^*, x_1^*, \ldots, x_m^*\}$ and $W^* = $ lin.sp.$\{w_1^*, \ldots, w_k^*\}$. Let $\lambda = c_1 w_1^* + \ldots + c_k w_k^* + d_1 x_1^* + \ldots + x_m^* \in V^*$. Then since all x_j^* annihilate W, $\lambda|W = (c_1 w_1^* + \ldots + c_k w_k^*)(W)$. It follows from this that the map from $V^* \longrightarrow W^*$ is surjective. Its kernel is evidently $W^{\perp}$. Hence $W^* \cong V^*/W^{\perp}$.

Therefore $\dim W^* = \dim V^* - \dim W^\perp$. But this tells us $\dim W^\perp = \dim V - \dim W$. This latter term is called the *co-dimension* of W in V. Since $\dim W^\perp = n - k$, $\dim W^{\perp\perp} = n - (n - k) = k$, that is $\dim W^\perp = \dim W$, and since $W \subseteq W^{\perp\perp}$ we see $W^{\perp\perp} = W$ proving 3. Evidently, 3 implies 4. We leave the proof of 5 to the reader. □

Corollary 3.5.9. *Let $W_1, \ldots, W_s$ be subspaces of V. Then*

1. $(\cap_i W_i)^\perp = \sum_i W_i^\perp$.

2. $(\sum_i W_i)^\perp = \cap_i (W_i^\perp)$.

Proof. By induction on s we may assume we have only 2 subspaces U and W of V. So we have to prove:

1. $(U \cap W)^\perp = U^\perp + W^\perp$.

2. $(U + W)^\perp = U^\perp \cap W^\perp$.

The proof of 2 is trivial since λ annihilates $U + W$ if and only if it does this to both U and W. As for 1, since $U \cap W$ is contained in both U and W, $(U \cap W)^\perp$ contains both $U^\perp$ and $W^\perp$. Since it is a subspace,

$$(U \cap W)^\perp \supseteq U^\perp + W^\perp.$$

To see these are actually equal, we take annihilators. $(U \cap W)^{\perp\perp} = U \cap V \subseteq (U^\perp + W^\perp)^\perp$. But by the first part this latter term is $U + V$. Therefore the annihilators of the terms in the equation are equal and so are the terms themselves. □

3.5.2   Systems of Linear Equations Revisited

We now apply these ideas to solving systems of homogeneous linear equations and get more insight into the particulars of the solution space. Let

$$a_{11}x_1 + \ldots + a_{1n}x_n = 0$$
$$\ldots\ldots\ldots\ldots\ldots\ldots\ldots$$
$$\ldots\ldots\ldots\ldots\ldots\ldots\ldots$$
$$a_{m1}x_1 + \ldots + a_{mn}x_n = 0$$

be a system of m homogeneous linear equations with coefficients a_{ij} from the field k in n unknowns. Here our objective is to quantify the non-trivial solutions. Since the set of all solutions S forms a vector space over k and therefore a subspace of k^n we are interested in its dimension, $\dim S$. To find it we regard each $a_{i1}x_1 + \ldots + a_{in}x_n$ as a linear functional on $k^n = V$. Namely, $\lambda_i(x_1, \ldots +, x_n) = a_{i1}x_1 + \ldots + a_{in}x_n$ $(\lambda_i \in V^*)$. When viewed in this way $S = \cap_{i=1,\ldots m} \operatorname{Ker} \lambda_i$. Let r be the number of linearly independent λ_i. That is, r is the dimension r of the subspace of V^* spanned by the λ_i. This is called the *rank* of the system. Notice that the solution space is the same as the solution space gotten by throwing away any of the dependent λ's.

To analyze this situation we first make an observation.

Proposition 3.5.10. *Let V be a vector space of dimension n, V^* be its dual and let $\lambda \neq 0 \in V^*$. Then $[\lambda] = \operatorname{Ker}(\lambda)^\perp$.*

Proof. If $\lambda \in V^*$ is not the zero functional, then since λ is surjective $\operatorname{Ker} \lambda$ has dimension $n - 1$. Taking annihilators, $\dim(\operatorname{Ker}(\lambda)^\perp = 1$. Let $v \in \operatorname{Ker} \lambda$ then $\lambda(v) = 0$, so $\lambda \in \operatorname{Ker}(\lambda)^\perp$ and by dimension $[\lambda] = \operatorname{Ker}(\lambda)^\perp$. $\qquad\square$

Theorem 3.5.11. *The set S of solutions is a subspace of V of dimension $n - r$.*

Proof. Since $\lambda_1, \ldots \lambda_r$ are linearly independent in V^*, this set can be extended to a basis $\lambda_1, \ldots \lambda_n$ of V^*. Now $S = \cap_{i=1}^r \operatorname{Ker} \lambda_i$. Hence $S^\perp = \sum_{i=1}^r (\operatorname{Ker} \lambda_i) = \sum_{i=1}^r [\lambda_i]$. Since $\dim S^\perp = r$, $\dim S = n - r$. $\quad\square$

Corollary 3.5.12. *The system has a non-trivial solution if and only if $r < n$, that is, if and only if the λ_i, $i = 1 \ldots r$ span a proper subspace of V^*. In particular, if $m < n$ there is a non-trivial solution.*

Proof. The system has a non-trivial solution if and only if $S \neq (0)$ i.e. if and only if $\dim S > 0$ that is if and only if $n - r > 0$, that is $n > r$. $\qquad\square$

We can extend these considerations to systems of non-homogeneous linear equations. Let

$$a_{11}x_1 + \ldots + a_{1n}x_n = b_1$$
$$\ldots\ldots\ldots\ldots\ldots\ldots\ldots\ldots\ldots\ldots\ldots\ldots\ldots\ldots\ldots$$
$$\ldots\ldots\ldots\ldots\ldots\ldots\ldots\ldots\ldots\ldots\ldots\ldots\ldots\ldots\ldots$$
$$a_{m1}x_1 + \ldots + a_{mn}x_n = b_m$$

be a system of m non-homogeneous linear equations with coefficients a_{ij} from a field k in n unknowns. Let us write this as $Ax = b$ where $x \in k^n$ is the unknown vector, $b \in k^m$ is a constant vector and A is a constant $m \times n$ matrix. Evidently if x is *any* solution to the associated homogeneous system $Ax = 0$ and y_0 is a particular solution to the non-homogeneous equation, then $A(x + y_0) = A(x) + A(y_0) = 0 + b = b$ so $x + y_0$ is a solution to the non-homogeneous equation. Conversely, suppose u_1 and u_2 are solutions to the non-homogeneous equation. Then $x = u_1 - u_2$ is a solution to the homogeneous system $Ax = A(u_1 - u_2) = b - b = 0$. Thus the solutions space to the non-homogeneous equation is a hyperplane $S + y_0$ in k^n of dimension equal to $\dim S$ i.e. of the homogeneous system as above. In order to find y_0 (in fact all of $S + y_0$) one can use Gaussian elimination (see above). When $n = m$ we give an alternative approach to this below using determinants.

3.5.3   The Adjoint of an Operator

Definition 3.5.13. Let $T : V \to W$ be a linear transformation. Its adjoint $T^* : W^* \to V^*$ is defined as follows (where we identify V and V^{**}).

$$T^*(w^*)(v) = w^*(T(v)).$$

Sometimes it is convenient to write this as,

$$\langle T^*(w^*), v \rangle = \langle w^*, T(v) \rangle.$$

The following proposition summarizes the properties of the adjoint whose proof we leave to the reader.

Proposition 3.5.14.

1. T^* *is k-linear.*

2. $T^{**} = T$ *(after identifying V with its double dual).*

3. $(T + S)^* = T^* + S^*$ *and* $(cT)^* = c \cdot T^*$.

4. $(TS)^* = S^* \cdot T^*$.

5. $\mathrm{Ker}(T^\star) = T(V)^\perp$. *Hence* $T^\star$ *is injective if and only if T is surjective. Hence by duality* $\mathrm{Im}(T^\star) = \mathrm{Ker}(T)$ *and T is injective if and only if* $T^\star$ *is surjective.*

6. *Let* $x_1, \ldots, x_n$ *be a basis of V and A be the matrix of* $T \in \mathrm{End}_k(V)$. *Then the matrix of* T^* *is* A^t.

3.6   Direct Sums

Let U and V be two vector spaces over the field k. Consider the cartesian product $W = U \times V$ as a set. Now, we define on W an operation $+$ and a scalar multiplication componentwise, i.e. by setting

$$(u_1, v_1) + (u_2 + v_2) := (u_1 + u_2, v_1 + v_2)$$

and

$$\lambda \cdot (u, v) := (\lambda u, \lambda v), \quad \text{for all } \lambda \in k.$$

It is easy to see that with these operations W becomes a vector space over k.

Definition 3.6.1. The vector space W defined above is called the *direct sum* of U and V and it is denoted by $U \oplus V$.

Example 3.6.2. Since a field k is a vector space over itself,

$$\underbrace{k \oplus k \oplus \cdots \oplus k}_{n-\text{times}} = k^n.$$

Now suppose U_1, U_2, V_1, and V_2 are vector spaces over k, and $\rho : U_1 \longrightarrow V_1$, $\sigma : U_2 \longrightarrow V_2$ are two linear maps. We then have the following definition.

Definition 3.6.3. We define the *direct sum of the linear maps ρ and σ* as the map

$$\rho \oplus \sigma : U_1 \oplus U_2 \longrightarrow V_1 \oplus V_2, \quad u_1 \oplus u_2 \mapsto \rho(u_1) \oplus \sigma(u_2).$$

Exercise 3.6.4. Show $\rho \oplus \sigma$ is a linear map.

Now, observe the vector space $U \oplus V$ comes with two natural, injective, linear maps, namely,

$$i_U : U \longrightarrow U \oplus V, \ \text{ such that } \ u \mapsto i_U(u) := (u,0);$$

and

$$i_V : V \longrightarrow U \oplus V, \ \text{ such that } \ v \mapsto i_V(v) := (0,v).$$

Exercise 3.6.5. Prove that if S and $T : U \oplus V \longrightarrow W$ are two linear maps such that $S \cdot i_U = T \cdot i_U$ and $S \cdot i_V = T \cdot i_V$, then $S = T$.

Definition 3.6.6. We say that the sequence

$$0 \longrightarrow U \overset{f}{\longrightarrow} W \overset{g}{\longrightarrow} V \longrightarrow 0$$

is a *short exact sequence* of k-vector spaces if and only if $\text{Im}(f) = \text{Ker}(g)$.

Let π be the map $V \to W$ in the sequence above. We say that the sequence *splits on the right* if there is a linear map $\sigma : W \to V$ called a *cross section*, such that $\pi \cdot \sigma = \text{Id}_W$. Similarly, let i be the map $U \to V$. We say that the sequence *splits on the left* if there is a linear map $\tau : V \to U$ such that $\tau \cdot i = \text{Id}_U$.

Exercise 3.6.7. Consider the short exact sequence

$$0 \to U \overset{f}{\to} V \overset{g}{\to} W \to 0.$$

Prove that the following statements are equivalent:

1. The sequence splits on the right.

2. The sequence splits on the left.

3. There is an isomorphism $j : V \longrightarrow U \oplus W$ such that the following diagram

$$
\begin{array}{ccccccccc}
0 & \longrightarrow & U & \xrightarrow{f} & V & \xrightarrow{g} & W & \longrightarrow & 0 \\
 & & \| & & \downarrow & & \| & & \\
0 & \longrightarrow & U & \xrightarrow{i_U} & U \oplus W & \xrightarrow{i_V} & W & \longrightarrow & 0
\end{array}
$$

commutes.

3.7   Tensor Products

We now define the tensor product of vector spaces.[3] For convenience we deal with just two. However, the reader will note that the construction of the tensor product and all its properties actually hold for the tensor product of any finite number of vector spaces. We will thoroughly explore this and related matters in Chapter 7.

Let U, V and W be vector spaces over the field k.

Definition 3.7.1. A *bilinear map* $\phi : V \times W \longrightarrow U$ is a map such that ϕ is linear on each factor, i.e. ϕ is linear in v for each fixed w and linear in w for each fixed v.

---

[3]The tensor product appears for the first time, in 1884 when Gibbs, [44], [45] in his study of the strain on a body introduced the tensor product of vectors in R^3 under the name *indeterminate product*, and extended it to n dimensions two years later. Voigt used tensors for a description of stress and strain on crystals in 1898 [112], and the term *tensor* first appeared with its modern meaning in his work. The word *tensor* comes from the Latin *tendere*, which in turn comes from the Greek $\tau\epsilon\iota\nu\omega$, meaning *to stretch*. Ricci applied tensors to differential geometry during the 1880s and 1890s. A paper of 1901, [102], written with Levi-Civita, was crucial in Einstein's work on general relativity. The widespread adoption of the term *tensor* both in physics and mathematics is due to Einstein's usage. Up to then, tensor products were built from vector spaces.

We already encountered bilinear maps when we studied the dual space. An example of a bilinear map occurs when $T : U \longrightarrow U^*$ is a linear map. Then the composite map $\phi \cdot T : V \times W \longrightarrow U^*$ is bilinear.

We now ask the following question: Is it possible to find the "biggest" vector space U' (and bilinear map $V \times W \longrightarrow U'$) such that for any vector space U, any bilinear map $V \times W \longrightarrow U$ factors through some linear map $U \longrightarrow U'$?

Definition 3.7.2. The *tensor product* $V \otimes_k W$ is a (finite dimensional) vector space over k, together with a bilinear map,

$$\otimes : V \times W \longrightarrow V \otimes_k W,$$

such that if $\beta : V \times W \longrightarrow U$ is any bilinear map, then there exists a *unique linear map* $\widehat{\beta} : V \otimes_k W \to U$ satisfying $\widehat{\beta} \circ \otimes = \beta$. In other words, the following diagram is commutative

$$
\begin{array}{ccc}
U \times W & \xrightarrow{\ \otimes\ } & V \otimes_k W \\
& \searrow{\scriptstyle \beta} & \downarrow{\scriptstyle \widehat{\beta}} \\
& & U
\end{array}
$$

For this reason it might be said that $(x, y) \mapsto x \otimes y$ is the "mother of all bilinear maps".

Theorem 3.7.3. *(Universal Property of the Tensor product) For any two k-vector spaces V and W, the tensor product $V \otimes_k W$ exists and it is unique (up to an isomorphism).*

Proof. We start by proving unicity. If the tensor product $V \otimes_k W$ exists, it must be unique. This is because of its universal mapping property. Indeed, since there is a unique linear map $\widehat{\beta} : V \otimes_k W \to U$ such that $\widehat{\beta} \circ \otimes = \beta$, if $(\mathfrak{T}, \phi)$ were another such pair, where $\mathfrak{T}$ is a k-vector space and

$$\phi : V \times W \to \mathfrak{T}$$

is a bilinear map, then $\widehat{\phi}$ and $\widehat{\otimes}$ would be mutually inverse vector space isomorphisms between $V \otimes W$ and $\mathfrak{T}$ interchanging $\otimes$ with ϕ. Thus they are isomorphic in the strongest possible sense.

We now give a construction of $V \otimes W$. Let $\mathcal{F}(V \times W)$ be the (infinite dimensional) vector space of all k-valued functions on $V \times W$ with finite support, that is, that vanish off a finite subset of $V \times W$. If f and g have finite support and $c \in k$, so do $f + g$ and cf. As noted earlier with pointwise operations $\mathcal{F}(V \times W)$ is a k-vector space. Let $v * w$ be the function which is 1 at (v, w) and zero elsewhere. Then each $f \in \mathcal{F}$ and f can be written $f = f(v_1, w_1)(u_1 * w_1) + \ldots + f(v_k, w_k)(u_k * w_k)$, where the $\{(v_i, w_i)\}$ is its support. In fact, as is easily checked, the coefficients here are uniquely determined so that the points of $V \times W$ are a basis for $\mathcal{F}(V \times W)$.

Consider the subspace $\mathcal{S}$ generated by $(v_1 + v_2) * w - v_1 * w - v_2 * w$, $v * (w_1 + w_2) - w * v_1 - w * v_2$, $c(v * w) - (cv) * w$ and $c(v * w) - cv * (cw)$. Defining $V \otimes W = \mathcal{F}/\mathcal{S}$ we get a k vector space with relations corresponding to setting the generators of $\mathcal{S}$ to be zero. We denote by $\otimes$ the corresponding projection. These relations then translate exactly to the statement that $\otimes$ is a bilinear function. Now let $\gamma : V \times W \to U$ be an arbitrary bilinear function.

We define $T_\gamma = T$ first on $\mathcal{F}$. Let $f = f(v_1, w_1)(u_1 * w_1) + \ldots + f(v_k, w_k)(u_k * w_k)$, as above (the coefficients are unique). Then take $Tf = \sum f(v_i, w_i)\gamma(v_i, w_i)$. If g is another such function with finite support and S is the union of the two supports, then $T(f + g) = \sum_S (f(v_i, w_i) + g(v_i, w - i))\gamma(v_i, w_i) = Tf + Tg$ and, similarly, $T(cf) = cT(f)$. Thus T is a linear function on $\mathcal{F}$ with values in U and since γ is bilinear, $T(\mathcal{S}) = (0)$. Therefore T induces a k linear map $\widetilde{T} : V \otimes W \to U$, where $\widetilde{T} \circ \pi = T$. For $(v, w) \in V \times W$, we have $\widetilde{T}(\pi(v, w)) = T(v * w) = \gamma(v, w)$. This proves the existence of $V \otimes_k W$. $\square$

Definition 3.7.4. For any pair $(u, v) \in U \times V$, its image $\otimes(u, v) \in U \otimes_k V$ is called a *pure* (or *indecomposable*) *tensor*, and it is denoted by $u \otimes v$.

Example 3.7.5. The vector space V over k can be regarded as the tensor product $k \otimes V$, with tensor multiplication $\lambda \otimes v = \lambda v$, $\lambda \in k$, $v \in V$.

Example 3.7.6. The vector space $\mathrm{Hom}(V_1, \ldots, V_n, \ k)$ can be regarded as the tensor product

$$\mathrm{Hom}(V_1, \ldots, V_n, \ m) \otimes \mathrm{Hom}(V_{m+1}, \ldots, V_n, \ k)$$

with tensor multiplication

$$(\phi \otimes \psi)(v_1, \ldots, v_n) = \phi(v_1, \ldots, v_m) \cdot \psi(v_{m+1}, \ldots, v_n).$$

Exercise 3.7.7. Prove $\mathrm{Hom}(V_1, \ldots, V_n, \ k)^\star \cong V_1^\star \otimes \cdots \otimes V_n^\star.$

Example 3.7.8. Calculate

$$(1,2) \otimes (1,3) - (-2,1) \otimes (1,-2).$$

Solution: Let $e_1 = (1,0)$ and $e_2 = (0,1)$. We rewrite the left hand side in terms of x and y and use bilinearity to expand.

Proposition 3.7.9. *If V, W are two vector spaces over the field k, then*

$$\dim V \otimes_k W = \dim V \cdot \dim W.$$

Proof. Now from the proof of Theorem 3.7.3, we see that each element of $V \otimes W$ can be expressed as a linear combination of $v \otimes w$ for $v \in V$ and $w \in W$. Hence, in the finite dimensional case, if $\{v_1, \ldots v_n\}$ is a basis of V and $\{w_1, \ldots w_m\}$ one of W, then, clearly $\{v_i \otimes w_j\}$ generates $V \otimes W$. Thus $V \otimes W$ is finite dimensional with dimension $\leq nm$. However, we do not yet know what its dimension is.

The equation $\dim V \otimes W = \dim V \cdot \dim W$ will follow if $\dim(V \otimes W) \geq nm$, since then the generators $\{v_i \otimes w_j\}$ are a basis. In particular, if $x \in V$ and $y \in W$ are each non-zero, then $x \otimes y$ is also non-zero. This is because each of them can be included in a basis for V and W respectively and since the tensor products of these basis vectors is a basis for $V \otimes W$, each is non-zero.

We now prove that $\dim(V \otimes W) \geq nm$. Let V^* be the dual space of V and (v, v^*) the dual pairing. For $(v, w) \in V \times W$ consider the linear transformation $V^* \to W$ given by $T_{(v,w)}(\alpha) = \alpha(v)w$. The bilinearity of the pairing makes $T_{(v,w)}$ a linear function of everything in sight. Here

$T_{(v,w)} \in \text{Hom}(V^*, W)$ and $(v, w) \mapsto T_{(v,w)}$ is a bilinear function β : $V \times W \to \text{Hom}(V^*, W)$. Hence there is a linear transformation L : $V \otimes W \to \text{Hom}(V^*, W)$ satisfying $L(v \otimes w) = T_{(v,w)}$. Since $\text{Hom}(V^*, W)$ has dimension nm it will be sufficient to show L is surjective.

Now if $\alpha_1, \ldots \alpha_n$ is the dual basis to $\{v_1, \ldots v_n\}$ of V^*, what is the map $\alpha_i \otimes w_j$? At v_k, $(\alpha_i \otimes w_j)(v_k) = \alpha_i(v_k)w_j = \delta_{i,k}w_j$. Thus $\alpha_i \otimes w_j = E_{i,j}$. Using linearity we see this bilinear map $V^* \times W \to V^* \otimes W$ is surjective. Hence L is also surjective and therefore $\dim(V \otimes W) \geq nm$. $\square$

The tensor product is a key construction in differential geometry and arises in the following way. Given a point p in a manifold M, one considers the tangent space $V = T_p(M)$ and its dual space V^*, called the cotangent space. One defines

$$V^{r,s} := \underbrace{V \otimes \cdots \otimes V}_{r-\text{times}} \otimes \underbrace{V^* \otimes \cdots \otimes V^*}_{s-\text{times}},$$

where the tangent space is taken r times and the cotangent space s times. Then one considers smooth vector fields on M such that at each p they take values in $V^{r,s}(p)$.

Some evident properties of the tensor product are:

$(x_1 + x_2 \otimes y) = (x_1 \otimes y) + (x_2 \otimes y)$, $(x \otimes y_1 + y_2) = (x \otimes y_1) + (x \otimes y_2)$, $(\lambda x \otimes y) = \lambda(x \otimes y) = (x \otimes \lambda y)$, where the x's and y's are in V and $\lambda \in k$.

Corollary 3.7.10. $V \otimes W^* \cong \text{Hom}(W, V)$.

Proof. Let $v \in V$ and $\alpha \in W^*$ and define the bilinear map

$$V \otimes W^* \to \text{Hom}(W, V) \text{ such that } (v \otimes \alpha)(w) = \alpha(w) \cdot v.$$

This map is injective (therefore bijective) because if $(v \otimes \alpha)(w) = 0$, then either v or α must be zero. Hence it is an isomorphism of these vector spaces. $\square$

3.7.1   Tensor Products of Linear Operators

We now turn to linear operators and their tensor products. Let $A \in \text{End}_k(V)$ and $B \in \text{End}_k(W)$. We define their tensor product to be the linear operator on $V \otimes W \to V \otimes W$ by, $A \otimes B(x \otimes y) = Ax \otimes By$. In particular, this is the definition for basis elements $x_i \otimes y_j$. We then extend by linearity. Thus $A \otimes B \in \text{End}_k(V \otimes W)$.

Choosing bases $x_1, \ldots, x_n$ for V and $y_1, \ldots, y_m$ for W one sees easily that the matrix representation of $A \otimes B$ with respect to $x_i \otimes y_j$, where $i = 1 \ldots n$ and $j = 1 \ldots m$ is given by the $nm \times nm$ matrix. The elements of this matrix are $m \times m$ blocks of the form $a_{k,l}B$, where $A = (a_{k,l})$ and B is the matrix of the linear transformation of B with respect to $y_1, \ldots, y_m$.

The matrix of the tensor product of these two matrices is

$$\begin{pmatrix} a_{11}B & \cdots & a_{1n}B \\ \vdots & \vdots & \vdots \\ a_{m1}B & \cdots & a_{mn}B \end{pmatrix}.$$

It follows that

$$\text{tr}(A \otimes B) = \sum_k \alpha_{kl} \, \text{tr} \, B = \text{tr} \, A \cdot \text{tr} \, B.$$

Example 3.7.11. Let $V = \mathbb{R}^2$ be a vector space over $\mathbb{R}$. Suppose $S : V \longrightarrow V$ and $T : V \longrightarrow V$ are linear maps with corresponding matrices

$$S = \begin{pmatrix} 0 & 1 \\ 1 & 0 \end{pmatrix}, \quad T = \begin{pmatrix} 1 & 2 \\ 0 & 3 \end{pmatrix}.$$

We want to compute the 4×4 matrix of $S \otimes T$ with respect to the ordered basis $e_1 \otimes e_1, e_1 \otimes e_2, e_2 \otimes e_1, e_2 \otimes e_2$ of $\mathbb{R}^2 \otimes \mathbb{R}^2$, where $e_1 = (1,0)$, and $e_2 = (0,1)$.

We know $S \otimes T$ is a linear map with the property $(S \otimes T)(v \otimes w) = S(v) \otimes T(w)$ for any tensor $v \otimes w \in V \otimes V$. Hence

$$(S \otimes T)(e_1 \otimes e_1) = S(e_1) \otimes T(e_1) = (0e_1 + e_2) \otimes e_1$$
$$= 0 \, e_1 \otimes e_1 + e_2 \otimes e_1 = e_2 \otimes e_1,$$

and

$$(S \otimes T)(e_1 \otimes e_2) = S(e_1) \otimes T(e_2) = (0e_1 + e_2) \otimes (2e_1 + 3e_2) = 2e_2 \otimes e_1 + 3e_2 \otimes e_2.$$

Also,

$$(S \otimes T)(e_2 \otimes e_1) = S(e_2) \otimes T(e_1) = e_1 \otimes e_1$$

and

$$(S \otimes T)(e_2 \otimes e_2) = S(e_2) \otimes T(e_2) = e_1 \otimes (2e_1 + 3e_2) = 2e_1 \otimes e_1 + 3e_1 \otimes e_2.$$

Therefore,

$$S \otimes T = \begin{pmatrix} 0 & 0 & 1 & 2 \\ 0 & 0 & 0 & 3 \\ 1 & 2 & 0 & 0 \\ 0 & 3 & 0 & 0 \end{pmatrix}.$$

Observe we can write the above matrix in the form

$$\begin{pmatrix} A_{11} & A_{12} \\ A_{21} & A_{22} \end{pmatrix},$$

where $A_{ij} = s_{ij}T$, and $(s_{ij}) = S$, i.e.

$$S \otimes T = \begin{pmatrix} 0 \cdot T & 1 \cdot T \\ 1 \cdot T & 0 \cdot T \end{pmatrix} = \begin{pmatrix} 0 & 0 & 1 & 2 \\ 0 & 0 & 0 & 3 \\ 1 & 2 & 0 & 0 \\ 0 & 3 & 0 & 0 \end{pmatrix}.$$

We summarize some straightforward but important features of the tensor product of operators whose verification we leave to the reader.

1. $A \otimes 0 = 0 = 0 \otimes B$, $I_n \otimes I_m = I_{nm}$.

2. $(A_1 + A_2) \otimes B = A_1 \otimes B + A_2 \otimes B$.

3. $A \otimes (B_1 + B_2) = A \otimes B_1 + A \otimes B_2$.

4. $aA \otimes bB = ab(A \otimes B)$.

5. $(A_1 A_2) \otimes (B_1 B_2) = (A_1 \otimes B_1)(A_2 \otimes B_2)$.

6. In particular, $(A \otimes I_m)(I_n \otimes B) = A \otimes B$ and if A and B are invertible, then so is $A \otimes B$ and $(A \otimes B)^{-1} = A^{-1} \otimes B^{-1}$.

Theorem 3.7.12. *Let* $A \in \text{End}_k(V)$ *and* $B \in \text{End}_k(W)$, *where* $n = \dim_k(V)$ *and* $m = \dim_k(W)$. *Let* λ_i *be the eigenvalues of* A *each counted according to multiplicity and, similarly, let* μ_j *be the eigenvalues of* B. *Then* $\{\lambda_i \mu_j\}$ *are the eigenvalues of* $A \otimes B$.

Proof. Let λ be an eigenvalue of A and μ an eigenvalue of B, with corresponding eigenvectors $x \in V$ and $y \in W$. Then

$$(A \otimes B)(x \otimes y) = Ax \otimes By = \lambda x \otimes \mu y = \lambda \mu x \otimes y.$$

Since x and y are each non-zero and, as we saw above, this makes $x \otimes y \neq 0$ it follows that each $\lambda \mu$ is an eigenvalue of $A \otimes B$. Thus we already have nm eigenvalues. But since $\dim(V \otimes W) = nm$ there are at most nm eigenvalues, completing the proof.   □

Remark 3.7.13. Let V be the space of velocity vectors in Newtonian 3-space and T be the vector space of time measurements. We know from Newtonian mechanics that "velocity times time equals displacement", so $V \otimes T$ is the space of displacement vectors in Newtonian 3-space. The point here is that physical quantities have units associated with them. Velocity is not a vector in $\mathbb{R}^3$ but lives in a totally different 3-dimensional vector space over $\mathbb{R}^3$ (the tangent space). To perform calculations, we identify the tangent space with $\mathbb{R}^3$ by choosing coordinate directions such as up, forward and right and units such as m/s, but these are artificial constructions. Displacement again lives in a different vector space, and the tensor product allows us to relate elements in these different physical spaces.

Corollary 3.7.14. *For* $A \in \text{End}_k(V)$ *and* $B \in \text{End}_k(W)$ *we have* $\det(A \otimes B) = (\det A)^n \cdot (\det B)^m$.

Here is a method of proof for dealing with this corollary which will be useful in other contexts.

Proof. To prove this we may assume k is algebraically closed. This is because all determinants involved are unchanged by field extensions. Hence we may pass to the algebraic closure of k. (See Appendix which proves the existence of the algebraic closure of a field.) We first assume the eigenvalues λ_i and μ_j are all distinct. In this case both A and B are diagonalizable and so PAP^{-1} and QBQ^{-1} are diagonal. Using 3.7.1 and the fact that $\det(PAP^{-1}) = \det(A)$ etc. to prove the formula in question for diagonalizable operators we may assume A and B are actually diagonal. In this case from the matrix representation above it follows that $A \otimes B$ is also diagonal with entries $\lambda_i \mu_j$. Thus

$$\det(A \otimes B) = \prod_i \lambda_i (\mu_1)^m \cdots \prod_i \lambda_i (\mu_m)^m$$

$$= \det A \mu_1^m \cdots \det A \mu_m^m = \overset{n}{\det} A \cdot \overset{m}{\det} B,$$

proving the statement whenever the eigenvalues of both A and B are distinct.

Now, in general, the subset D_V of $\operatorname{End}_k(V)$ where the eigenvalues of all elements are distinct is a Zariski dense set; similarly for $\operatorname{End}_k(W)$. Thus $\det(A \otimes B) = (\det A)^n \cdot (\det B)^m$ on the Zariski dense subset $D_V \times D_W$ of $\operatorname{End}_k(V) \times \operatorname{End}_k(W)$. Since these are both polynomial and therefore Zariski continuous functions, they agree everywhere. (For all this see the section on the Zariski topology in Chapter 9.) $\square$

However for the reader who is not familiar with the Zariski topology, since polynomials are Euclidean continuous, one can use exactly the same argument in the case of real or complex operators by substituting the Euclidean topology for the Zariski topology. (Similarly, one shows the diagonalizable operators are Euclidean dense in $\operatorname{End}_k(V)$, where $k = \mathbb{C}$ at the end of Chapter 9 as well).

Exercise 3.7.15. Prove that the definition we gave of $A \otimes B$ yields the tensor product $\operatorname{End}_k(V) \otimes \operatorname{End}_k(W)$.

3.8   Complexification of a Real Vector Space

Why would we want to complexify real vector spaces? One reason is to solve equations. If we want to prove theorems about real solutions to a system of real linear equations or a system of real linear differential equations, it can be convenient as a first step to examine the complex solution space. Then we would try to use our knowledge of the complex solution space (for instance, its dimension) to get information about the real solution space. In the other direction, we may want to know if a subspace of $\mathbb{C}^n$ which is given to us as the solution set to a system of complex linear equations is also the solution set to a system of real linear equations. In what follows we will find a convenient way to describe such subspaces once we understand the different ways that a complex vector space can occur as the complexification of a real subspace.

Let V be a real vector space. To pass from V to a naturally associated complex vector space, we need to give a meaning to $(a + bi)v$, where $a + bi \in \mathbb{C}$ and $v \in V$.

Definition 3.8.1. The *complexification* $V_{\mathbb{C}}$ of the real vector space V is defined to be

$$V_{\mathbb{C}} = V \oplus V,$$

with scalar multiplication

$$(a + bi)(v_1, v_2) = (av_1 - bv_2, \ bv_1 + av_2), \quad a, \ b \in \mathbb{R}.$$

This scalar multiplication is natural if one identifies the pair $(v_1, v_2) \in V \oplus V$ with the (formal) sum $v_1 + iv_2$. Indeed,

$$(a + bi)(v_1 + iv_2) = av_1 + aiv_2 + biv_1 - bv_2 = (av_1 - bv_2) + i(bv_1 + av_2).$$

In particular,

$$i(v_1, v_2) = (-v_2, v_1).$$

It is left to the reader to check that $V_{\mathbb{C}}$ equipped with the above scalar multiplication is a complex vector space e.g., $z(z_0(v_1, v_2)) = zz_0(v_1, v_2)$. Since $i(v, 0) = (0, v)$

$$(v_1, v_2) = (v_1, 0) + (0, v_2) = (v_1, 0) + i(v_2, 0). \tag{$*$}$$

Therefore, the elements of $V_\mathbb{C}$ look like $v_1 + iv_2$, except iv_2 has no meaning while $i(v_2, 0)$ does: it is $(0, v_2)$.

The two real subspaces $V \oplus \{0\}$ and $\{0\} \oplus V$ of $V_\mathbb{C}$ behave like V, since the addition is componentwise and $a(v, 0) = (av, 0)$ as well as $a(0, v) = (0, av)$, $a \in \mathbb{R}$.

Definition 3.8.2. The $\mathbb{R}$-linear function

$$V \hookrightarrow V \oplus \{0\} \quad \text{such that} \quad v \mapsto (v, 0)$$

will be called the *standard embedding* of V into $V \oplus \{0\}$.

So, we identify $V \oplus \{0\}$ with a copy of V inside $V_\mathbb{C}$ and under this identification one sees the complexification $V_\mathbb{C}$ as $V + iV$ (using ($*$)).

Example 3.8.3. In the particular case where $V = \mathbb{R}$, its complexification $\mathbb{R}_\mathbb{C}$ is the set of ordered pairs of real numbers (x, y) with scalar multiplication $(a + bi)(x, y) = (ax - by, bx + ay)$.

Since $(a + bi)(x + yi) = (ax - by) + (bx + ay)i$, the complex vector space $\mathbb{R}_\mathbb{C}$ is isomorphic to $\mathbb{C}$ (as $\mathbb{C}$-vector spaces) under the isomorphism $(x, y) \mapsto x + yi$.

Remark 3.8.4. The space $\mathbb{C}$ as a real vector space is 2 dimensional. Its complexification $\mathbb{C}_\mathbb{C}$ is not the vector space over $\mathbb{C}$, since $\dim_\mathbb{C} \mathbb{C} = 1$ has to have complex dimension 2 because $\mathbb{C}$ has real dimension 2. Rather, $\mathbb{C}_\mathbb{C} = \mathbb{C} \oplus \mathbb{C}$ with scalar multiplication

$$(a + bi)(z_1, z_2) = (az_1 - bz_2, bz_1 + az_2).$$

We leave to the reader the proof of the following proposition.

Proposition 3.8.5. *If* $V = \{0\}$ *then,* $V_\mathbb{C} = \{0\}$. *If* $V \neq \{0\}$ *and* $\{e_1, \ldots, e_n\}$ *is a basis of the real vector space* V, *then* $\{(e_1, 0), \ldots, (e_n, 0)\}$ *is a* $\mathbb{C}$-*basis of* $V_\mathbb{C}$. *In particular,*

$$\dim_\mathbb{C}(V_\mathbb{C}) = \dim_\mathbb{R}(V)$$

for all real vector spaces V.

Proposition 3.8.6. *Every R-linear map* $\phi : V \longrightarrow W$ *of real vector spaces extends in a unique way to a* $\mathbb{C}$-*linear map of the complexifications: there is a unique* $\mathbb{C}$-*linear map* $\phi_{\mathbb{C}} : V_{\mathbb{C}} \longrightarrow W_{\mathbb{C}}$ *making the diagram*

$$
\begin{array}{ccc}
V & \xrightarrow{\ \phi\ } & W \\[2pt]
id_V \downarrow & & \uparrow id_W \\[2pt]
V_{\mathbb{C}} & \xrightarrow[\ \phi_{\mathbb{C}}\ ]{} & W_{\mathbb{C}}
\end{array}
$$

commute, where the vertical maps are the standard embeddings of real vector spaces into their complexifications.

Proof. If the $\mathbb{C}$-linear map $\phi_{\mathbb{C}}$ exists, then the commutativity of the diagram says $\phi_{\mathbb{C}}(v, 0) = (\phi(v), 0)$ for all $v \in V$. Therefore, if (v_1, v_2) is any element of $V_{\mathbb{C}}$, we get

$$
\begin{aligned}
\phi_{\mathbb{C}}(v_1, v_2) &= \phi_{\mathbb{C}}(v_1, 0) + \phi_{\mathbb{C}}(0, v_2) \\
&= \phi_{\mathbb{C}}(v_1, 0) + \phi_{\mathbb{C}}\big(i(v_2, 0)\big) \\
&= \phi_{\mathbb{C}}(v_1, 0) + i\phi_{\mathbb{C}}(v_2, 0) \\
&= (\phi(v_1), 0) + i\big(\phi(v_2), 0\big) \\
&= (\phi(v_1), 0) + \big(0, \phi(v_2)\big) \\
&= (\phi(v_1), \phi(v_2)).
\end{aligned}
$$

This forces us to define the map $\phi_{\mathbb{C}} : V_{\mathbb{C}} \longrightarrow W_{\mathbb{C}}$ by

$$
\phi_{\mathbb{C}}(v_1, v_2) := \big(\phi(v_1), \phi(v_2)\big).
$$

We leave to the reader to check that $\phi_{\mathbb{C}}$ is $\mathbb{C}$-linear and that it commutes with the multiplication by i. $\square$

Definition 3.8.7. We call $\phi_{\mathbb{C}}$ the *complexification* of ϕ.

Example 3.8.8. Let A be a 2×2 real matrix. Its complexification

$$
A_{\mathbb{C}} : (\mathbb{R}^2)_{\mathbb{C}} \longrightarrow (\mathbb{R}^2)_{\mathbb{C}}
$$

is given by

$$
A_{\mathbb{C}}(w_1, w_2) = (Aw_1, Aw_2).
$$

The isomorphism $\psi : (\mathbb{R}^2)_{\mathbb{C}} \longrightarrow \mathbb{C}^2$ given by $\psi(w_1, w_2) = w_1 + iw_2$ identifies $A_{\mathbb{C}}$ with the map

$$\mathbb{C}^2 \longrightarrow \mathbb{C}^2, \quad \text{such that} \quad w_1 + iw_2 \mapsto Aw_1 + iAw_2.$$

Since A as a matrix acting on complex vectors satisfies $Aw_1 + iAw_2 = A(w_1 + iw_2)$, the diagram

$$
\begin{array}{ccc}
(\mathbb{R}^2)_{\mathbb{C}} & \xrightarrow{\ A_{\mathbb{C}}\ } & (\mathbb{R}^2)_{\mathbb{C}} \\
\psi \downarrow & & \uparrow \psi \\
\mathbb{C}^2 & \xrightarrow[\ A\ ]{} & \mathbb{C}^2
\end{array}
$$

commutes. This is the sense in which the complexification $A_{\mathbb{C}}$ is just the matrix A acting on $\mathbb{C}^2$.

It is straightforward to check using the definitions that if $\phi : V \longrightarrow W$ and $\chi : W \longrightarrow Z$ are $\mathbb{R}$-linear transformations, then

$$(\chi \circ \phi)_{\mathbb{C}} = \chi_{\mathbb{C}} \circ \phi_{\mathbb{C}}, \quad \text{and} \quad (id_V)_{\mathbb{C}} = id_{V_{\mathbb{C}}}.$$

Also, if U is a real subspace of V, then $U_{\mathbb{C}}$ is a complex subspace of $V_{\mathbb{C}}$.

3.8.1   Complexifying with Tensor Products

While the definition of $V_{\mathbb{C}}$ does not depend on the choice of a basis of V, implicitly it depends on a choice of the real basis $\{1, i\}$ of $\mathbb{C}$. It is also possible to give a completely basis-free construction of the complexification using tensor products. The idea is that $V_{\mathbb{C}}$ behaves like $\mathbb{C} \otimes_{\mathbb{R}} V$ and the complexification

$$\phi_{\mathbb{R}} : V_{\mathbb{C}} \longrightarrow W_{\mathbb{C}}$$

of an $\mathbb{R}$-linear map $\phi : V \longrightarrow W$ behaves like the $\mathbb{C}$-linear map

$$1 \otimes \phi : \mathbb{C} \otimes_{\mathbb{R}} V \longrightarrow \mathbb{C} \otimes_{\mathbb{R}} W.$$

Indeed, we have

Theorem 3.8.9. *For every real vector space V, there is a unique iso-morphism of $\mathbb{C}$-vector spaces*

$$f_V : V_{\mathbb{C}} \longrightarrow \mathbb{C} \otimes_{\mathbb{R}} V, \quad such \ that \ f_V(v_1, v_2) = 1 \otimes v_1 + i \otimes v_2$$

making the following diagram

$$
\begin{array}{ccc}
V & \xrightarrow{\ i\ } & V_{\mathbb{C}} \\
& {}_{i}\searrow & \downarrow {}^{f_V} \\
& & \mathbb{C} \otimes_{\mathbb{R}} V
\end{array}
$$

commute. In addition, if $\phi : V \longrightarrow W$ is any $\mathbb{R}$-linear map of real vector spaces, then the following diagram of $\mathbb{C}$-linear maps

$$
\begin{array}{ccc}
V_{\mathbb{C}} & \xrightarrow{\ \phi_{\mathbb{C}}\ } & W_{\mathbb{C}} \\
{}^{f_V}\downarrow & & \downarrow {}^{f_W} \\
\mathbb{C} \otimes_{\mathbb{R}} V & \xrightarrow[1 \otimes \phi]{} & \mathbb{C} \otimes_{\mathbb{R}} W
\end{array}
$$

commutes.

Proof. If f_V exists, then

$$
\begin{aligned}
f_V(v_1, v_2) = f_V((v_1, 0) + i(v_2, 0)) \quad &= f_V(v_1, 0) + i f_V(v_2, 0) \\
= 1 \otimes v_1 + i(1 \otimes v_2) \quad\quad &= 1 \otimes v_1 + i \otimes v_2.
\end{aligned}
$$

So, we define f_V by the above expression, and one can easily check that f_V is a $\mathbb{R}$-linear map. To see that f_V is in fact a $\mathbb{C}$-linear map, we compute

$$f_V\big(i(v_1, v_2)\big) = f_V(-v_2, v_1) = 1 \otimes (-v_2) + i \otimes v_1 = -1 \otimes v_2 + i \otimes v_1,$$

and

$$i f_V(v_1, v_2) = i(1 \otimes v_1 + i \otimes v_2) = i \otimes v_1 + (-1) \otimes v_2 = -1 \otimes v_2 + i \otimes v_1.$$

To show that f_V is an isomorphism, we calculate the inverse map $\mathbb{C} \otimes_{\mathbb{R}} V \longrightarrow V_{\mathbb{C}}$. Since $1 \otimes v$ corresponds to $(v, 0)$, one expects the element $z \otimes v = z(1 \otimes v)$ to be sent to $z(v, 0)$. Hence, we construct such a map by first considering the map

$$\mathbb{C} \times V \longrightarrow V_{\mathbb{C}}, \quad : \quad (z, v) \mapsto z(v, 0).$$

This is an $\mathbb{R}$-bilinear map, so it induces an $\mathbb{R}$-linear map

$$g_V : \mathbb{C} \otimes_{\mathbb{R}} \longrightarrow V_{\mathbb{C}}, \quad : \quad g_V(z, v) = z(v, 0).$$

To prove that g_V is $\mathbb{C}$-linear, it suffices to check that $g_V(zt) = z g_V(t)$ when t is an elementary tensor, say $t = z_1 \otimes v$. We have

$$g_V\big(z(z_1 \otimes v)\big) = g_V(zz_1 \otimes v) = zz_1(v, 0)$$

and

$$z g_V(z_1 \otimes v) = z\big(z_1(v, 0)\big) = zz_1(v, 0).$$

Finally, to show that f_V and g_V are inverses of each other, we have

$$\begin{aligned}
g_V\big(f_V(v_1, v_2)\big) &= g_V(1 \otimes v_1 + i \otimes v_2) \\
&= (v_1, 0) + i(v_2, 0) = (v_1, 0) + (0, v_2) \\
&= (v_1, v_2).
\end{aligned}$$

Now, it is easy to verify the commutativity of the diagram.   □

Remark 3.8.10. Since $V_{\mathbb{C}}$ and $\mathbb{R} \otimes_{\mathbb{R}} V$ are both the direct sum of subspaces V and iV (using the standard embeddings of V into $V_{\mathbb{C}}$ and $\mathbb{C} \otimes_{\mathbb{R}} V$.). This suggests calling a complex vector space V the complexification of a real subspace W when $V = W + iW$ and $W \cap iW = \{0\}$. For example, in this sense $\mathbb{C}^n$ is the complexification of $\mathbb{R}^n$. The distinction from the preceding meaning of complexification is that now we are talking about how a pre-existing complex vector space can be a complexification of a real subspace of it, rather than starting with a real vector space and creating a complexification out of it.

Proposition 3.8.11. *Let V be a complex vector space and W be a real subspace. The following are equivalent:*

1. V is the complexification of W.

2. *Any $\mathbb{R}$-basis of W is a $\mathbb{C}$-basis of V.*

3. *Some $\mathbb{R}$-basis of W is a $\mathbb{C}$-basis of V.*

4. *the $\mathbb{C}$-linear map $\mathbb{C} \otimes_{\mathbb{R}} W \to V$ given by $z \otimes w \mapsto zw$ on elementary tensors is an isomorphism of complex vector spaces.*

Proof. (1) $\implies$ (2) Let $(e_k)_{k=1,\dots,m}$ be a real basis of W. Since $V = W + iW$, V is spanned over $\mathbb{R}$ by $\{e_k, ie_k \mid k = 1, \dots m\}$, so V is spanned over $\mathbb{C}$ by $(e_k)_{k=1,\dots,m}$. To show linear independence of $(e_k)_{k=1,\dots,m}$ over $\mathbb{C}$, suppose

$$\sum (a_k + ib_k)e_k = 0.$$

Then

$$\sum a_k e_k = i \sum (-b_k)e_k \in W \cap iW = \{0\},$$

and therefore, $a_k = b_k = 0$ for all k.

That (2) $\implies$ (3) is obvious.

(3) $\implies$ (4). If some real basis $(e_k)_{k=1,\dots,m}$ of W is a $\mathbb{C}$-basis of V, then the $\mathbb{C}$-linear map

$$\mathbb{C} \otimes_{\mathbb{R}} W \quad \text{such that } z \otimes w \mapsto zw$$

sends $1 \otimes e_k$ to e_k, and so it identifies $\mathbb{C}$-bases of two complex vector spaces. Therefore this map is an isomorphism of complex vector spaces.

(4) $\implies$ (1). Since

$$\mathbb{C} \otimes_{\mathbb{R}} W = 1 \otimes W + i(1 \otimes W)$$

and

$$(1 \otimes W) \cap i(1 \otimes W) = 0,$$

the isomorphism with V shows that V is a complexification of W. $\square$

Example 3.8.12. $M_2(\mathbb{C})$ is the complexification of two real subspaces, namely, the subspace $M_2(\mathbb{R})$ and the subspace

$$\left\{ \begin{pmatrix} a & b + ci \\ b - ci & d \end{pmatrix} : a, b, c, d \in \mathbb{R} \right\}.$$

One can easily check that both are 4-dimensional real subspaces containing a $\mathbb{C}$-basis of $M_2(\mathbb{C})$.

3.8.2   Real Forms and Complex Conjugation

Real subspaces with a given complexification are defined as follows:

Definition 3.8.13. If V is a complex vector space, a *real form* (or $\mathbb{R}$-*form*) of V is a real subspace W of V having V as its complexification.

Definition 3.8.14. A *conjugation* on a complex vector space V is a map $j : V \longrightarrow V$ such that the following holds.

1. $j(v_1 + v_2) = j(v_1) + j(v_2)$, for any $v_1, v_2 \in V$.

2. $j(zv) = zj(v)$, for all $z \in \mathbb{C}$ and $v \in V$.

3. $j(j(v)) = v$ for all $v \in V$.

We observe that a conjugation is $\mathbb{R}$-linear, but is not $\mathbb{C}$-linear, since $j(iv) = -ij(v)$.

The complexification $V_{\mathbb{C}}$ of the real vector space V has more structure than an ordinary complex vector space. It comes with a *canonical complex conjugation* map

$$\chi : V_{\mathbb{C}} \longrightarrow \overline{V_{\mathbb{C}}}, \quad \text{defined by } \chi(v \otimes z) = v \otimes \bar{z}.$$

The map χ can be seen either as a conjugate-linear map from $V_{\mathbb{C}}$ to itself, or as a complex linear isomorphism from $V_{\mathbb{C}}$ to its complex conjugate $\overline{V_{\mathbb{C}}}$.

Chapter 4

Inner Product Spaces

Let V be a real or complex vector space, not necessarily of finite dimension. We begin with the complex case.

Definition 4.0.1. A *Hermitian form* on a complex vector space V is a map

$$\langle \cdot, \cdot \rangle : V \times V \longrightarrow \mathbb{C}$$

satisfying:

1. $\langle \cdot, \cdot$ is linear in the first variable.

2. $\langle \cdot, \cdot$ is conjugate linear in the second variable.

3. $\langle x, y \rangle = \overline{\langle y, x \rangle}$.

4. $\langle \cdot, \cdot$ is positive definite, that is $\langle x, x \rangle \geq 0$ and can only be 0 if $x = 0$.

If V is a real vector space, we just replace "conjugate linear" by linear in 2 and condition 3 by symmetry. In either case $(V, \langle \cdot, \cdot \rangle)$ will be called a complex (respectively real) *inner product space*, or just an inner product space.

Example 4.0.2. The typical example of a finite dimensional real inner product space is: $V = \mathbb{R}^n$ with the usual inner product

$$\langle x, y \rangle = \sum_{i=1}^{n} x_i y_i,$$

while in the complex case $V = \mathbb{C}^n$ with the inner product

$$\langle x, y \rangle = \sum_{i}^{n} x_i \bar{y}_i.$$

We ask the reader to check that these are indeed real or complex inner product spaces, respectively. These examples bear the same relationship to general finite dimensional inner product spaces as finite dimensional vector spaces have to choosing the standard basis rather than having an arbitrary basis.

Exercise 4.0.3. Given a real inner product space, show one obtains a Hermitian inner product space upon complexification.

Exercise 4.0.4. We ask the reader to think about what *all possible* finite dimensional real or complex inner product spaces might be.

Before turning to more exotic examples of real or complex inner product spaces, we formulate an important theorem. Here we write $\| x \|^2$ for $\langle x, x \rangle$.

Theorem 4.0.5. *(Cauchy-Schwarz Inequality) Let $(V, \langle \cdot, \cdot \rangle)$ be a real or complex inner product space (finite or infinite dimensional). For each $x, y \in V$,*

$$|\langle x, y \rangle|^2 \leq \| x \|^2 \| y \|^2 .$$

In particular in the examples above, in the complex case

$$\left| \sum_{i=1}^{n} x_i \bar{y}_i \right|^2 \leq \sum_{i=1}^{n} |x_i|^2 |y_i|^2$$

while in the real case,

$$\left| \sum_{i=1}^{n} x_i y_i \right|^2 \leq \sum_{i=1}^{n} |x_i|^2 |y_i|^2.$$

Proof. We first deal with the more difficult complex case. Since the right hand side of the Cauchy-Schwarz[1] inequality is non-negative, to prove it we may certainly assume $\langle x, y \rangle \neq 0$ and in particular, both x and $y \neq 0$. Let $c = \frac{|\langle y, x \rangle|}{\langle y, x \rangle} \in \mathbb{C}$. Then $|c| = 1$ and

$$0 \leq \left| \frac{cx}{\| x \|} - \frac{y}{\| y \|} \right|^2 = \frac{|c|^2 \| x \|^2}{\| x \|^2} - 2 \frac{\Re(\langle cy, x \rangle)}{\| y \| \| x \|} + \frac{\| y \|^2}{\| y \|^2}.$$

Thus,

$$0 \leq 2 \| y \| \| x \| - 2 \Re(c \langle y, x \rangle),$$

and so,

$$\Re(c \langle y, x \rangle) = \Re(|\langle y, x \rangle|) \leq \| y \| \| x \| .$$

$\square$

The real case follows directly from this, or it can be proved directly as follows. Let $V, \langle , \rangle$ be a real inner product space. For $x, y \in V$ and $t \in \mathbb{R}$,

$$0 \leq \langle x + ty, x + ty \rangle = \langle x, x \rangle + 2t \langle x, y \rangle + t^2 \langle y, y \rangle,$$

which for fixed x and y is a non-negative quadratic function q of t with real coefficients. Since $q \geq 0$ it must have discriminant $\Delta \geq 0$. Thus,

$$(2 \langle x, y \rangle)^2 \leq 4 \| x \|^2 \| y \|^2 .$$

Later we will obtain a generalization as well as a simpler proof of the Cauchy-Schwarz inequality using Gramians (see Proposition 4.8.2).

---

[1] Hermann A. Schwarz (1843-1921) a German Jewish mathematician was influenced by Kummer and Weierstrass. While still a graduate student, Schwarz made an important contribution by discovering a new minimal surface. Schwarz also gave the first rigorous proof of the Riemann mapping theorem and gave the alternating method for solving the Dirichlet problem which soon became the standard technique. He also answered the question of whether a given minimal surface really yields a minimal area which used the idea of constructing a function by successive approximations, led Émile Picard to his existence proof for solutions of differential equations. The same paper contains the inequality for integrals now known as the "Schwarz inequality". The Cauchy-Schwarz inequality appears pretty much everywhere in mathematics. His work on what is now known as the "Schwarz function" led to the development of the theory of automorphic functions.

More interesting examples of real or complex inner product spaces whose dimension is not finite are $l_2(\mathbb{R})$, respectively $l_2(\mathbb{C})$. Here we shall treat the more difficult complex case, the real case following immediately from this.

Example 4.0.6. Let V be the set of all sequences of complex numbers $(x_n)_{n\in\mathbb{N}}$ satisfying

$$\sum_{n=1}^{\infty} |x_n|^2 < \infty.$$

Such sequences are called *square-summable*. An inner product on $l_2(\mathbb{C})$ is defined by

$$\langle (x_n)_{n\in\mathbb{N}}, (y_n)_{n\in\mathbb{N}} \rangle = \sum_{n=1}^{\infty} x_n y_n, \quad \text{respectively} \quad \sum_{n}^{\infty} x_n \bar{y}_n$$

We must show $l_2(\mathbb{C})$ is a complex inner product space. In particular, that $\langle x, y \rangle$ is well defined and possesses the properties of an inner product. Suppose $(x_n)_{n\in\mathbb{N}}$ and $(y_n)_{n\in\mathbb{N}}$ are two such sequences. By the *finite* dimensional version of Cauchy-Schwarz we know

$$|\sum_{i=1}^{n} x_i \bar{y}_i|^2 \le \sum_{i=1}^{n} |x_i|^2 |y_i|^2.$$

Let x and y be square summable sequences and $x|n$ and $y|n$ be their truncation at the n-th coordinate. Then, as above,

$$(\langle x|n, y|n \rangle)^2 \le \| x|n \|^2 \| y|n \|^2 \le \| x \|^2 \| y \|^2 .$$

Therefore $\lim_{n\to\infty}((\langle x|n, y|n \rangle)^2$ exists. Hence, so does

$$\lim_{n\to\infty} (\langle x|n, y|n \rangle) = \langle x, y \rangle.$$

We leave to the reader to check the other details making these spaces real or complex inner product spaces.

Now, given an inner product space $(V, \langle, \rangle)$, we define:

Definition 4.0.7. The *norm* (or *the length*) of the vector $x \in V$ is defined as

$$\| x \| = \sqrt{\langle x, x \rangle}.$$

Since the inner product is positive definite $\| x \| = 0$ if and only if $x = 0$. Therefore if $x \neq 0$, we can *normalize* it by letting $y = \frac{x}{\|x\|}$ and then $\| y \| = 1$.

Definition 4.0.8. A set $\{x_1, ..., x_m, ...\} \subset V$ is called *orthonormal* if

$$\langle x_i, x_j \rangle = \delta_{ij}, \quad \forall\, i, j.$$

Proposition 4.0.9. *An orthonormal set is linearly independent.*

Proof. Suppose there is a dependence relation between the (finite set of) vectors

$$\sum_{i=1}^{m} c_i x_i = 0, \quad \text{with} \quad c_i \neq 0, \quad \forall\, i = 1, \ldots, m,$$

where the c_i are in $\mathbb{R}$ or $\mathbb{C}$.

Since $\{x_1, ..., x_m\}$ is an orthonormal set, for each $j = 1, ..., m$

$$\left\langle \sum_{i=1}^{m} c_i x_i, \ x_j \right\rangle = \langle 0, \ x_j \rangle = 0 = \sum_{i=1}^{m} c_i \langle x_i, \ x_j \rangle$$

$$= \sum_{i=1}^{m} c_i \delta_{ij} = c_j.$$

Hence, $c_j = 0$ for every $j = 1, ..., m$. $\qquad\qquad\square$

Definition 4.0.10. The vectors $(x_i)_{i \in I}$ form an *orthonormal basis* for V if it is orthonormal and generates V.

In an inner product space, what served as a basis in a vector space will be replaced by an orthonormal basis. The coefficients of a vector in an orthonormal basis are easy to determine.

Definition 4.0.11. If $\{x_i : i \in I\}$ is an orthonormal basis of V and $x = \sum_{i=1}^{n} c_i x_i$. Then the coefficients, c_i, are given by:

$$c_j = \langle x, \, x_j \rangle, \quad \forall \, j = 1, ..., n.$$

They are called the *Fourier coefficients*[2] of x.

We close this section with the question, when is the Cauchy-Schwarz inequality an equality? As usual, we deal with the more difficult case by working over $\mathbb{C}$.

Corollary 4.0.12. *The Cauchy-Schwarz inequality is an equality if and only if x and y are linearly dependent.*

Proof. If $y = cx$, then $\langle cx, x \rangle = c \parallel x \parallel^2$ and so $|\langle cx, x \rangle|^2 = |c|^2 \parallel x \parallel^4$. On the other hand, this is just $\langle cx, cx \rangle \langle x, x \rangle$.

Conversely, suppose $|\langle x, y \rangle|^2 = \parallel x \parallel^2 \cdot \parallel y \parallel^2$. We will show x and y are linearly dependent. To do so we may assume both x and $y \neq 0$. Replace y by the vector $y/ \parallel y \parallel$ of norm 1. Then the equality becomes

$$\parallel x \parallel^2 = |\langle x, \frac{y}{\parallel y \parallel} \rangle|^2.$$

Now, consider the vector $x - \langle x, y \rangle y$. It is clear that it belongs to the linear span of $\{x, y\}$ and since we have normalized y,

$$\left\langle x - \langle x, y \rangle y \, , \, y \right\rangle = \langle x, y \rangle - \langle x, y \rangle \parallel y \parallel^2 = 0$$

---

[2]Joseph Fourier (1768-1830) Italian-French mathematician and physicist was the son of a tailor. Commissions in the scientific corps of the army were reserved for those of "good birth", and being thus ineligible, he accepted a military lectureship in mathematics. Although a supporter of the French Revolution, he was briefly imprisoned during the Terror but in 1795 was appointed to the École Normale, and subsequently succeeded Lagrange at the École Polytechnique. Fourier accompanied Napoleon on his Egyptian expedition in 1798, as scientific adviser, and was appointed secretary of the Institut d'Égypte. In 1801 Napoleon appointed Fourier Prefect (Governor) of the Department of Isère in Grenoble, where he oversaw projects such as road construction. It was there he began to experiment with heat flow and wrote the ground breaking "The Analytic Theory of Heat" and developed what we now call Fourier series.

and therefore

$$x - \langle x, y \rangle y \perp y. \qquad (\star)$$

Now there are two possibilities:

1. Either $x - \langle x, y \rangle y = 0$, and then, x is a multiple of y and we are done.

2. Or, $x - \langle x, y \rangle y \neq 0$. In this case

$$\left\langle x - \langle x, y \rangle y, x \right\rangle = \| x \|^2 - \langle x, y \rangle \langle y, x \rangle$$
$$= \| x \|^2 - |\langle x, y \rangle|^2 = 0,$$

by assumption. Therefore,

$$x - \langle x, y \rangle y \perp x. \qquad (\star\star)$$

The relations $(\star)$ and $(\star\star)$ give a contradiction since, as we saw, $x - \langle x, y \rangle y$ belongs to the linear span of $\{x, y\}$. Therefore, this vector cannot be perpendicular to both x and y. $\square$

4.0.1   Gram-Schmidt Orthogonalization

Let $(V, \langle \cdot, \cdot \rangle)$ be an inner product space over $k = \mathbb{R}$ or $\mathbb{C}$.

Theorem 4.0.13. *(Gram-Schmidt orthogonalization[3])* *Let* $\{v_1, v_2, ..., \}$ *be a linearly independent, finite (or possibly countable) set of vectors in V. Then, there is an orthonormal family of vectors $\{u_1, u_2..., \}$ such that $lin.sp_k(\{u_1, ..., u_i\}) = lin.sp_k(\{v_1, ..., v_i\})$, for every $i \geq 1$.*

---

[3]The orthogonalization method described here is named for the Danish mathematician Jurgen Pedersen Gram (1850-1916) and the German-Estonian mathematician Erhard Schmidt (1876-1959), both students of David Hilbert. However, this appeared earlier in the work of Laplace and Cauchy.

Proof. Since $v_1, ..., v_n ...$ are linearly independent, $v_1 \neq 0$, and so we replace it by the normalized vector $u_1 = \frac{v_1}{\|v_1\|}$. Hence, $\| u_1 \| = 1$.

Now consider the vector $\widetilde{u}_2 = v_2 - \langle v_2, u_1 \rangle u_1$. Obviously, $\widetilde{u}_2 \neq 0$. For if not, $v_2 - \langle v_2, u_1 \rangle u_1 = 0$. Hence $v_2 = $ multiple of v_1, which is a contradiction since v_1, v_2 are linearly independent. Now let $k \geq 2$ and reason by induction.

$$u_k = \frac{v_k - \sum_{i=1}^{k-1} \langle v_k, u_i \rangle u_i}{\| v_k - \sum_{i=1}^{k-1} \langle v_k, u_i \rangle u_i \|}.$$

□

Corollary 4.0.14. *In a countable dimensional inner product space, any orthonormal set S can extended to an orthonormal basis.*

Proof. By proposition 4.0.9 S is linearly independent. Therefore it can be extended to a basis, B, of V. Applying Gram-Schmidt to $B \supseteq S$ by first running through S completes the proof. □

As we shall see shortly, Gram-Schmidt orthogonalization can be used in Lie groups to prove the *Iwasawa decomposition* (see 4.7.13).

4.0.1.1   Legendre Polynomials

Here we present a final example of an infinite dimensional real inner product space, as well as Gram-Schmidt orthogonalization on it namely the **Legendre polynomials**.[4] Here V is the real Hilbert space $L_2[-1, 1]$ defined just below.

---

[4]These polynomials were first studied by the French mathematician Adrien-Marie Legendre (1752-1833), who made important contributions to special functions, elliptic integrals, number theory, probability and the calculus of variations. We encounter Legendre polynomials in the solution of many problems particularly in the representation of the rotation group acting on the sphere S^2 by spherical harmonics. This is connected with solving Laplace equation $\nabla^2 F(x) = 0$ in $\mathbb{R}^3$ and problems in potential theory in Physics.

Consider the real inner product space $C[-1,1]$ of continuous real valued functions on $[-1,1]$ equipped with the inner product

$$\langle f, g \rangle = \int_{-1}^{1} f(x)g(x)dx.$$

$L_2[-1,1]$ consists of all real valued measurable functions f defined on $[-1,1]$ such that $\int_{-1}^{1} |f(x)|^2 dx = \| f \|^2 < \infty$. $L_2[-1,1]$ contains $C[-1,1]$ (as a dense set in the norm).

In particular, $L_2[-1,1]$ contains the set $\{1, x, x^2, x^3, ...\}$, $x \in [-1,1]$, of linearly independent functions (vectors). Their linear span contains all polynomials. By the Stone Weierstrass theorem, these polynomials are uniformly dense in $C[-1,1]$ and therefore are norm-dense in $L_2[-1,1]$.

We leave to the reader to check that $L_2[-1,1]$ is an inner product space. (It has a countable dense subset, namely all the polynomials on $[-1,1]$ with rational coefficients). We use Gram-Schmidt orthogonalization to construct an orthonormal basis, $\{\phi_1, \phi_2, ...\}$. First, set $h_1(x) := 1$, so $\| h_1(x) \| = \left(\int_{-1}^{1} dx \right)^{\frac{1}{2}} = \sqrt{2}$. Hence,

$$\phi_1(x) = \frac{h_1(x)}{\| h_1(x) \|} = \frac{1}{\sqrt{2}}.$$

Choose h_2 by setting $h_2(x) = a + bx$. Then h_1, and h_2 are linearly independent. Since we want $h_2 \perp h_1$, we must have

$$0 = \int_{-1}^{1} h_1(x)h_2(x)dx = \int_{-1}^{1} (a + bx)dx = \left[ax + \frac{b}{2}x^2 \right]_{-1}^{1} = 2a.$$

Therefore, $a = 0$ and $h_2(x) = \frac{b}{2}x$. In addition,

$$\| h_2 \| = \int_{-1}^{1} \frac{b}{2}x^2 dx = \frac{b}{2}\sqrt{\frac{2}{3}}.$$

Thus

$$\phi_2(x) = \frac{h_2(x)}{\| h_2(x) \|} = \sqrt{\frac{3}{2}}\, x.$$

Now, we take $h_3(x) = c + dx + ex^2$, and since we want $h_3 \perp h_2$ and $h_3 \perp h_1$ we must have $\int_{-1}^{1}(c + dx + ex^2)dx = 0$ and $\int_{-1}^{1} \frac{b}{2}x(c + dx + ex^2)dx = 0$. Calculating these integrals we find that $d = 0$, and $e = -3c$ and so $h_3(x) = c(x - 3x^2)$. Normalizing $h_3(x)$, we get

$$\phi_3(x) = \frac{c(x - 3x^2)}{\| c(x - 3x^2) \|} = \sqrt{\frac{5}{8}} \, (3x^2 - 1).$$

Continuing, in general we find,

$$\phi_n(x) = \sqrt{\frac{2n - 1}{2}} \, p_{n-1}(x), \quad \text{where} \quad p_n(x) = \frac{1}{2^n n!} \frac{d^n}{dx^n}(x^2 - 1)^n.$$

The polynomials $\phi_n(x)$ are called the *Legendre polynomials*.

4.0.2   Bessel's Inequality and Parseval's Equation

In this subsection, we will consider only finite-dimensional vector spaces (although all this works in the infinite dimensional case as well).

Let V be an inner product space of finite dimension and $\{v_1, ..., v_n\}$ be an orthonormal family of vectors, with $n \leq \dim V$.

Theorem 4.0.15. *For each $v \in V$,*

$$\| v \|^2 \geq \sum_{i=1}^{n} |\langle v, v_i \rangle|^2 \quad \text{(Bessel's inequality)}.$$

If $n = \dim V$, we get equality (Parseval's equation).

Proof. For $v \in V$, $\| v - \sum_{i=1}^n \langle v, v_i \rangle v_i \|^2 \geq 0$. Now,

$$
\| v - \sum_{i=1}^n \langle v, v_i \rangle v_i \|^2 = \left\langle v - \sum_{i=1}^n \langle v, v_i \rangle v_i , \ v - \sum_{i=1}^n \langle v, v_i \rangle v_i \right\rangle
$$

$$
= \| v \|^2 - \left\langle v, \sum_{i=1}^n \langle v, v_i \rangle v_i \right\rangle - \left\langle \sum_{i=1}^n \langle v, v_i \rangle v_i, v \right\rangle
$$

$$
+ \left\langle \sum_{i=1}^n \langle v, v_i \rangle v_i , \ \sum_{i=1}^n \langle v, v_i \rangle v_i \right\rangle
$$

$$
= \| v \|^2 - \sum_{i=1}^n \overline{\langle v, v_i \rangle} \langle v, v_j \rangle - \sum_{i=1}^n \langle v, v_i \rangle \langle v_i, v \rangle
$$

$$
+ \left\langle \sum_{\substack{i,j \\ i=j}}^n \langle v, v_i \rangle v_i , \ \sum_{\substack{i,j \\ i=j}}^n \langle v, v_i \rangle v_i \right\rangle
$$

$$
= \| v \|^2 - \sum_{i=1}^n |\langle v, v_i \rangle|^2 \geq 0
$$

where we used the fact that $\langle v_i, v \rangle = \overline{\langle v, v_i \rangle}$ and also that $\langle v_i, v_j \rangle = \delta_{ij}$.
So,

$$
\| v \|^2 \geq \sum_{i=1}^n |\langle v, v_i \rangle|^2.
$$

When $n = \dim V$, then $v = \sum_{i=1}^n \langle v, v_i \rangle v_i$. Therefore,

$$
\| v \|^2 - \sum_{i=1}^n |\langle v, v_i \rangle|^2 = 0.
$$

□

We note in the case l_2, Bessel's inequality and Parseval's equation are respectively,

$$
\| v \|^2 \geq \sum_{i=1}^\infty |\langle v, v_i \rangle|^2
$$

and

$$\| v \|^2 = \sum_{i=1}^{\infty} |\langle v, v_i \rangle|^2.$$

Polarizing, the Parseval equation takes the form (see 4.10.1 of Chapter I, Vol. II).

Corollary 4.0.16. *Let* $\{v_1, ..., v_n\}$ *be an orthonormal basis in* V, *and* x, y *in* V, *then*

$$\langle x, y \rangle = \sum_{i=1}^{n} \langle x, v_i \rangle \langle v_i, y \rangle \quad (Parseval's \ equation).$$

Exercise 4.0.17. Use Bessel's inequality to give a different proof of the Cauchy-Schwarz Inequality.

4.1   Subspaces and their Orthocomplements

For the remainder of this chapter we will deal only with finite dimensional real or complex inner product spaces, $(V, \langle \cdot, \cdot \rangle)$. Let $k = \mathbb{R}$, or $\mathbb{C}$, and let W be a subspace of V.

Definition 4.1.1. We call the subspace $W^\perp := \{v \in V \ : \ \langle v, W \rangle = 0\}$ the *orthogonal complement* of W.

Evidently, $W^\perp$ is a subspace of V and $W \cap W^\perp = (0)$, since if $w \in W \cap W^\perp$, then $\langle w, w \rangle = 0$, and so $w = 0$. The following theorem sums up the various properties of the orthocomplement of a subspace.

Theorem 4.1.2. *The subspace* $W^\perp$ *satisfies the following properties:*

1. $V = W \oplus W^\perp$.

2. $\dim_k W^\perp = \dim_k V - \dim_k W$.

3. $\dim_k W^{\perp\perp} = \dim_k W$.

4. *If* $W_1 \subset W_2$, *then* $W_2^\perp \subset W_1^\perp$.

5. $\left(lin.sp(W_1 + ... + W_k) \right)^{\perp} = \bigcap_{i=1}^{k} W_i^{\perp}.$

6. $\left(\bigcap_i^k W_i \right)^{\perp} = \sum_{i=1}^{k} W_i^{\perp}.$

Proof. Indeed,

1. By Gram-Schmidt choose an orthonormal basis $\{x_1, ..., x_k\}$ of W and extend it to an orthonormal basis $\{x_1, ..., x_k, ..., x_n\}$ of V. Hence $\{x_{k+1}, ..., x_n\} \subset W^{\perp}$. We will prove they also span $W^{\perp}$. For this, let $x \in W^{\perp}$. Since x is in V, $x = \sum_{i=1}^{n} c_i x_i = \sum_{i=1}^{n} \langle x, x_i \rangle x_i$. On the other hand, since $x \in W^{\perp}$, $x \perp \{x_1, ..., x_k\}$. Thus $\langle x, x_j \rangle = 0$, for all $j = 1, ..., k$. Therefore

$$x = \sum_{i=k+1}^{n} \langle x, x_i \rangle x_i,$$

   which means $\{x_{k+1}, ..., x_n\}$ spans $W^{\perp}$. Hence they form an orthonormal basis of $W^{\perp}$. So, $V = W \oplus W^{\perp}$ and $\dim W^{\perp} = \dim V - \dim W^{\perp}$.

2. $\dim W^{\perp\perp} = \dim V - \dim W^{\perp} = \dim V - (\dim V - \dim W) = \dim W.$

3. Let $w \in W$. Then, $\langle w, w^{\perp} \rangle = 0$ for every $w^{\perp} \in W^{\perp}$. That means that $w \in W^{\perp\perp}$. Hence $W \subset W^{\perp\perp}$. But the two subspaces W and $W^{\perp}$ have the same dimension. Therefore, $W = W^{\perp\perp}$.

4. It is obvious.

5. Suppose $v \in (W_1 + \cdots + W_k)^{\perp}$. This is equivalent to $v \perp W_i$ for every $i = 1, ..., k$, which in turn is equivalent to $v \in \bigcap_{i=1}^{k} W_i^{\perp}$.

6. Let $v \in \sum_{i=1}^{k} W_i^{\perp}$. This means $v = w_1^{\perp} + \cdots + w_k^{\perp}$, where $w_i^{\perp} \in W_i^{\perp}$. Now, let $w \in \bigcap_{i=1}^{k} W_i$. Consider $\langle v, w \rangle$. Obviously $\langle v, w \rangle = 0$, which means

$$v \in \left(\bigcap_{i=1}^{k} W_i \right)^{\perp}.$$

Therefore,

$$\sum_{i=1}^{k} W_i^{\perp} \subset \left(\bigcap_{i=1}^{k} W_i \right)^{\perp}.$$

By 4,

$$\left(\left(\bigcap_{i=1}^{k} W_i \right)^{\perp} \right)^{\perp} \subset \left(\sum_{I=1}^{k} W_i^{\perp} \right)^{\perp}.$$

The left side is equal to $\bigcap_{i=1}^{k} W_i$ while the right side, by 5, is equal to $\bigcap_{i=1}^{k} W_i^{\perp}$. Hence,

$$\bigcap_{i=1}^{k} W_i \subset \bigcap_{i=1}^{k} W_i^{\perp} = \left(\sum_{i=1}^{k} W_i \right)^{\perp}.$$

$\square$

4.2   The Adjoint Operator

Let $(V, \langle \cdot \, , \, \cdot \rangle)$ be a real or complex inner product space and T be a linear map from V to itself.

Definition 4.2.1. We shall say the map $T^{\star} : V \to V$ is *the adjoint* of T if for all x, y in V

$$\langle T(x) \, , \, y \rangle = \langle x, T^{\star}(y) \rangle.$$

Clearly, if there is such a map, it is linear.

To find $T^{\star}$ we have to find its values at y for each $y \in V$. To do so, let $y \in V$, and $\{e_1, ..., e_n\}$ be an orthonormal basis of V. We consider the Fourier coefficients of $T^{\star}(y)$ defined by

$$T^{\star}(y) = \sum_{i=1}^{n} \langle T^{\star}(y), e_i \rangle e_i.$$

Thus if we define $\langle T^{\star}(y), e_i \rangle := \langle y, T(e_i) \rangle$, we get

$$T^{\star}(y) = \sum_{i=1}^{n} \langle y, T(e_i) \rangle e_i.$$

One checks easily that $T^\star$, defined in this way, satisfies

$$\langle T^\star(y), x \rangle = \langle y, T(x) \rangle,$$

and taking conjugates we see,

$$\langle T(x), y \rangle = \langle x, T^\star(y) \rangle,$$

for every $x \in V$.

Exercise 4.2.2. Prove that the definition of the adjoint is independent of the choice of the orthonormal basis. Prove given T, that T^* is unique. That is, * is a function.

Exercise 4.2.3. Prove the following for linear operators T and S:

1. $(TS)^\star = S^\star T^\star$.

2. $(T + S)^\star = T^\star + S^\star$.

3. $(\lambda T)^\star = \overline{\lambda} T^\star$, for every $\lambda \in k$.

4. $(T^\star)^\star = T$.

5. $\mathrm{tr}(T^\star) = \overline{\mathrm{tr}(T)}$.

6. $\det(T^\star) = \overline{\det(T)}$.

7. $T^\star = \overline{T^t}$.

4.3 Unitary and Orthogonal Operators

In this and the next section we introduce the cast of characters, that is, the various types of linear operators pertinent to real or complex inner product spaces.

Definition 4.3.1. Let $(V, \langle \cdot, \cdot \rangle)$ be a complex inner product space. We shall say $U : V \longrightarrow V$ is *unitary* if U preserves $\langle \cdot, \cdot \rangle$. That is, $\langle U(x), U(y) \rangle = \langle x, y \rangle$. Similarly, if $(V, \langle, \rangle)$ is a real inner product space, we shall say $O : V \longrightarrow V$ is *orthogonal* if O preserves $\langle \cdot, \cdot \rangle$.

Unitary operators can be characterized in various ways.

Proposition 4.3.2. *The following statements are equivalent:*

1. U is a unitary operator.

2. U takes an orthonormal basis to an orthonormal basis.

3. U preserves the norm, i.e. $\| U(x) \| = \| x \|$, for all $x \in V$.

4. $U^{\star} = U^{-1}$.

Proof. Evidently 1 and 2 are equivalent. Taking $x = y$ in the definition of a unitary operator proves 3. To see the converse, we use *polarization*. That is over $\mathbb{C}$,

$$\langle x, y \rangle = \frac{1}{4}\left(\| x + y \|^2 - \| x - y \|^2 \right) + \frac{1}{4i}\left(\| x + y \|^2 - \| x - iy \|^2 \right).$$

If $\| U(x) \| = \| x \|$, then U preserves the squares of the norms, preserves the signs $(+)$ or $(-)$ of the $\| \cdot \|^2$, preserves $\frac{1}{4}$, $\frac{1}{2}$, or $\frac{1}{4i}$ of the $\| \cdot \|^2$, so taking into consideration the polarization identity, $\langle U(x), U(y) \rangle = \langle x, y \rangle$. See Chapter 1 of Vol.II 4.10.1, for a general treatment of the concept of polarization.

If U is unitary, then

$$\langle U(x), U(y) \rangle = \langle x, U^{\star}U(y) \rangle = \text{by (3)} = \langle x, y \rangle.$$

Since this holds for all x and y by the uniqueness of the adjoint, $U^{\star}U = I$, i.e. $U^{\star} = U^{-1}$. Here all steps are reversible so this in turn proves U is unitary.   □

Exercise 4.3.3. Prove the polarization identities on both the real and complex cases. Notice in the real case the polarization identity is simpler. State and prove the analogous result to the one just above for orthogonal operators on a real inner product space.

Exercise 4.3.4. If $U_n(\mathbb{C})$ stands for the set of all unitary operators on a complex inner product space of complex dimension n and $O_n(\mathbb{R})$ stands for the set of all orthogonal operators on a real inner product space of real dimension n, prove $U_n(\mathbb{C})$ and $O_n(\mathbb{R})$ are groups.

Since det is a homogeneous polynomial in its matrix coordinates $\det(\bar{A}) = \overline{\det(A)}$. Moreover, $\det(A^t) = \det(A)$. Therefore, $\det(A^*) = \overline{\det(A)}$. As a consequence, if U is unitary, then, $\det(UU^{-1}) = \det(I) = 1$. Hence, $1 = \det(U)\det(U^*) = \det(U)\overline{\det(U)}$. Thus $|\det(U)|^2 = 1$ and so $\det(U)$ lies on the unit circle.

Corollary 4.3.5. *If U is a unitary operator, $|\det(U)| = 1$. In particular, if O is an orthogonal operator, $\det(O) = \pm 1$.*

4.3.1 Eigenvalues of Orthogonal and Unitary Operators

As we shall see, the nature of the eigenvalues of many of the operators we encounter in this chapter will be of great import. Here is a case in point.

Theorem 4.3.6. *The eigenvalues of a unitary (respectively orthogonal) operator all lie on the unit circle.*

Proof. If $x \neq 0$ is an eigenvector, $U(x) = \lambda x$, where $\lambda \in \mathbb{C}$. Notice first of all $\lambda \neq 0$. For if it were 0, then $U(x)$ would be zero and since $x \neq 0$ this means U is singular, which is impossible since U^{-1} exists.

Next we show x is also an eigenvector of U^{-1} with eigenvalue λ^{-1}. This is because since $Ux = \lambda x$, $U^{-1}Ux = x = U^{-1}(\lambda x)$. Now we calculate $\langle Ux, x \rangle$ two ways. On the one hand, it is $\lambda \parallel x \parallel^2$, but on the other it is

$$\langle x, U^{-1}x \rangle = \langle x, \lambda^{-1}x \rangle = \overline{\lambda^{-1}} \parallel x \parallel^2 .$$

Thus $\lambda^{-1} = \overline{\lambda}$. $\qquad\qquad\square$

4.4 Symmetric and Hermitian Operators

Definition 4.4.1. When $(V, \langle \cdot, \cdot \rangle)$ is a complex inner product space, an operator H is called *Hermitian* if $H = H^\star$. Similarly, in the real case an operator S is called *symmetric* if $S = S^t$.

In other words, an operator H is Hermitian if and only if $\langle H(x), y \rangle = \langle x, H(y) \rangle$ and S is symmetric if and only if $\langle S(x), y \rangle = \langle x, S(y) \rangle$, both for all x, y in V. For this reason sometimes one uses unified terminology and just calls such operators *self-adjoint*.

Proposition 4.4.2. *For any operator T, the operator $T^\star T$ is self-adjoint.*

This follows immediately from the properties of the adjoint.

Definition 4.4.3. A Hermitian (respectively symmetric) operator T is called *positive definite* if all its eigenvalues are strictly positive. We write $\mathcal{H}^+$ for the positive definite Hermitian operators and $\mathcal{S}^+$ for the positive definite symmetric ones.

Proposition 4.4.4. *The operator $T^\star T$ is positive definite if and only if T is invertible.*

Proof. Suppose that $T^\star T$ is positive definite. Then $T^\star T(x) = \lambda x$ with $\lambda > 0$ and $x \neq 0$. Since we know that $T^\star T$ is a self-adjoint operator, we get

$$\langle T^\star T(x), x \rangle = \langle \lambda x, x \rangle = \lambda x, x \rangle = \langle T(x), T(x) \rangle$$

and since $\lambda > 0$ and we see $T(x) \neq 0$. Hence T is invertible. Since all steps are reversible the converse also follows. $\square$

Proposition 4.4.5. *If U is unitary and H Hermitian, then UHU^{-1} is Hermitian. Similarly, if O is orthogonal and S symmetric, then OSO^{-1} is symmetric.*

Proof. Indeed, $(UHU^{-1})^\star = (U^{-1})^\star H^\star U^\star = UHU^{-1}$. $\square$

Another example of a self adjoint operator is orthogonal projection onto a subspace. Let $(V, \langle , \rangle)$ be an inner product space and W be a subspace of V. We now define P_W, the projection of V onto W. Write $V = W \oplus W^\perp$ the orthogonal direct sum. For $v \in V$, $v = w + w^\perp$ (uniquely) and define $P_W(v) = w$. Then $P_W : V \to W$ is a linear operator taking V onto W. Ker $P_W = W^\perp$. We leave the easy verification of these facts to the reader. We also ask the reader to prove P_W is self-adjoint.

Proposition 4.4.6. *The operator P_W is a self-adjoint operator.*

Proof. We show
$$\langle P_W(v), v_1 \rangle = \langle v, P_W(v_1) \rangle$$
for all v, v_1 in V. Write $v_1 = w_1 + w_1^\perp$. Then $P_W(v) = w$, and

$$\langle P_W(v), v_1 \rangle = \langle w, v_1 \rangle = \langle w, w_1 + w_1^\perp \rangle = \langle v, w_1 \rangle.$$

On the other hand,

$$\langle v, P_W(v_1) \rangle = \langle v, w_1 \rangle = \langle w + w^\perp, w_1 \rangle = \langle w, w_1 \rangle.$$

Therefore,
$$\langle P_W(v), v_1 \rangle = \langle v, P_W(v_1) \rangle.$$

$\square$

4.4.1   Skew-Symmetric and Skew-Hermitian Operators

These are defined quite analogously to the Hermitian and Symmetric operators.

Definition 4.4.7. When $(V, \langle \cdot, \cdot \rangle)$ is a complex inner product space, an operator H is called *skew-Hermitian* if $H^\star = -H$. Similarly, in the real case an operator S is called *skew-symmetric* if $S^t = -S$.

In other words, an operator H is skew-Hermitian if and only if for all x, y in V, $\langle H(x), y \rangle = -\langle x, H(y) \rangle$ and S is skew-symmetric if and only if $\langle S(x), y \rangle = -\langle x, S(y) \rangle$.

The proof of the following proposition is identical to the analogous one for Hermitian and symmetric operators, so we omit it.

Proposition 4.4.8. *If U is unitary and H skew-Hermitian, then UHU^{-1} is skew-Hermitian. Similarly, if O is orthogonal and S skew-symmetric, then OSO^{-1} is skew-symmetric.*

Theorem 4.4.9. *The eigenvalues of a Hermitian or symmetric operator are real. The eigenvalues of a skew-Hermitian or skew-symmetric operator are purely imaginary.*

Proof. Suppose H is Hermitian and $x \neq 0$ is an eigenvector with eigenvalue λ. Then

$$\langle Hx, x \rangle = \lambda \parallel x \parallel^2 .$$

But since H is Hermitian this is $\langle x, Hx \rangle = \bar{\lambda} \parallel x \parallel^2$. It follows that $\lambda = \bar{\lambda}$ so λ is real. Similarly if we have a symmetric operator, then it is self-adjoint on the real vectors on which it operates (but $\lambda \in \mathbb{C}$). Nonetheless, the same argument works and shows the eigenvalues are real.

Now if H were skew-Hermitian, then $\langle Hx, x \rangle = \lambda \parallel x \parallel^2$. But this is $\langle x, -Hx \rangle = -\bar{\lambda} \parallel x \parallel^2$. Hence in this case $\lambda = -\bar{\lambda}$ and is purely imaginary. Similarly, if we have a skew symmetric operator, then it is skew adjoint on the real vectors on which it operates. Hence the same argument works and shows its eigenvalues are purely imaginary.   □

We denote by $\mathcal{H}$ the set of all Hermitian operators, by $\mathcal{S}$ all symmetric operators, by $\mathcal{A}$ all skew Hermitian operators and by $\mathcal{B}$ all skew symmetric operators. Also $\mathcal{H}^+$ stands for the positive definite Hermitian operators and $\mathcal{S}^+$ for the positive definite symmetric ones, $\mathcal{H}$ and $\mathcal{A}$ being subsets of $\mathrm{End}_{\mathbb{C}}(V)$ and $\mathcal{S}$ and $\mathcal{B}$ being subsets of $\mathrm{End}_{\mathbb{R}}(V)$, where $(V, \langle \cdot, \cdot \rangle)$ is a complex, respectively real, inner product space. In either case we let $\dim V = n$.

Proposition 4.4.10. $\mathcal{H}$ *and* $\mathcal{A}$ *are* real *vector subspaces of the complex vector space* $\mathrm{End}_{\mathbb{C}}(V)$. $\dim_{\mathbb{R}} \mathcal{H} = n^2$ *and* $\dim_{\mathbb{R}} \mathcal{A} = n^2$.

We ask the reader to verify that these are not complex vector spaces.

Proof. That these are real subspaces follows immediately from the properties of * above. As to the dimensions, the strictly lower triangular terms of a Hermitian operator are determined by the strictly upper triangular terms and there are $n(n-1)/2$ *complex* terms making $n(n-1)$ real terms. On the diagonal there are precisely n real terms. Thus the total is $n^2 - n + n = n^2$. A similar calculation shows $\dim_{\mathbb{R}} \mathcal{A} = n^2$.   □

Proposition 4.4.11. $\mathcal{S}$ *and* $\mathcal{B}$ *are* real *vector subspaces of the real vector space* $\mathrm{End}_{\mathbb{R}}(V)$. *Also,* $\dim_{\mathbb{R}} \mathcal{S} = n(n+1)/2$ *and* $\dim_{\mathbb{R}} \mathcal{B} = n(n-1)/2$.

Proof. Here we also leave the proof that $\mathcal{S}$ and $\mathcal{B}$ are subspaces to the reader. As to the dimensions, the strictly lower triangular terms of a symmetric operator are determined by the strictly upper triangular terms and there are $n(n-1)/2$ terms. On the diagonal there are precisely n real terms. Thus the total is $n(n+1)/2$. A similar argument applies to $\mathcal{B}$ except that on the diagonal everything must be zero. Hence $\dim_\mathbb{R} \mathcal{B} = n(n-1)/2$. $\square$

Exercise 4.4.12. Let $T \in \mathrm{M}(n, \mathbb{C})$. Then $\frac{T+T^*}{2}$ is Hermitian and $\frac{T-T^*}{2}$ is skew-Hermitian. Also

$$T = \frac{T+T^*}{2} + \frac{T-T^*}{2}.$$

This gives an $\mathbb{R}$ linear direct sum decomposition of

$$\mathrm{M}(n, \mathbb{C}) = \mathcal{H} \oplus \mathcal{A}.$$

The next exercise is similar.

Exercise 4.4.13. Let $T \in \mathrm{M}(n, \mathbb{R})$. Then $\frac{T+T^t}{2}$ is symmetric and $\frac{T-T^t}{2}$ is skew symmetric. Also

$$T = \frac{T+T^t}{2} + \frac{T-T^t}{2}.$$

This gives an $\mathbb{R}$ linear direct sum decomposition of

$$\mathrm{M}(n, \mathbb{R}) = \mathcal{S} \oplus \mathcal{B}.$$

Exercise 4.4.14. Prove T is skew-Hermitian if and only if iT is Hermitian.

Proposition 4.4.15. *Let A be a skew-Hermitian operator. Then, if $\dim_\mathbb{C} V$ is odd, $\det(A)$ is pure imaginary. If $\dim_\mathbb{C} V$ is even, $\det(A)$ is real. Similarly, let B be a skew-symmetric operator. If $\dim V$ is odd, $\det(B) = 0$. If $\dim V$ is even, we can draw no conclusion.*

Proof. Suppose A is a skew-Hermitian operator. Then, since $\det(A^\star) = \overline{\det(A)}$ and $\det(-A) = \det(A^\star) = \overline{\det(A)}$, then $(-1)^n \det(A) = \overline{\det(A)}$, proving the first statement. Alternatively, since the determinant of A is the product of all its eigenvalues with multiplicity, which we know to be purely imaginary. Hence, if the dimension of V is odd the result is a pure imaginary number while if it is even the result is a real number.

In the case S is a skew-symmetric operator, then $\det(S^\star) = \det(S) = \det(-S) = (-1)^n \det(S)$. So, if $\dim V$ is odd, $\det(S) = 0$, and if it is even number no conclusion is possible. $\qquad\square$

4.5   The Cayley Transform

The classical Cayley transform is the biholomorphic map $z \mapsto \frac{z-i}{z+i}$ taking the upper half plane onto the interior of the unit disk all in $\mathbb{C}$. Here we will prove an analogue of this. Namely, in a real inner product space, if X is skew-symmetric, then $I+X$ is invertible and $O = (I-X)(I+X)^{-1}$ is orthogonal. Conversely, if $X = (I-O)(I+O)^{-1}$, where $O \in \mathrm{SO}(n,\mathbb{R})$, then X is skew-symmetric and $I + X$ is invertible.

Let $(V, \langle\ ,\ \rangle)$ be a finite dimensional real inner product space. In order to state and prove the properties of the Cayley transform we need the following two lemmas.

Lemma 4.5.1. *If X is skew-symmetric, then $\det(I + X) \neq 0$.*

Proof. If $\det(I + X) = 0$, then -1 is an eigenvalue of X. Hence there exists an eigenvector $v \neq 0$ so that $Xv = -v$. But then

$$\langle Xv, v \rangle = - \parallel v \parallel^2 .$$

On the other hand,

$$\langle Xv, v \rangle = \langle v, X^t v \rangle = \langle v, -Xv \rangle = -\langle v, -v \rangle = \parallel v \parallel^2 .$$

Thus $\parallel v \parallel^2 = 0$ and so $v = 0$, a contradiction. $\qquad\square$

Lemma 4.5.2. *If X is any matrix, then $I + X$ and $I - X$ commute. Hence, also if $I-X$ is invertible, then $I+X$ and $(I-X)^{-1}$ also commute.*

Proof. $(I + X)(I - X) = I - X^2$. Computing in the opposite order obviously gives the same result. Hence if $I - X$ is invertible,

$$(I - X)(I + X)(I - X)^{-1} = I + X.$$

Thus $I + X$ and $(I - X)^{-1}$ also commute.   □

We can now state the theorem.

Theorem 4.5.3. *If X is skew-symmetric, then $I + X$ is invertible and $O = (I - X)(I + X)^{-1}$ is orthogonal. Conversely, if*

$$X = (I - O)(I + O)^{-1},$$

where $O \in \mathrm{SO}(n, \mathbb{R})$, then X is skew-symmetric and $\det(I + X) \neq 0$.

Proof. $I + X$ is invertible by Lemma 4.5.1. Taking the transpose of both sides of the first equation we get

$$O^t = ((I + X)^t)^{-1}(I - X)^t = (I + X^t)^{-1}(I - X^t) = (I - X)^{-1}(I + X).$$

Therefore, $OO^t = (I - X)(I + X)^{-1}(I - X)^{-1}(I + X)$. By Lemma 4.5.2 this is $(I - X)(I - X)^{-1}(I + X)^{-1}(I + X) = I$. Thus O is orthogonal.

For the converse, $(I + O)^{-1}$ exists because O does not have -1 as an eigenvalue. For if it did, we could split off that eigenvector and get an orthogonal decomposition of V with a complementary invariant subspace of co-dimension 1 on which the restriction of O acts as something in $\mathrm{SO}(n - 1, \mathbb{R})$. Hence we would get $\det O = -1$, a contradiction.

Multiplying the second equation by $I + O$ on the right tells us

$$X(I + O) = I - O.$$

Taking transposes and account of the fact that O is orthogonal shows $(I + O)^t X^t = (I - O)^t$ and so $(I + O^{-1})X^t = (I - O^{-1})$. Now multiply this equation on the left by O and get

$$(O + I)X^t = O - I.$$

Thus $X^t = (O + I)^{-1}(O - I)$, whereas $X = (I - O)(I + O)^{-1}$. Adding these we get

$$X + X^t = (I - O)(I + O)^{-1} + (O + I)^{-1}(O - I)$$
$$= (I + O)^{-1}(I - O) + (O + I)^{-1}(O - I).$$

But this latter term is just

$$((I + O)^{-1} - (O + I)^{-1})(I - O) = 0(I - O) = 0.$$

Hence X is skew-symmetric. $\qquad\square$

This argument extends directly to the complex situation as follows. We remind the reader that $\mathrm{SU}(n, \mathbb{C})$ *stands for the unitary operators of determinant 1.*

Corollary 4.5.4. *If* X *is skew Hermitian, then* $I + X$ *is invertible and* $U = (I - X)(I + X)^{-1}$ *is unitary. Conversely, if* $X = (I - U)(I + U)^{-1}$, *where* $U \in \mathrm{SU}(n, \mathbb{C})$, *then* X *is skew-Hermitian and* $\det(I + X) \neq 0$.

4.6 Normal Operators

In order to unify the treatment of the various classes of operators we have encountered in the chapter we now turn to Normal *operators. These are defined s follows:*

Definition 4.6.1. In the complex case an operator T is called *normal* means it commutes with its adjoint T^*.

$$T^\star T = TT^\star.$$

In the real case, S is *normal* means $TT^t = TT^t$. Evidently Hermitian, skew-Hermitian, unitary, symmetric and skew-symmetric and orthogonal operators are all normal.

The following shows that normal operators are polynomially closed.

Proposition 4.6.2. *Let* T *be a normal operator and* $p(x)$ *be a polynomial with real or complex coefficients. Then* $p(T)$ *is a normal operator.*

Proof. Let $p(x) = a_n x^n + \cdots + a_1 x + a_0$ with $a_i \in \mathbb{C}$, or $\mathbb{R}$. Then, using the properties of the adjoint operator (see Exercise 4.2.3), we get

$$(p(T))^\star = (a_n T^n + \cdots + a_1 T + a_0)^\star$$
$$= \bar{a}_n (T^\star)^n + \cdots + \bar{a}_1 T^\star + \bar{a}_0 = \bar{p}(T^\star)$$

where $\bar{p}(T^\star)$ denotes the polynomial with coefficients the conjugates of those of p. Clearly, in the real case $\bar{p}(T^\star) = p(T^\star)$. Now, the two polynomials $p(T)$ and $\bar{p}(T^\star)$ commute, since T and $T^\star$ commute. Hence $p(T)$ is a normal operator. □

4.6.1   The Spectral Theorem

We now establish the diagonalization of normal operators. What we mean by this is given a normal operator T on a real or complex inner product space, V, there exists an orthonormal basis of V such that in that basis T is a diagonal matrix. (Of course the diagonal elements are eigenvalues and the corresponding vectors are eigenvectors.) However, there is one proviso we must make here and that is in the skew-symmetric case: since the eigenvalues are not real but purely imaginary, although the operator is diagonalizable over $\mathbb{C}$ it will only be block diagonaizable, with 2×2 blocks, over $\mathbb{R}$.

Theorem 4.6.3. *A a normal operator T can be diagonalized with respect to an orthonormal basis of eigenvectors.*

We will prove this in the complex case; the real case being completely analogous. However, before embarking on the proof of this important result, we mention an alternative way to understand what diagonalization means and leave its proof to the reader.

Proposition 4.6.4. *There exists an orthonormal basis of V such that in that basis T is a diagonal matrix if and only if there exists a unitary operator U on V so that UTU^{-1} is diagonal. Similarly, in the real case there exists an orthonormal basis of V such that in that basis T complexifies to a diagonal matrix if and only if there exists an orthogonal operator O on V so that OTO^{-1} is diagonal in the complexification.*

The last statement means

$$OTO^{-1} = \begin{pmatrix} \begin{pmatrix} a_1 & b_1 \\ -b_1 & a_1 \end{pmatrix} & & & & & & \\ & \ddots & & & & & \\ & & \begin{pmatrix} a_r & b_r \\ -b_r & a_r \end{pmatrix} & & & & \\ & & & \lambda_1 & & & \\ & & & & \lambda_2 & & \\ & & & & & \ddots & \\ & & & & & & \lambda_n \end{pmatrix}$$

where the λ_i are the real eigenvalues of T and the 2×2 blocks correspond to the complex conjugate eigenvalues of T.

The proof of the Spectral theorem requires the following lemmas, the first of which has nothing to do with normality.

Lemma 4.6.5. *If T is any linear operator on an inner product space and λ and μ are two distinct eigenvalues of T, then $V_\lambda \perp V_\mu$.*

Proof. Let $x \in V_\lambda$ and $y \in V_\mu$. Then $T(x) = \lambda x$ and $T(y) = \mu y$. Now,

$$\langle T(x), y \rangle = \lambda \langle x, y \rangle, \quad \text{and} \quad \langle x, T^\star(y) \rangle = \langle x, \overline{\mu} y \rangle.$$

But $\langle T(x), y \rangle = \langle x, T^\star(y) \rangle$. Therefore, either $\lambda \langle x, y \rangle = \mu \langle x, y \rangle$ or $(\lambda - \mu)\langle x, y \rangle = 0$. But, by assumption $\lambda \neq \mu$, so $\langle x, y \rangle = 0$, i.e. $x \perp y$. $\square$

Lemma 4.6.6. *Let T be a normal operator and W be an invariant subspace under both T and $T^\star$. Then, $S = T|_W$ is a normal operator on W.*

Proof. We know that $\langle T(x), y \rangle = \langle x.T^\star(y) \rangle$ for every x and y in V, a fortiori for every x and y in W. Now, $\langle S(x), y \rangle = \langle x, S^\star(y) \rangle$ and $S^\star = T^\star|_W$, and since T and $T^\star$ commute, so do S and $S^\star$. $\square$

Lemma 4.6.7. *Let T be a normal operator, and $x \neq 0 \in V$. Then*

$$T(x) = \lambda x \quad \Leftrightarrow \quad T^{\star}(x) = \overline{\lambda} x.$$

In other words, $V_\lambda(T) = V_{\overline{\lambda}}(T^{\star})$.

Proof. First, we prove that $\| T(x) \| = \| T^{\star}(x) \|$. Indeed, since T is a normal operator,

$$\| T(x) \|^2 = \langle x, T^{\star}T(x) \rangle = \langle x, TT^{\star}(x) \rangle = \langle T^{\star}(x), T^{\star}(x) \rangle = \| T^{\star}(x) \|^2.$$

Now, since T is normal, so is $T - \lambda I$ for any $\lambda \in \mathbb{C}$. Moreover,

$$(T - \lambda I)^{\star} = T^{\star} - \overline{\lambda} I.$$

We now check normality of $T - \lambda I$.

$$(T - \lambda I)(T^{\star} - \overline{\lambda} I) = TT^{\star} - \overline{\lambda} T - \lambda T^{\star} + |\lambda|^2 I \quad (1),$$

and

$$(T^{\star} - \overline{\lambda} I)(T - \lambda I) = T^{\star}T - \overline{\lambda} T - \lambda T^{\star} + |\lambda|^2 I \quad (2)$$

are equal. But, since T is a normal operator $TT^{\star} = T^{\star}T$. Therefore $(1) = (2)$. In other words,

$$(T - \lambda I)(T - \lambda I)^{\star} = (T - \lambda I)^{\star}(T - \lambda I),$$

i.e, $T - \lambda I$ is a normal operator. Hence

$$\| (T - \lambda I)(x) \| = \| (T^{\star} - \overline{\lambda} I)(x) \|. \quad (3).$$

By (3) it follows that if λ is an eigenvalue of T then $\overline{\lambda}$ is an eigenvalue of $T^{\star}$ and conversely. $\qquad \square$

Corollary 4.6.8. *If T is normal, then $V_\lambda(T)$ is $T^{\star}$-invariant.*

Corollary 4.6.9. *If T is a normal operator, then $V = \mathrm{Ker}(T) \oplus Im(T)$ is an orthogonal direct sum.*

Proof. Let $x \in \mathrm{Ker}(T)$ and $y \in \mathrm{Im}(T)$. Then, $T(x) = 0$, and by lemma 4.6.7, $T^\star(x) = 0$. Also, $y = T(z)$ for some $z \in V$,

$$\langle x, y \rangle = \langle x, T(z) \rangle = \langle T^\star(x), z \rangle = \langle 0, z \rangle = 0.$$

Therefore $x \perp y$, and since $\dim(V) = \dim(\mathrm{Ker}(T)) + \dim(\mathrm{Im}(T))$, the proof is complete. $\square$

Lemma 4.6.10. *Let T be a normal operator and V_λ a geometric eigenspace. Then V_λ is T-invariant and $V_\lambda^\perp$ is $T^\star$-invariant.*

Proof. Since T is normal, V_λ is T-invariant (see corollary 4.6.8). So what we have to prove is that $V_\lambda^\perp$ is $T^\star$-invariant. Take a vector $v_\lambda^\perp \in V_\lambda^\star$. Then, since $T(v_\lambda) \in V_\lambda$ and $v_\lambda^\perp \in V_\lambda^\perp$,

$$\langle T^\star(v_\lambda^\perp), v_\lambda \rangle = \langle v_\lambda^\perp, T(v_\lambda) \rangle = 0.$$

Hence, $T^\star(v_\lambda^\perp) \in V_\lambda^\perp$, and this for any $v_\lambda^\perp$. Therefore,

$$T^\star(V_\lambda^\perp) \subset V_\lambda^\perp.$$

Let $v_\lambda \in V_\lambda$, then $v_\lambda^\perp \in V_\lambda^\perp$. It follows that,

$$\langle T(v_\lambda^\perp), v_\lambda \rangle = \langle v_\lambda^\perp, T^\star(v_\lambda) \rangle = 0 \quad \Rightarrow \quad T(v_\lambda^\perp) \in V_\lambda^\perp.$$

$\square$

We are now ready to prove the Spectral theorem for normal operators.

Theorem 4.6.11. (Spectral Theorem For Normal Operators) *Let T be a normal operator on a finite dimensional inner product vector space $(V, \langle \cdot, \cdot \rangle)$. Then,*

$$V = \bigoplus_{\lambda \in Spec(T)} V_\lambda.$$

Proof. Let $\lambda \in \mathrm{Spec}(T)$. As we saw in the lemmas just above both V_λ and $V_\lambda^\perp$ are T-invariant. So, in an appropriate orthonormal basis, T can be written as,

$$T = \left(\begin{array}{c|c} T|_{V_\lambda} & 0 \\ \hline 0 & T|_{V_\lambda^\perp} \end{array} \right).$$

We complete the proof by induction on $T|_{V_\lambda}$ and $T|_{V_\lambda}^\perp$.   □

   Here is a short alternative proof of the diagonalizability part of Spectral theorem in the complex case.[5] Notice that, although it gets diagonalizability, it does not get *ortho* diagonalizability!

Proof. **(2nd Proof of the Spectral Theorem)** It suffices to prove that the minimal polynomial $m_T(x)$ of T has no multiple roots (for this see Chapter 6, Proposition 6.6.7). Suppose $m_T(x) = (x - \lambda)^2 g(x)$ for some $\lambda \in \mathbb{C}$ and some polynomial $g(x)$ of degree strictly smaller than the degree of $m_T(x)$. Consider $T - \lambda I$. This is a normal operator. Now, pick a $v \in V$ and put $(T - \lambda I)(g(T)(v)) = w$. Obviously, $w \in \mathrm{Im}(T - \lambda I)$ and also

$$(T - \lambda I)(w) = (T - \lambda I)^2 g(T)(v) = m_T(T)(v) = 0,$$

which implies $w \in \mathrm{Ker}(T - \lambda I)$. Therefore, by Corollary 4.6.9, $w = 0$. This means T is annihilated by the polynomial $(x - \lambda)g(x)$, which is of degree strictly smaller than $m_T(x)$, a contradiction. Since the minimal polynomial has simple roots, T is diagonalizable.   □

Corollary 4.6.12. *Hermitian, skew-Hermitian, unitary, symmetric, skew-symmetric and orthogonal operators are all orthogonally diagonalizable over* $\mathbb{C}$. *Since the eigenvalues of a symmetric operator are real it is actually ortho-diagonalizable over* $\mathbb{R}$. *However, the skew-symmetric and orthogonal operators are only 2×2 block diagonalizable over* $\mathbb{R}$. *In all cases everything is determined by the nature of the eigenvalues.*

---

[5]For a treatment of the Spectral theorem on Hilbert space, see for example Paul Halmos, [54].

When T is orthogonal,

$$OTO^{-1} = \begin{pmatrix} \begin{pmatrix} \cos\theta_1 & \sin\theta_1 \\ -\sin\theta_1 & \cos\theta_1 \end{pmatrix} & & & & & & \\ & \ddots & & & & & \\ & & \begin{pmatrix} \cos\theta_k & \sin\theta_k \\ -\sin\theta_k & \cos\theta_k \end{pmatrix} & & & & \\ & & & 1 & & & \\ & & & & \ddots & & \\ & & & & & 1 & \\ & & & & & & -1 \\ & & & & & & & \ddots \\ & & & & & & & & \end{pmatrix}$$

Evidently, if $T \in \mathrm{SO}(n, \mathbb{R})$, the number of -1's in the above matrix is even.

Finally, if T is a skew-symmetric,

$$OTO^{-1} = \begin{pmatrix} \begin{pmatrix} 0 & a_1 \\ -a_1 & 0 \end{pmatrix} & & & & & \\ & \ddots & & & & \\ & & \begin{pmatrix} 0 & a_r \\ -a_r & 0 \end{pmatrix} & & & \\ & & & 0 & & \\ & & & & 0 & \\ & & & & & \ddots \\ & & & & & & 0 \end{pmatrix}.$$

We remind the reader that, in general, for an operator T, when λ an eigenvalue of T and χ_T is the characteristic polynomial, the algebraic multiplicity of λ is its multiplicity as a root of χ_T, while its geometric multiplicity is the dimension of the eigenspace V_λ. As we know, in general, the algebraic multiplicity is $\geq$ the geometric multiplicity. However, when T is a normal operator the two coincide and this is why we get diagonalizability!

As we shall see in Chapter 6, by the Jordan form

$$PTP^{-1} = \begin{pmatrix} \square & & & \\ & \square & & \\ & & \ddots & \\ & & & \square \end{pmatrix}.$$

But when T is a normal operator, each block is diagonal.

Corollary 4.6.13. *If T is a normal operator, and $(T - \lambda I)^m x = 0$ for some positive integer m, then $(T - \lambda I)x = 0$.*

Proof. As we know (see Lemma 4.6.7) if T is normal, so is $T - \lambda I$. Hence, it is sufficient to prove if T is normal and $T^m(x) = 0$, then $T(x) = 0$.

Choose a number j such that $2^j > m$. Then $T^{2^j}(x) = 0$. We will show $T^{2^{j-1}}(x) = 0$. Then by induction $T(x)$ itself is 0. To do so, set

$$H = T^{2^{m-1}}.$$

Then,

$$H^\star = \left(T^{2^{m-1}} \right)^\star.$$

First, we will prove the claim when T is Hermitian. Then

$$H^\star = T^{2^{m-1}}.$$

This means that H also is Hermitian. Hence,

$$H^\star H = T^{2^{m-1}} T^{2^{m-1}} = T^{2^m}.$$

On the other hand, $H^\star(H(x)) = 0$ by hypothesis and so we obtain $\langle H^\star H(x), x \rangle = 0$. But

$$\langle H^\star H(x), x \rangle = \langle H(x), H(x) \rangle = \| H(x) \|^2 .$$

Thus $H(x) = 0$, which implies $T^{2^{m-1}}$ is Hermitian (since $H = H^\star$). Continuing in this way, we get the conclusion when T is Hermitian.

Now, let T be an *arbitrary* normal operator. Consider the operator $H = TT^\star$ which we know is Hermitian. Then

$$H^k = (TT^\star)^k = T^k(T^\star)^k = (T^\star)^k T^k$$

since T is normal. Suppose that $T^k = 0$. Then $H^k(x) = 0$. So, by the above, $H(x) = 0$. This means

$$\langle TT^\star(x), x \rangle = \langle T^\star(x), T^\star(x) \rangle = \| T^\star(x) \|^2 .$$

But $\langle TT^\star(x), x \rangle = \langle H(x), x \rangle = 0$. Hence, $T^\star(x) = 0$ and so $T(x) = 0$. $\qquad\qquad\qquad\qquad\qquad\qquad\qquad\qquad\qquad\qquad\qquad\qquad\square$

Exercise 4.6.14. Using the fact that the eigenvalues are real, give an independent proof that symmetric operators are orthogonally diagonalizable over $\mathbb{R}$.

Hint: Use orthocomplements and induction.

4.7   Some Applications to Lie Groups

4.7.1   Positive Definite Operators

Let V be a real or complex inner product space, and T a linear operator on V.

Definition 4.7.1. We call *square root* of T, and we denote it by $\sqrt{T}$, any operator $A = \sqrt{T}$ such that $A^2 = T$.

Not all square matrices have a square root (see Gallier, [39]). If T is an invertible operator over $\mathbb{C}$, then it always has a square root. This is because the exponential map $\mathrm{M}(n, \mathbb{C}) \to \mathrm{GL}(n, \mathbb{C})$ is surjective (see exercise 6.6.11). Hence $T = \mathrm{Exp}(X)$ for some $X \in \mathrm{M}(n, \mathbb{C})$. Hence $\mathrm{Exp}(1/2X)^2 = \mathrm{Exp}(X) = T$. Over $\mathbb{R}$ unipotent operators, or more generally any operator in the range of the exponential map, has all its k-th roots, $k \geq 2$. But if T is a singular operator, as for example the nilpotent operator,

$$T = \begin{pmatrix} 0 & 1 \\ 0 & 0 \end{pmatrix},$$

it has no square root (check).

Furthermore, the square root of a matrix, if it exists, is not necessarily unique. For example, when $V = \mathbb{R}^2$ and T is

$$T = \begin{pmatrix} 33 & 24 \\ 48 & 57 \end{pmatrix}.$$

A simple calculation shows both

$$\pm \begin{pmatrix} 1 & 4 \\ 8 & 5 \end{pmatrix} \quad \text{and} \quad \pm \begin{pmatrix} 5 & 2 \\ 4 & 7 \end{pmatrix}$$

are square roots of T.

The point here is that positive definite Hermitian or symmetric operators have unique square roots.

Proposition 4.7.2. *Any positive definite Hermitian or positive definite symmetric operator has a unique square root which is also positive definite Hermitian or symmetric.*

Proof. We will prove it in the case of Hermitian operators, the symmetric case being similar. If H is a Hermitian operator, by the spectral theorem that there is a unitary matrix U such that

$$UHU^{-1} = \begin{pmatrix} \lambda_1 & & & & \\ & \ddots & & 0 & \\ & & \lambda_k & & \\ & 0 & & \ddots & \\ & & & & \lambda_n \end{pmatrix}, \qquad \lambda_i \in \mathbb{R},$$

and, because H is positive definite, $\lambda_i > 0$ for any $i = 1, 2, ..., n$. Now, consider the matrix,

$$P = U^{-1} \begin{pmatrix} \sqrt{\lambda_1} & & & & \\ & \ddots & & 0 & \\ & & \sqrt{\lambda_k} & & \\ & 0 & & \ddots & \\ & & & & \sqrt{\lambda_n} \end{pmatrix} U.$$

Obviously, P is Hermitian since the diagonal matrix is. Then

$$P^2 = U^{-1}\left(\sqrt{\lambda_i}\right)_i UU^{-1}\left(\sqrt{\lambda_i}\right)_i U = U^{-1}\left(\sqrt{\lambda_i}\right)_i U = H.$$

So $P^2 = P$ and P is unique. $\qquad\square$

We now explain what the interest is in positive definite operators and answer a question raised at the very beginning of this chapter.

Proposition 4.7.3. *Let $(V, \langle,\rangle)$ be a complex inner product space. Then V comes with a positive definite Hermitian form, β given by*

$$\beta(x, y) = \langle Bx, y\rangle, \quad x, \ y \in V,$$

where B is some positive definite Hermitian matrix and $\langle \cdot, \cdot \rangle$ is the standard positive definite Hermitian form on V. An analogous statement holds over $\mathbb{R}$, for real symmetric forms.

Proof. If $\beta : V \times V \to \mathbb{C}$ is a positive definite Hermitian form on V, let $\{v_1, \dots, v_n\}$ be an orthonormal basis of V. Then for $x = \sum_{i=1}^n x_i v_i$ and $y = \sum_{j=1}^n y_j v_j$ in V, let $\beta(v_i, v_j) = b_{i,j} = B$, the matrix of β relative to the orthonormal basis $\{v_1, \dots, v_n\}$. Then $\beta(x, y) = \sum_{i,j} x_i y_j b_{i,j}$. Thus β and B each uniquely determine the other. In matrix terms $\beta(x, y) = y^* Bx$. Now since β is positive definite and Hermitian the matrix B must clearly be Hermitian and $x^* Bx > 0$ if $x \neq 0$. That is B is positive definite and $\beta(x, y) = \langle Bx, y\rangle$, where $\langle \cdot, \cdot\rangle$ is the standard Hermitian form on V. $\qquad\square$

Exercise 4.7.4.

1. Let $\langle \cdot, \cdot \rangle$ be the standard Hermitian form on V and B be a positive definite Hermitian operator on V. Show the map defined by $\beta_B(x, y) = \langle Bx, y\rangle$ is a positive definite Hermitian form by using square roots.

2. Show the map $B \mapsto \beta_B$ is injective.

3. What happens to B when we change the orthonormal basis to $\{w_1, \dots, w_n\}$?

Using this proposition and the fact that each positive definite Hermitian operator B has a unique square root $\sqrt{B}$, we see

$$\beta(x, y) = \langle \sqrt{B}x, \sqrt{B}y \rangle$$

and renaming $\sqrt{B} = C$, $\beta(x, y) = \langle Cx, Cy \rangle$, where C is positive definite Hermitian.

Now, let β_0 be such a Hermitian positive definite form. We can consider the unitary group defined by such β_0 which we denote by $U_0(n, \mathbb{C})$, while $U(n, \mathbb{C})$ denotes the unitary group defined by the standard form, $\langle \cdot, \cdot \rangle$.

Theorem 4.7.5. *The groups $U_0(n, \mathbb{C})$ and $U(n, \mathbb{C})$ are conjugate in $GL(n, \mathbb{C})$.*

Thus for any two positive definite Hermitian (respectively symmetric) forms the corresponding unitary (respetively orthogonal) groups are conjugate. In other words we will have proved the conjugacy of maximal compact subgroups for $GL(n, \mathbb{C})$ (respectively $GL(n, \mathbb{R})$).

Proof. We first see what these groups actually are. By definition

$$U(n, \mathbb{C}) = \{u \in GL(n, \mathbb{C}) \mid uu^\star = I\},$$

while

$$\langle Au_0x, Au_0y \rangle = \langle Ax, Ay \rangle$$

for some positive definite Hermitian matrix A. But

$$\langle Au_0x, Au_0y \rangle = \langle x, (Au_0)^\star Au_0y \rangle$$

and

$$\langle Ax, Ay \rangle = \langle x, A^\star Ay \rangle.$$

Therefore,

$$\langle x, (Au_0)^\star Au_0y \rangle = \langle x, AA^\star y \rangle.$$

Since this holds for all $(u_0)^\star Au_0 = AA^\star$, that is,

$$u_0^\star A^\star Au_0 = AA^\star.$$

Letting $AA^\star = T$, we see

$$u_0^\star T u_0 = T.$$

Now, let $u \in U(n, \mathbb{C})$ and consider $u_0 = (\sqrt{T})^{-1} u \sqrt{T}$. It follows that $u_0 \in U_0(n)$. Hence $(\sqrt{T})^{-1} U \sqrt{T} \subseteq U_0$ and by dimension and connectedness considerations we would conclude,

$$(\sqrt{T})^{-1} U \sqrt{T} = U_0.$$

However, since $O(n, \mathbb{R})$ is not connected, this type of argument fails in $GL(n, \mathbb{R})$. So we just observe that the same argument works *symmetrically*, i.e.

$$\sqrt{T} U_0 (\sqrt{T})^{-1} \subseteq U.$$

But $\sqrt{T} U_0 (\sqrt{T})^{-1} \supseteq U$, so they are equal.
   It remains just to show $(\sqrt{T})^{-1} u \sqrt{T} \in U_0$, i.e.

$$((\sqrt{T})^{-1} u \sqrt{T})^\star T (\sqrt{T})^{-1} u \sqrt{T}$$

is equal to T. Indeed,

$$((\sqrt{T})^{-1} u \sqrt{T})^\star T (\sqrt{T})^{-1} u \sqrt{T} = \sqrt{T} u^\star (\sqrt{T})^{-1} T (\sqrt{T})^{-1} u \sqrt{T}$$
$$= \sqrt{T} u^\star u \sqrt{T}$$

which, since $u^\star u = I$, is equal to $\sqrt{T}\sqrt{T} = T$.   □

4.7.1.1   Exercises

Exercise 4.7.6. Let $v_1, v_2, \ldots, v_n$ be vectors in a real inner product space. Define A to be the $n \times n$ matrix whose (i, j)-entry is $\langle v_i, v_j \rangle$. Prove that A is positive definite if and only if the vectors are linearly independent.

Exercise 4.7.7. Let $A \in U(n, \mathbb{C})$. Suppose that A has only nonnegative eigenvalues. Prove that $A = I$.

4.7.2   The Topology on $\mathcal{H}^+$ and $\mathcal{P}^+$

In order to understand the topology on $\mathcal{H}^+$ and $\mathcal{P}^+$ we will need the following theorem of complex analysis.

Theorem 4.7.8. (Rouché's Theorem[6]) *Let Ω be a simply connected domain, γ be a positively oriented, piecewise smooth, simple closed curve in Ω and f and g be meromorphic functions defined on Ω with the property that, on γ,*

$$|f(z)| > |g(z)|,$$

and neither f nor g has zeros or poles on γ. Then,

$$Z_{f+g} - P_{f+g} = Z_f - P_g,$$

where P_f is the number of poles of f, and Z_f the number of zeros of f all within γ.

Notice that if f and g are polynomials, they do not have poles, therefore Rouché's theorem simply says that $Z_{f+g} = Z_f$.

We now turn to the topology of these positive definite forms.

Theorem 4.7.9. *Both $\mathcal{S}^+$ and $\mathcal{H}^+$ are non-trivial open subsets of $\mathcal{S}$ and $\mathcal{H}$ respectively. As such, an open set in a real vector space is, in a natural way, a real analytic manifold of the same dimension.*

Proof. Consider the following polynomials,

$$p(z) = \sum_{i=1}^{n} p_i z^i \quad \text{and} \quad q(z) = \sum_{i=1}^{n} q_i z^i$$

with complex coefficients, of degree n, and let $z_1, ..., z_n$ and $w_1, ..., w_n$ be their respective roots, counted with multiplicities. From Rouché's theorem it follows that if $|p_i - q_i| < \delta$, then (after possible reordering of the w_i's), $|z_i - w_i| < \epsilon$ for all $i = 1, ..., n$. Now suppose that $\mathcal{H}^+$ is not open in $\mathcal{H}$. Then its complement can not be closed. In other words,

---

[6]For a proof see e.g. Martin Moskowitz, *A Course in Complex Analysis in one Variable*, World Scientific Publishing Co. Singapore, 2002, pp. 63-65.

there must exist an h not in $\mathcal{H}^+$ and a sequence $(x_j)_j \in \mathcal{H} - \mathcal{H}^+$ with $x_j \longrightarrow h$, in $\mathrm{M}(n, \mathbb{C})$. Since h is positive definite, all its eigenvalues are positive.

Choose an $\epsilon > 0$ so that the ϵ-balls about the eigenvalues of h satisfy

$$\bigcup_{i=1}^{n} B_{\lambda_i}(\epsilon) \subseteq \mathcal{H}_+,$$

where $\mathcal{H}_+$ denotes the right half plane (the right of the axis). Since the coefficients of the characteristic polynomial of an operator are polynomials and so continuous functions of the matrix entries and $x_j \longrightarrow h$, for j sufficiently large, the coefficients of the characteristic polynomial of x_j are within δ of the corresponding coefficient of the characteristic polynomial of h. Hence, all the eigenvalues of such an x_j are positive. Therefore x_j is in $\mathcal{H}^+$, which contradicts the fact that none of the x_j is supposed to be in $\mathcal{H}^+$. Thus $\mathcal{H}^+$ is open in $\mathcal{H}$. Finally, intersecting everything in sight with $\mathrm{M}(n, \mathbb{R})$ shows $\mathcal{S}^+$ is also open in $\mathcal{S}$.   □

We remind the reader that when we apply the exponential map or the logarithm to an operator we shall write Exp or Log and when these functions apply to a complex number we write exp or log. Upon restriction, the exponential map of $\mathrm{M}(n, \mathbb{C})$ is a real analytic map from $\mathcal{H}$ onto $\mathcal{H}^+$. As we shall see, its inverse, is given by

$$\mathrm{Log}\, h = \log(\mathrm{tr}\, h)I - \sum_{i=1}^{\infty}(I - \frac{h}{\mathrm{tr}\, h})^i,$$

which is an analytic function on $\mathcal{H}^+$. Thus $\mathrm{Exp} : \mathcal{H} \to \mathcal{H}^+$ will be an analytic diffeomorphism. Hence restriction of Exp to any real subspace of $\mathcal{H}$ we get a real analytic diffeomorphism of the subspace with its image. Upon restriction, Exp is a real analytic diffeomorphism between $\mathcal{P}$ and $\mathcal{P}^+$.

Theorem 4.7.10. *The exponential map* $\mathrm{Exp} : \mathcal{H} \longrightarrow \mathcal{H}^+$ *is a diffeomorphism. The same holds for* $\mathrm{Exp} : \mathcal{S} \longrightarrow \mathcal{S}^+$. *In particular,* Exp *is bijective* $\mathcal{H} \longrightarrow \mathcal{H}^+$.

Proof. We shall do this for $\mathcal{H}$, the real case being completely analogous. Suppose $h \in \mathcal{H}^+$ is diagonal with eigenvalues $h_i > 0$. Then $\text{tr}(h) > 0$ and $0 < \frac{h_i}{\text{tr}(h)}$. Therefore $\log(\text{tr}(h))$ is well defined and $\log(\frac{h_i}{\text{tr}(h)})$ is defined for all i. But since $0 < \frac{h_i}{\text{tr}(h)} < 1$, we see that $0 < (1 - \frac{h_i}{\text{tr}(h)})^k < 1$ for all positive integers k. Hence $\text{Log}(\frac{I-h}{\text{tr}(h)})$ is given by an absolutely convergent power series $-\sum_{i=1}^{\infty}(I - \frac{h}{\text{tr}\,h})^i / i$. If u is a unitary operator so that uhu^{-1} is diagonal, then $\text{tr}(uhu^{-1}) = \text{tr}(h)$ and since conjugation by u commutes with any convergent power series, this series actually converges for all $h \in \mathcal{H}^+$ and is a real analytic function Log on $\mathcal{H}^+$. Because on the diagonal part of $\mathcal{H}^+$ this function inverts Exp, and both Exp and this power series commute with conjugation, it inverts Exp everywhere on $\mathcal{H}^+$. Finally, $\log(\text{tr}(h))I$ and $\text{Log}(\frac{h}{\text{tr}\,h})$ commute and Exp, as a sum of commuting matrices, is the product of the Exp's. Since Log inverts Exp on the diagonal part of $\mathcal{H}^+$ it follows that

$$\text{Log}(h) = \log(\text{tr}(h))I + \text{Log}(\frac{h}{\text{tr}\,h}) = \log(\text{tr}(h))I - \sum_{i=1}^{\infty}(I - \frac{h}{\text{tr}\,h})^i / i.$$

$\square$

4.7.3   The Polar Decomposition

Here we prove the Polar Decomposition Theorem.

Recall that for any $g \in \text{GL}(n, \mathbb{C})$, $g^*g \in \mathcal{H}^+$. It follows that for all $g \in \text{GL}(n, \mathbb{C})$, $\text{Log}(g^*g) \in \mathcal{H}$ and since this is a real linear space, also $\frac{1}{2}\text{Log}(g^*g) \in \mathcal{H}$. This means we can apply Exp and conclude the following:

Corollary 4.7.11. $h(g) = \text{Exp}(\frac{1}{2}\text{Log}(g^*g)) \in \mathcal{H}^+$ *is a real analytic function from* $\text{GL}(n, \mathbb{C}) \to \mathcal{H}^+$.

Proof. Since $h(g)^2 = g^*g$, and $h(g)^{-1} \in \mathcal{H}^+$, $g(g^*g)^{-1}g^* = I$. Thus, $gh(g)^{-1} = u(g)$ is unitary for each $g \in \text{GL}(n, \mathbb{C})$. Since group multiplication and inversion are analytic functions (in $\text{GL}(n\mathbb{C})$, $u(g)$ is also a real analytic function on $\text{GL}(n, \mathbb{C})$ (as is $h(g)$). Now this decomposition $g = uh$, where $u \in U$ and $h \in \mathcal{H}^+$ is actually unique. For suppose

$u_1 h_1 = g = u_2 h_2$. Then $u_2^{-1} u_1 = h_2 h_1^{-1}$ so that $h_2 h_1^{-1}$ is unitary. This means $(h_2 h_1^{-1})^* = (h_2 h_1^{-1})^{-1}$ and hence $h_1^2 = h_2^2$. But since h_1 and $h_2 \in \mathcal{H}^+$, each is an exponential of something in $\mathcal{H}$; $h_i = \operatorname{Exp} x_i$. But then $h_i^2 = \operatorname{Exp} 2x_i$ and since Exp is injective on $\mathcal{H}$, we get $2x_1 = 2x_2$ so $x_1 = x_2$. Therefore $h_1 = h_2$ and $u_1 = u_2$.

The upshot of all this is that we have a real analytic map

$$\operatorname{GL}(n, \mathbb{C}) \to \operatorname{U}(n, \mathbb{C}) \times \mathcal{H}^+$$

given by $g \mapsto u(g)h(g)$. Since $g = u(g)h(g)$ for every g (multiplication in $\operatorname{GL}(n, \mathbb{C})$), this map is surjective and has a real analytic inverse.

We summarize these facts as the following Polar Decomposition Theorem.

Theorem 4.7.12. *The map $g \mapsto u(g)h(g)$ gives a real analytic diffeomorphism $\operatorname{GL}(n, \mathbb{C}) \to \operatorname{U}(n, \mathbb{C}) \times \mathcal{H}^+$. Identical reasoning also shows that as a real analytic manifold $\operatorname{GL}(n, \mathbb{R})$ is, in the same way, diffeomorphic to $O(n, \mathbb{R}) \times \mathcal{P}^+$.*

From this it follows that, since $\mathcal{H}^+$ and $\mathcal{P}^+$ are each diffeomorphic with a Euclidean space, and are therefore topologically trivial, in each case the topology of the non-compact group is completely determined by that of the compact one. In this situation, one calls the compact group a *deformation retract* of the non-compact group. Since $\mathcal{P}^+$ and $\mathcal{H}^+$ are diffeomorphic images under Exp of some Euclidean space, one calls them *exponential submanifolds*. For example, connectedness, the number of components, simple connectedness and the fundamental group of the non-compact group are each the same as that of the compact one. Thus for all $n \geq 1$, $\operatorname{GL}(n, \mathbb{C})$ is connected and its fundamental group is $\mathbb{Z}$, while for all n, $\operatorname{GL}(n, \mathbb{R})$ has 2 components and the fundamental group of its identity component is $\mathbb{Z}_2$ for $n \geq 3$ and $\mathbb{Z}$ for $n = 2$. These facts follow from the long exact homotopy sequence for a fibration and are explained in C. Chevalley [24].

Although what we have done in this subsection may seem rather special, it leads to a very general result known as the E. Cartan Decomposition theorem which however we cannot pursue as it would take us too far afield. For further details on this see [1].

4.7.4   The Iwasawa Decomposition

Here we deal with the *Iwasawa decomposition theorem* and the *Bruhat lemma* both in the case of the general linear group G. As before, G stands for $\mathrm{GL}(n,\mathbb{R})$ or $\mathrm{GL}(n,\mathbb{C})$. $\mathrm{O}(n,\mathbb{R})$ denotes the orthogonal group and $\mathrm{U}(n,\mathbb{R})$ the unitary group.[7]

The Iwasawa decomposition theorem states that

$$G = K \cdot A \cdot N,$$

where

$$\underline{\mathrm{GL}(n,\mathbb{R})} \qquad\qquad\qquad\qquad \underline{\mathrm{GL}(n,\mathbb{C})}$$

$$K = \mathrm{O}(\mathrm{n},\mathbb{R}) \qquad\qquad\qquad\qquad K = \mathrm{U}(n,\mathbb{C})$$

$$a = \begin{pmatrix} a_{11} & & 0 \\ & \ddots & \\ 0 & & a_{nn} \end{pmatrix} \qquad a_{ii} > 0, \quad \forall\, i, \qquad a = \begin{pmatrix} a_{11} & & 0 \\ & \ddots & \\ 0 & & a_{nn} \end{pmatrix}$$

$$n = \begin{pmatrix} 1 & & 0 \\ & \ddots & \\ \star_{\mathbb{R}} & & 1 \end{pmatrix} \qquad\qquad\qquad n = \begin{pmatrix} 1 & & 0 \\ & \ddots & \\ \star_{\mathbb{C}} & & 1 \end{pmatrix}$$

For varieties sake here we shall work over $\mathbb{R}$, the case of $\mathbb{C}$ being analogous. Thus, A is the subgroup of diagonal matrices in $G = \mathrm{GL}(n,\mathbb{R})$

---

[7]Kenkichi Iwasawa (1917-1998) Japanese mathematician spent the early and difficult years of his career during the war in Japan. In 1949 he published his famous paper *On some types of topological groups* where among other things contains what is now known as the Iwasawa decomposition of a real sem-isimple Lie group. He then returned to algebraic number theory which was the subject of his dissertation. The ideas he introduced there have had a fundamental impact on the development of mathematics in the second half of the 20th century. Iwasawa remained as Fine Professor of mathematics at Princeton when he published his book, Local class field theory upon his retirement in 1986. He then returned to Tokyo where he spent his last years.

with *positive* entries, N, the subgroup of uni-triangular matrices n in G and $T = AN$, the group of real triangular matrices with positive diagonal entries. As an exercise we propose the reader formulate and prove the analogous statements for $\mathrm{GL}(n, \mathbb{C})$.

Theorem 4.7.13. *Every invertible linear transformation g in G can be uniquely written as $g = kan$.*

Thus we have another decomposition $G = KAN$, where AN is a connected simply connected open set in Euclidean space. Sometimes the Iwasawa theorem is more useful while sometimes the E. Cartan theorem is.

Proof. We write V for $\mathbb{R}^n$. Let $\{e_1, ..., e_n\}$ be the standard orthonormal basis of V and choose $g \in G$ so that

$$g^{-1}(e_i) := v_i.$$

In this way we obtain a basis $\{v_1, ..., v_n\}$ of V, but perhaps not an orthonormal basis.

We now apply the Gram-Schmidt orthogonalization process to this basis and get an orthonormal basis, $\{u_1, ..., u_n\}$. On the other hand, since $v_1, ..., v_n$ is a basis,

$$u_i = \sum_{j \le i} a_{ji} v_j, \quad \text{with} \quad a_{ii} > 0, \quad (1)$$

and this for all $i \le n$. But, by the Gram-Schmidt construction, each u_i is $\perp$ to all v_j with $j \le i$, and since $a_{ji} = \langle u_i, v_j \rangle$ for any j, it follows that $a_{ji} = 0$ for all $j \le i$. Therefore, equation (1) tells us

$$\begin{pmatrix} u_1 \\ u_2 \\ \vdots \\ u_n \end{pmatrix} = \begin{pmatrix} a_{11} & & & \\ a_{12} & a_{22} & & 0 \\ \vdots & & \ddots & \\ a_{1n} & a_{2n} & \cdots & a_{nn} \end{pmatrix} \begin{pmatrix} v_1 \\ v_2 \\ \vdots \\ v_{nn} \end{pmatrix},$$

where a_{ji}, $j \leq i$ are the Fourier coefficients of u_i in the Gram-Schmidt orthogonalization. But,

$$
\begin{pmatrix} a_{11} & & 0 \\ & a_{22} & \\ & & \ddots \\ \bigstar & & a_{nn} \end{pmatrix} = \begin{pmatrix} a_{11} & & 0 \\ & a_{22} & \\ & & \ddots \\ 0 & & a_{nn} \end{pmatrix} \begin{pmatrix} 1 & & \\ & 1 & 0 \\ & & \ddots \\ \bigstar & & 1 \end{pmatrix}.
$$

Therefore,

$$
\begin{pmatrix} u_1 \\ u_2 \\ \vdots \\ u_n \end{pmatrix} = \begin{pmatrix} a_{11} & & 0 \\ & a_{22} & \\ & & \ddots \\ 0 & & a_{nn} \end{pmatrix} \begin{pmatrix} 1 & & \\ & 1 & 0 \\ & & \ddots \\ \bigstar & & 1 \end{pmatrix} \begin{pmatrix} v_1 \\ v_2 \\ \vdots \\ v_n \end{pmatrix}. \tag{2}
$$

Now, consider the two orthonormal bases, $e_1, ..., e_n$, and $u_1, ..., u_n$ we used. We can pass from one to the other by a unique orthogonal (resp. unitary) operator k. In other words, there is a $k \in \mathrm{O}(n, \mathbb{R})$ (respectively $\mathrm{U}(n, \mathbb{C})$) so that $k(u_i) = e_i = g(v_i)$ since $g^{-1}(e_i) = v_i$. Hence, using (2) we get:

$$
k \begin{pmatrix} u_1 \\ u_2 \\ \vdots \\ u_n \end{pmatrix} = k \begin{pmatrix} a_{11} & & 0 \\ 0 & a_{22} & 0 \\ & & \ddots \\ 0 & & a_{nn} \end{pmatrix} \begin{pmatrix} 1 & & \\ & 1 & 0 \\ & & \ddots \\ \bigstar_{\mathbb{R},\mathbb{C}} & & 1 \end{pmatrix} \begin{pmatrix} v_1 \\ v_2 \\ \vdots \\ v_n \end{pmatrix}
$$

$$
= \begin{pmatrix} e_1 \\ e_2 \\ \vdots \\ e_n \end{pmatrix} = g \begin{pmatrix} u_1 \\ u_2 \\ \vdots \\ u_n \end{pmatrix}.
$$

From this it follows that

$$
k \begin{pmatrix} a_{11} & & 0 \\ & a_{22} & \\ & & \ddots \\ 0 & & a_{nn} \end{pmatrix} \begin{pmatrix} 1 & & \\ & 1 & 0 \\ & & \ddots \\ \bigstar_{\mathbb{R},\mathbb{C}} & & 1 \end{pmatrix} = g, \quad \forall \ g \in \mathrm{GL}(n, \mathbb{R}), \ \mathrm{GL}(n, \mathbb{C}),
$$

where k is an orthogonal (resp. unitary) operator. In other words $g = kan$.

Turning to the uniqueness of the components, suppose $kan = k_1 a_1 n_1$. Write $T = AN$. Then, letting $t = an$ and $t_1 = a_1 n_1$ we get $k_1^{-1} k = t_1 t^{-1}$. Both K and T are groups (T being a group because A normalizes N). Hence, if $k_2 = k_1^{-1} k$ and $t_2 = t_1 t^{-1}$ we see that $k_2 \in K \cap T$. But since $k_2 \in K$ all its eigenvalues are of absolute value 1, while elements of T have all their eigenvalues positive. It follows that all the eigenvalues of k_2 are equal to 1 and since k_2 is diagonalizable, it must be I. Thus $k = k_1$ and therefore also $t = t_1$.

Using a similar argument $a_1^{-1} a = n_1 n^{-1}$ so that $a_2 \in A \cap N$. As these are also groups, we see that the eigenvalues of a_2 are all equal to 1. Because A is diagonal, a_2 is the identity. Hence $a = a_1$ and therefore also $n = n_1$.   □

The subgroup AN of G is connected and simply connected since it is homeomorphic to Euclidean space. Moreover, the multiplication map

$$K \times A \times N \longrightarrow G \qquad (k, a, n) \mapsto k \cdot a \cdot n = g$$

is a homeomorphism (actually a real analytic homeomorphism). Therefore, since the topology of AN is trivial, we see, just as before, that the topology of G (i.e. the topology of $\mathrm{GL}(n, \mathbb{R})$), is completely determined by that of the compact group K and K is a *deformation retract* of G.

Exercise 4.7.14. Show that $Z_K(A)$, the centralizer of A in K, is just (1). *Hint*: Use the above proof of uniqueness.

Exercise 4.7.15. Show that the Iwasawa decomposition theorem extends to the groups, $G = \mathrm{SL}(n, \mathbb{R})$ and $G = \mathrm{SL}(n, \mathbb{C})$.
    Hint: Here one must take the compact group K to be $\mathrm{SO}(n, \mathbb{R})$ (respectively $\mathrm{SU}(n, \mathbb{C})$) and as A, the diagonal matrices with positive entries. Notice N is automatically a subgroup of G.

4.7.4.1 $|\det|$ **and Volume in** $\mathbb{R}^n$.

We now give an alternative approach to det versus vol based on the Iwasawa decomposition theorem.

Theorem 4.7.16. *If $U \neq \varnothing$ is an open set in $\mathbb{R}^n$ and $T : \mathbb{R}^n \longrightarrow \mathbb{R}^n$ is a linear transformation, then*

$$\mathrm{vol}(T(U)) = |\det(T)|\,\mathrm{vol}(U) \quad (1).$$

(This is independent of the set U and depends only on T.)

Proof. First, we notice that if $\det T = 0$ then T as a singular matrix must send U to a lower dimension variety, thus $\mathrm{vol}(T(U)) = 0$, and our statement is true since $0 = 0 \cdot \mathrm{vol}(U)$. Next, suppose that $T \in \mathrm{GL}(n, \mathbb{R})$. According to Iwasawa decomposition we have $T = k \cdot a \cdot n$.

Now, assuming equation (1) holds for every linear transformation S and for any open sets $U \subset \mathbb{R}^n$, $U \neq \varnothing$, we get

$$\mathrm{vol}(S(U)) = |\det(S)|\,\mathrm{vol}(U).$$

Hence, if $S \in \mathrm{GL}(n, \mathbb{R})$, then $TS \in \mathrm{GL}(n, \mathbb{R})$, and

$$\mathrm{vol}(TS(U)) = |\det(TS)|\,\mathrm{vol}(U).$$

On the other hand,

$$\mathrm{vol}(TS(U)) = |\det(T)|\,\mathrm{vol}(S(U)) = |\det(T)| \cdot |\det(S)|\,\mathrm{vol}(U).$$

So, if we prove that (1) is true for k, a, and n, then it will be true for $k \cdot a \cdot n$, i.e. for any $T \in \mathrm{GL}(n, \mathbb{R})$.

First, let us check what happens for k. Since $k \in \mathrm{O}(n, \mathbb{R})$, it is an isometry, so it preserves distances and therefore volumes. So $\mathrm{vol}(k(U)) = \mathrm{vol}(U)|\det(k)| = \mathrm{vol}(U)$, since $|\det(k)| = 1$. Turning to a. The matrix a acts by scaling along the coordinate axis by a_{ii} for x_i, for any i. Hence, $\mathrm{vol}(a(U)) = |a_{11}a_{22}...a_{nn}|\,\mathrm{vol}(U) = |\det(a)|\,\mathrm{vol}(U)$. Finally, for n. We can see that n acts by shearing volumes along an axis, preserving the base and the height, therefore it preserves volume. This follows since $\det n = 1$. To see this visually take as space $\mathbb{R}^2$ and

$$n = \begin{pmatrix} 1 & 0 \\ x & 1 \end{pmatrix}$$

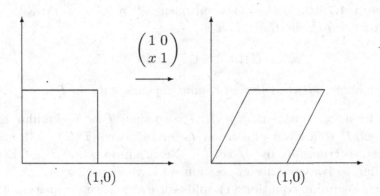

Then (see image above) we have

$$\begin{pmatrix} 1 & 0 \\ x & 1 \end{pmatrix} \begin{pmatrix} 1 \\ 0 \end{pmatrix} = \begin{pmatrix} 1 \\ x \end{pmatrix}, \quad \begin{pmatrix} 1 & 0 \\ x & 1 \end{pmatrix} \begin{pmatrix} 0 \\ 1 \end{pmatrix} = \begin{pmatrix} 0 \\ 1 \end{pmatrix}.$$

Therefore, $\mathrm{vol}(n(U)) = \mathrm{vol}(U)$. $\square$

4.7.5   The Bruhat Decomposition

We now turn to the *Bruhat decomposition*. Here again the real and complex cases are completely analogous and as above we shall treat only the real case. Writing $\mathbb{R}^n = V$ and $\mathrm{GL}(n, \mathbb{R}) = G$, we let Ω denote the (finite group) of all permutations of n objects and, for $\omega \in \Omega$, denote by $\bar{\omega}$ the associated permutation matrix, i.e. the permutation induced by ω on some fixed basis of V. Notice that this is a $1:1$ correspondence, so that the order of Ω is the same as that of $\bar{\Omega}$, which is $n!$, T is the full group of triangular matrices.

Theorem 4.7.17. $G = \bigcup_{\omega \in \Omega} T\bar{\omega}T$. *That is to say, the mapping* $\bar{\omega} \mapsto T\bar{\omega}T$ *is* onto *the set of all double cosets,* TgT, $g \in G$. *Moreover, this map is* $1:1$ *and these double cosets are pairwise disjoint.*

On the right we have a *double coset space* sometimes written $T\bar{\Omega}T$. Since $\bar{\omega} = I$ is always included, one of these double cosets is just T. This is called the trivial double coset; it has lower dimension than G.

To see the map is $1 : 1$, first observe that if $g = kt$, then $TgT = TkT$. Hence the question is if $t_1 k_1 t_1' = t_2 k_2 t_2'$, where k_1 and $k_2 \in K$ and all the t's are in T, does this force $k_1 = k_2$? Pulling all the t's together we see that $tk_1 = k_2 t'$, where t and $t' \in T$. Thus $k_1 k_2^{-1} \in K \cap T$ and arguing as before, we see that $k_1 = k_2$. This same argument shows the double cosets involved are pairwise disjoint (check).

In the case of GL(2) one can actually verify the Bruhat decomposition explicitly as follows: For $g \in \mathrm{GL}(2)$,

$$g = \begin{pmatrix} a & b \\ c & d \end{pmatrix}$$

with $\det(g) \neq 0$. If $c = 0$ then g is already triangular so $g \in T$, the trivial double coset. On the other hand, if $c \neq 0$, then we write $g = uvw$, where

$$u = \begin{pmatrix} 1 & \frac{a}{c} \\ 0 & 1 \end{pmatrix}$$

$$v = \begin{pmatrix} 0 & 1 \\ 1 & 0 \end{pmatrix}$$

and

$$w = \begin{pmatrix} c & d \\ 0 & -\frac{\det(g)}{c} \end{pmatrix}.$$

Exercise 4.7.18. The reader should check $g = uvw$.

Here the first and third matrices are in T and the second is a permutation of order 2 (actually a reflection). Thus G is a disjoint union of two double cosets, one of which is trivial while the other is a true double coset. Notice this last double coset is a space of dimension 4, as is G.

Exercise 4.7.19. Let
$$\omega_0 = \begin{pmatrix} 0 & 1 \\ 1 & 0 \end{pmatrix}.$$

Show $T\omega_0 T$ is both open and Euclidean dense in G.

To verify the Bruhat decomposition in general for $GL(n)$ we shall require three lemmas. We first consider the n^2 dimensional real vector space $End(V)$ on which we have a positive definite symmetric form β given by[8] $\beta(A, B) = tr(AB^t)$.

Lemma 4.7.20. *The map* $\beta(A, B) = tr(AB^t)$ *is symmetric bilinear and positive definite.*

Proof. The bilinearity is clear. Since $(AB^t)^t = BA^t$ and because we are over the reals, $tr(X) = tr(X^t)$ for all X. It follows that β is symmetric. Let $A = (a_{i,j})$ be any $n \times n$ real matrix. Then $\beta(A, A) = tr(AA^t)$. A direct calculation shows $tr(AA^t) = \sum_{i,j=1}^{n} a_{i,j}^2$, proving β is positive definite. $\square$

In order to proceed we shall need to make a few remarks about Lie algebras. (In our case only finite dimensional real Lie algebras will matter so we restrict ourselves to these). Since $End(V)$ is an associative algebra under matrix multiplication, it becomes a Lie algebra under the operation commutator, $[x, y] = xy - yx$, called the bracket. (This is a type of non-associative algebra). Since in our case only finite dimensional real Lie algebras will matter, we restrict ourselves to these. Indeed, we can restrict ourselves to subalgebras of $End(V)$, that is, vector subspaces of $End(V)$, which are closed under bracketing.

An automorphism α of such a Lie algebra is an $\mathbb{R}$ linear map which is bijective and preserves the bracket. That is, $\alpha[x, y] = [\alpha(x), \alpha(y)]$. Now let $g \in GL(V)$ and $\alpha_g(x) = gxg^{-1}$, for $x \in End(V)$ (This is called an inner automorphism of $End(V)$). As is easily checked, each α_g is an automorphism of $End(V)$. So when we have a subalgebra of $End(V)$,

---

[8]β is usually called the *Hilbert Schmidt norm* on operators. In the complex case it takes the form $\beta(A, B) = tr(AB^*)$. This norm works for certain operators even in the case of Hilbert spaces of infinite dimension.

to see if α_g is an automorphism of it we only have to check that it is invariant under α_g.

We let $\mathfrak{t}$, $\mathfrak{n}$ and $\mathfrak{d}$ denote, respectively, the upper triangular matrices, strictly upper triangular and diagonal matrices of $\mathrm{End}(V)$. As the reader can check, these are Lie subalgebras of the Lie algebra $\mathrm{End}(V)$.

Lemma 4.7.21. *1. The algebra,* $\mathrm{End}(V)$ *is the orthogonal direct sum of subalgebras,* $\mathfrak{n}^t \oplus \mathfrak{d} \oplus \mathfrak{n}$. *In particular,* $\dim(\mathfrak{t}) + \dim(\mathfrak{n}) = n^2$.

2. Conjugation by $k \in O(n)$ *is an algebra automorphism of* $\mathrm{End}(V)$ *which commutes with the transpose.*

Proof. That these are subalgebras and the sum is direct is clear. We leave the verification of orthogonality to the reader as an exercise. As we explained just above, conjugation by $k \in O(n)$ is an algebra automorphism of $\mathrm{End}(V)$. It commutes with ^{t} because $k^{-1} = k^t$.   □

Lemma 4.7.22. *For each* $g \in G$, $\mathfrak{t} = \mathfrak{n} + (gtg^{-1} \cap \mathfrak{t})$.

Proof. Using the Iwasawa decomposition theorem 4.7.13 write $g = kt$, where $t = an \in T$. Suppose the lemma were true for $g = k$. Then since $gtg^{-1} = kttt^{-1}k^{-1} = ktk^{-1}$, we would have

$$\mathfrak{n} + (gtg^{-1} \cap \mathfrak{t}) = \mathfrak{n} + (ktk^{-1} \cap \mathfrak{t}) = \mathfrak{t}.$$

Hence to prove the lemma we may assume $g \in K$.

Next, we show $\mathfrak{n} \cap (gtg^{-1} \cap \mathfrak{t}) = \mathfrak{n} \cap (g\mathfrak{n}g^{-1})$. This is because

$$\mathfrak{n} \cap (gtg^{-1} \cap \mathfrak{t}) \subseteq \mathfrak{n} \cap (gtg^{-1}) \subseteq \mathfrak{n} \cap (g\mathfrak{n}g^{-1}) \subseteq \mathfrak{n} \cap (gtg^{-1} \cap \mathfrak{t}).$$

Calculating orthocomplements,

$$(gtg^{-1} \cap \mathfrak{t})^{\perp} = \mathfrak{t}^{\perp} + gt(g^{-1})^{\perp}$$

and because K preserves the inner product the last term is $g\mathfrak{t}^{\perp}g^{-1}$. Since $g \in K$ and lemma 4.7.21 tells us we have an orthogonal decomposition, finally we get $\mathfrak{n}^t + g\mathfrak{n}^tg^{-1} = (\mathfrak{n} + g\mathfrak{n}g^{-1})^t$. The dimension of this last space is the same as $\dim(\mathfrak{n} + g\mathfrak{n}g^{-1})$. On the other hand $\dim(gtg^{-1} \cap \mathfrak{t})^{\perp} = n^2 - \dim(gtg^{-1} \cap \mathfrak{t})$. Thus

$$n^2 = \dim(\mathfrak{t}) + \dim(\mathfrak{n}) = \dim(gtg^{-1} \cap \mathfrak{t}) + \dim(\mathfrak{n} + g\mathfrak{n}g^{-1}).$$

But $\dim(\mathfrak{n} + g\mathfrak{n}g^{-1}) = 2\dim(\mathfrak{n}) - \dim(\mathfrak{n} \cap g\mathfrak{n}g^{-1})$. Hence

$$\dim(\mathfrak{t}) + \dim(\mathfrak{n}) = \dim(gtg^{-1} \cap \mathfrak{t}) + 2\dim(\mathfrak{n}) - \dim(\mathfrak{n} \cap g\mathfrak{n}g^{-1}).$$

Therefore,

$$\dim(\mathfrak{t}) = \dim(\mathfrak{n}) + \dim(gtg^{-1} \cap \mathfrak{t}) - \dim(\mathfrak{n} \cap g\mathfrak{n}g^{-1}).$$

In order to show $\mathfrak{t} = \mathfrak{n} + (gtg^{-1} \cap \mathfrak{t})$, since the right side is a subset of the left it is sufficient to prove their dimensions are the same and this is exactly what the above equation does. $\square$

Lemma 4.7.23. *Let* $t \in T$ *and have* n *distinct eigenvalues. Then there exists a (unique)* $u \in N$ *so that* utu^{-1} *is diagonal.*

Proof. We prove this by induction on n. Let

$$t = \begin{pmatrix} \lambda_1 & v_1 \\ 0 & t_1 \end{pmatrix},$$

where v_1 is an $n - 1$ vector and t_1 is a triangular matrix of order $n - 1$ with distinct eigenvalues. By induction there is a unique unipotent u_1 with $u_1 t_1 u_1^{-1} = d_1$, with d_1 diagonal. Let

$$u = \begin{pmatrix} 1 & w_1 \\ 0 & u_1 \end{pmatrix},$$

where w_1 is some $n - 1$ vector to be determined. Then u is unipotent and is unique if w_1 is. Moreover, since block matrix multiplication is exactly semi-direct product multiplication (check!) we see that

$$utu^{-1} = \begin{pmatrix} 1 & w_1 \\ 0 & u_1 \end{pmatrix} \cdot \begin{pmatrix} \lambda_1 & v_1 \\ 0 & t_1 \end{pmatrix} \cdot \begin{pmatrix} 1 & -w_1(u_1)^{-1} \\ 0 & -u_1^{-1} \end{pmatrix}.$$

This is

$$\begin{pmatrix} \lambda_1 & -\lambda_1 w_1 u_1^{-1} + v_1 u_1^{-1} + w_1 t_1 u_1^{-1} \\ 0 & d_1 \end{pmatrix}.$$

Thus it remains to see that we can choose a unique w_1 for which $(\lambda_1 I - d_1)(w_1) = u_1(v_1)$. This comes down to whether $\lambda_1 I - d_1$ is non-singular. It is since t_1 is triangular and has distinct eigenvalues. ☐

We now turn to the proof of Theorem 4.7.17.

Proof. Let $g \in G$ and $d \in \mathfrak{d}$ be the matrix with diagonal entries $1, 2. \ldots n$. By Lemma 4.7.22 there is some $n \in \mathfrak{n}$ so that $d = n + gtg^{-1}$ where $gtg^{-1} \in \mathfrak{t}$. Because $d - n = gtg^{-1}$ has the same spectrum as d, it follows that gtg^{-1} and therefore also t has as its spectrum the first n integers. Thus $g^{-1}(d - n)g = t$ is a triangular matrix with eigenvalues some permutation of the first n integers. On the other hand, since these eigenvalues are distinct, by lemma 4.7.23 there is some unipotent operator $u_1 \in N$ so that $u_1 t u_1^{-1} = d_1$ with d_1 diagonal. Thus $u_1 g^{-1}(d - n)gu_1^{-1} = d_1$ and, of course, there is a permutation matrix W such that $W d_1 W^{-1} = d$. This tells us $u_1 g^{-1}(d-n)gu_1^{-1} = W^{-1}dW$. Hence $d - n = u_1^{-1} g W^{-1} dW g^{-1} u_1$. However, since $d - n$ is also triangular with distinct eigenvalues, applying lemma 4.7.23 again shows us there is a $u_2 \in N$ with $d - n = u_2 d u_2^{-1}$. Thus

$$d = u_2^{-1} u_1^{-1} g W^{-1} dW g^{-1} u_1 u_2,$$

so that $W g^{-1} u_1 u_2$ commutes with d. Since the eigenvalues of d are distinct $W u_1 g^{-1} u_2 = d_2$ some diagonal matrix. Hence $g = u_2 d_2^{-1} w u_1 \in T w N \subseteq T w T$. ☐

We conclude our treatment of the full linear group with the following proposition. Since this result is of secondary importance we only sketch the proof.

Proposition 4.7.24. *For $G = \mathrm{GL}(n, \mathbb{R})$ one of the double cosets is open and dense in G (Euclidean topology).*

Proof. We choose the double coset determined by the cyclic permutation ω of order n (the largest order of all cyclic permutations). Thus

$$
\bar{\omega} = \begin{pmatrix} 0 & \cdots & \cdots & 1 \\ 0 & \cdots & 1 & 0 \\ \vdots & \vdots & \vdots & \vdots \\ 1 & \cdots\cdots & 0 & 0 \end{pmatrix}.
$$

Let $\bar{\omega}$ be as above. To see that $T\bar{\omega}T$ is open and dense it suffices to see that $\bar{\omega}T\bar{\omega}T$ is itself open and dense. But $\bar{\omega}T\bar{\omega} = T^t$. (Prove!) Hence $\bar{\omega}T\bar{\omega}T = T^tT$. Now write out this matrix multiplication. The remainder of the proof is left to the reader as an exercise.   □

When we have a group such as $G = \mathrm{SL}(n, \mathbb{R})$ or $\mathrm{SL}(n, \mathbb{C})$, just as with the Iwasawa decomposition theorem we shall have to redefine some of the ingredients of the Bruhat lemma. We replace the A above by $A \cap G$. Then $B = AN$ is called a *Borel subgroup* of G. It is a maximal connected solvable subgroup of G. B just as in the Iwasawa decomposition theorem will take the place of T in the Bruhat lemma. But what will take the place of S_n, the symmetric group? This will be $\mathcal{W}(G, A)$, the *Weyl group* which is defined by

$$
\mathcal{W}(G, A) = N_G(A)/Z_G(A).
$$

In terms of the Weyl group the Bruhat lemma is formulated as

$$
G = \bigcup_{w \in \mathcal{W}} B\bar{w}B,
$$

where for $w \in \mathcal{W}$ we write $\bar{w}$ for any pre-image in $N_G(A)$ of w.

Exercise 4.7.25.

1. Show that T is a maximal connected solvable subgroup of $\mathrm{GL}(n, \mathbb{R})$, respectively $\mathrm{GL}(n, \mathbb{C})$.

2. Show that if $G = \mathrm{GL}(n, \mathbb{R})$ or $\mathrm{GL}(n, \mathbb{C})$, then $N_G(A)/Z_G(A) = S_n$.

3. When $G = \mathrm{SL}(2, \mathbb{R})$, or $\mathrm{SL}(2, \mathbb{C})$ find $N_G(A)$, $Z_G(A)$ and representatives of $N_G(A)/Z_G(A)$.

Exercise 4.7.26. State and prove the Bruhat lemma for $SL(2, \mathbb{R})$ and $SL(2, \mathbb{C})$.

4.8 Gramians

Let $(V, \langle \cdot, \cdot \rangle)$ be a finite dimensional real inner product space of dimension n, and $\{a_1, \ldots, a_k\}$ be a set of $k \leq n$ vectors. We consider the $k \times k$ (symmetric) matrix, $M = (\langle a_i, a_j \rangle)$.

Definition 4.8.1. The *Gramian* of $a_1, \ldots, a_k$ is defined as

$$G(a_1, \ldots, a_k) = \det(\langle a_i, a_j \rangle).$$

Proposition 4.8.2. $G(a_1, \ldots, a_k) \geq 0$ *and is* 0 *if and only if* $\{a_1, \ldots, a_k\}$ *are linearly dependent.*

This result is a *very significant generalization* of the Cauchy-Schwarz inequality. For example, when $k = 3$ it says for any 3 vectors, $\{a_1, a_2, a_3\}$ in V,

$$\| a_1 \|^2 + \| a_2 \|^2 + \| a_3 \|^2 + \langle a_1, a_2 \rangle \langle a_1, a_3 \rangle \langle a_2, a_3 \rangle$$

$$\geq \| a_1 \|^2 \langle a_2, a_3 \rangle + \| a_2 \|^2 \langle a_1, a_3 \rangle + \| a_3 \|^2 \langle a_2, a_3 \rangle,$$

with equality if and only if $\{a_1, a_2, a_3\}$, are linearly dependent.

Proof. First suppose $\{a_1, \ldots, a_k\}$ are linearly dependent. Then there are λ_i, not all zero, so that $\sum \lambda_i a_i = 0$. Hence for each $j = 1 \ldots k$, $\sum \lambda_i \langle a_i, a_j \rangle = 0$. This says that the column vectors of M are themselves linearly dependent. Hence $G = 0$.

To complete the proof we must show that if $\{a_1, \ldots, a_k\}$ are linearly independent, then $G > 0$. For this let $W = \text{lin.sp.}\{a_1, \ldots a_k\}$. Then W is a subspace of V and $\{a_1, \ldots, a_k\}$ is a basis for W and we may replace W by V (making $k = n$). Suppose $\{a_1, \ldots, a_n\}$ is a basis for V. Let $\{e_1, \ldots, e_n\}$ be an orthonormal basis for V. Then there exists a unique linear operator A so that $A(e_i) = a_i$, for all i. Hence for all i, j, $\langle Ae_i, Ae_j \rangle = \langle a_i, a_j \rangle$. But $\langle Ae_i, Ae_j \rangle = \langle A^t Ae_i, e_j \rangle$. Let $B = A^t A$.

Then $\det B = \det^2 A$, and since A is invertible $\det A \neq 0$. Therefore $\det B > 0$. On the other hand, since $\langle Be_i, e_j \rangle = \langle a_i, a_j \rangle$ and the e_i are orthonormal $\langle Be_i, e_j \rangle = b_{i,j}$ and so $\det B = G$. $\qquad\qquad \square$

We remark that, appropriately modified, this argument works equally well in the complex Hermitian case. In particular, this gives an alternative proof of the more difficult Schwarz inequality in the complex Hermitian case and, at the same time, unifies the real symmetric and complex Hermitian cases.

For our next theorem we need two lemmas.

Lemma 4.8.3. *Let $\{a_1, \ldots, a_k\}$ be a basis for a space W, $\{e_1, \ldots, e_k\}$ be an orthonormal basis of W and $A(e_i) = a_i$ for all $i = 1 \ldots, k$. Assume the matrix of $B = A^t A$ with respect to $\{e_1, \ldots, e_k\}$ is diagonal. Then for $x \in W$,*

$$\| x \|^2 = \sum_{1=1}^{k} (\langle a_i, x \rangle)^2.$$

Proof. Apply the polar decomposition theorem given above to $A = OP$, where O is orthogonal and P is positive definite symmetric. Then $P(e_i) = O^{-1}(a_i)$. Replacing $\{a_1, \ldots, a_k\}$ by $\{O^{-1}a_1, \ldots, O^{-1}a_k\}$ and x by $O^{-1}x$ we see that we can assume $A = P$. That is, A is positive definite and symmetric. Hence the original equation is,

$$\| x \|^2 = \frac{\sum_{i=1}^{k} (\langle e_i, A(x) \rangle)^2}{a_{i,i}^2}.$$

Now the e_i's form an orthonormal basis of eigenvectors of A. Hence,

$$\| x \|^2 = \sum_{i=1}^{k} \langle e_i, x \rangle^2 = \sum_{i=1}^{k} x_i^2.$$

On the other hand,

$$\frac{\sum_{i=1}^{k} (\langle e_i, A(x) \rangle)^2}{a_{i,i}^2} = \frac{\sum_{i=1}^{k} (a_{i,i}^2 x_i^2)}{a_{i,i}^2}$$

is also $\sum_{i=1}^{k} x_i^2$. $\qquad\qquad \square$

Lemma 4.8.4. *Let W be a subspace of V and P_W be the orthogonal projection of V onto W. Then*

$$\| x - P_W(x) \|^2 = \langle x, x - P_W(x)\rangle.$$

Proof.

$$\| x - P_W(x) \|^2 = \langle x - P_W(x), x - P_W(x)\rangle$$
$$= \| x \|^2 - 2\langle x, P_W(x)\rangle + \langle P_W(x), P_W(x)\rangle.$$

Since P_W is a projection, $\langle x - P_W(x), P_W(x)\rangle = 0$ and so $\langle x, P_W(x)\rangle = \langle P_W(x), P_W(x)\rangle$ and therefore $\| x - P_W(x) \|^2 = \langle x, P_W(x) - x\rangle$.   □

We shall now find a formula for the distance of a point $x \in V$ to a subspace W of V in terms of Gramians. We denote this distance by $d(x, W)$.

Theorem 4.8.5. *Let $\{a_1, \ldots, a_k\}$ be a basis of W. Then*

$$d(x, W)^2 = \frac{G(a_1, \ldots, a_k, x)}{G(a_1, \ldots, a_k)}.$$

Proof. If x lies in W, then $\{a_1, \ldots, a_k, x\}$ are linearly dependent. Therefore the right side is 0. But then so is the left. Hence we can assume $x \in V - W$.

Let P_W be the orthogonal projection of V onto W. Then by Lemma 4.8.4 $d(x, W)^2 = \| x - P_W(x) \|^2 = \langle x, x - P_W(x)\rangle$.

We first consider the case when the a_i form an orthonormal basis of W. Then $G(a_1, \ldots, a_k) = 1$ and, as is easily seen,

$$G(a_1, \ldots, a_k, x) = \langle x, x\rangle - \sum_{i=1}^{k} x_i^2.$$

On the other hand, $\langle x, x - P_W(x)\rangle = \langle x, x\rangle - \langle x, P_W(x)\rangle$. But, as one checks easily, $\langle x, P_W(x)\rangle = \sum_{i=1}^{k} x_i^2$, proving the theorem in this case.

Now, $d(x, W)^2$ is an *intrinsic* notion independent of the basis chosen for W. Let e_i be an orthonormal basis and the a_i just be a basis of W.

As above, choose A so that $Ae_i = a_i$ for all i. Then, $A \in \mathrm{GL}(W)$ and $\langle Ae_i, Ae_j \rangle = \langle a_i, a_j \rangle = \langle Be_i, e_j \rangle$, where $B = A^t A$. Then $\langle Be_i, e_j \rangle = \sum_{l=1}^{k} b_{i,l} \langle e_l, e_j \rangle = b_{i,j}$. So here the matrix M for the e_i is just B and, of course, $G(a_1, \ldots, a_k) = \det B > 0$.

However, we can do slightly better by choosing a (perhaps different) orthonormal basis that makes B diagonal. This is because if O is orthogonal, it takes e_i to another orthonormal basis and we get all such as O varies over the orthogonal group. Hence, we can choose A so that $AOe_i = a_i$ for all i. But then $(AO)^t AO = O^t A^t AO = O^{-1} BO$ and since B is self adjoint, it is diagonal in a well chosen orthonormal basis. Using the fact that B is diagonal, a direct calculation using induction on k shows,

$$G(a_1, \ldots, a_k, x) = \det B \langle x, x \rangle - \sum_{i=1}^{k} b_{1,1} \ldots \widehat{b_{i,i}} \ldots b_{k,k} \langle a_i, x \rangle^2.$$

Hence,

$$\frac{G(a_1, \ldots, a_k, x)}{G(a_1, \ldots, a_k)} = \langle x, x \rangle - \sum_{i=1}^{k} \frac{\langle a_i, x \rangle^2}{b_{i,i}}.$$

Since $P_W(x)$ lies in W, by Lemma 4.8.3

$$\| P_W(x) \|^2 = \sum_{1=1}^{k} (\langle a_i, x \rangle)^2.$$

Hence, $\frac{G(a_1,\ldots,a_k,x)}{G(a_1,\ldots,a_k)}$ equals the same expression with the a_i replaced by the e_i and therefore, by what we already know, this proves the theorem.
□

Now let $\{a_1, \ldots, a_k\}$ be a basis of W. We define the *polyhedron* determined by $\{a_1, \ldots, a_k\}$ to be

$$\mathcal{P}(a_1, \ldots, a_k) = \left\{ \sum_{i=1}^{k} \lambda_i a_i : 0 \le \lambda_i \le 1, i = 1 \ldots, k \right\}.$$

We conclude this section with the *volume formula* for polyhedra.

Corollary 4.8.6. *Let* $\{a_1, \ldots, a_k\}$ *be a basis of* W. *Then*

$$\text{vol}(\mathcal{P})^2(a_1, \ldots, a_k) = G(a_1, \ldots, a_k).$$

Proof. We prove this by induction on k. When $k = 1$ it just says $\| a_1 \|^2 = \langle a_1, a_1 \rangle$. Now suppose the result holds for $k - 1$ and consider $\mathcal{P}(a_1, \ldots, a_k)$. Its volume is the product of the perpendicular distance from a_k to the base times the $\text{vol}_{k-1} \mathcal{P}(a_1, \ldots, a_{k-1})$. But by induction $\text{vol}_{k-1} \mathcal{P}(a_1, \ldots, a_{k-1})^2 = G(a_1, \ldots, a_{k-1})$. Hence $\text{vol}(\mathcal{P})(a_1, \ldots, a_k) = d(a_k, W)\sqrt{G(a_1, \ldots, a_{k-1})}$. But by the previous theorem,

$$d(a_k, W) = \sqrt{\frac{G(a_1, \ldots, a_{k-1}, a_k)}{G(a_1, \ldots, a_{k-1})}}.$$

Hence $\text{vol}(\mathcal{P})(a_1, \ldots, a_k) = \sqrt{G(a_1, \ldots, a_k)}$. $\qquad\square$

Exercise 4.8.7. The complex (Hermitian) case of the Gramian is similar to the real one and we leave as an exercise the formulation and proofs of the analogous results to the reader.

4.9 Schur's Theorems and Eigenvalue Estimates

In this section (which shows the Jordan canonical form of Chapter 6 is not the be all and end all) we prove *Schur's triangularization theorem* and *Schur's inequality*, each of these yielding alternative characterizations of a *normal operator* and the finally we get estimates for $\sum_{i=1}^{n} |\lambda_i|^2$ and $|\lambda_i|$, where the λ_i are the eigenvalues of an $n \times n$ complex matrix A.

Notice however, that in the theorem below because A is arbitrary, although the conjugator is unitary, we do not conjugate A to a diagonal matrix, but merely a triangular one.

Theorem 4.9.1. *(Schur's triangularization theorem) Let A be an $n \times n$ complex matrix (with eigenvalues λ_i). Then there exists a unitary matrix U with $UAU^{-1} = T$, where T is upper triangular. (If A is a real*

*matrix and the λ_i are also real, then U can be chosen real orthogonal).
A is normal if and only if T is diagonal.*

Proof. We prove this by induction on n, the case of $n = 1$ being trivially true. Let v be an eigenvector of A with eigenvalue λ_1, which we can normalize i.e. $\| v \| = 1$. By Gram-Schmidt extend this to an orthonormal basis of the space and let V be any unitary matrix with v as its first column. Then

$$\begin{vmatrix} \lambda_1 & Y \\ 0 & A_1 \end{vmatrix} = V^{-1}AV,$$

where $A_1 \in M_{n-1}(\mathbb{C})$ and Y is a row vector in $\mathbb{C}^{n-1}$.

By induction, choose a unitary matrix U_1 of order $n - 1$ so that $U_1 A_1 U_1^{-1} = T_1$, where T_1 is upper triangular. Let W be the n-th order matrix of the form

$$\begin{vmatrix} 1 & 0 \\ 0 & U_1 \end{vmatrix},$$

and $U = V \cdot W$. Then, U is clearly unitary and

$$\begin{vmatrix} \lambda_1 & Y_1 \\ 0 & T_1 \end{vmatrix} = UAU^{-1}.$$

As this matrix is upper triangular this proves the first statement.

When A is normal, the Spectral Theorem 4.6.11 shows UAU^{-1} is diagonal for some unitary U. If T is diagonal, then $A = U^{-1}TU$, which is normal. Evidently, if A and all the λ_i are real, the argument shows U can be taken orthogonal.   □

Of course if A is normal, since we are over $\mathbb{C}$ this block triangular matrix is block diagonal.

We now turn to Schur's inequality. Here, as above, $A = (a_{i,j}) \in M_n(\mathbb{C})$.

Corollary 4.9.2. *(Schur's inequality)*

$$\sum_{i=1}^{n} |\lambda_i|^2 \leq \sum_{i,j=1}^{n} |a_{i,j}|^2,$$

with equality if and only if A is normal.

Proof. By Theorem 4.9.1 choose a unitary matrix U with $UAU^* = T$. Then, $T^* = UA^*U^*$, so that $TT^* = UAA^*U^{-1}$, so $\operatorname{tr}(TT^*) = \operatorname{tr}(AA^*)$. Now, for any matrix $X = (x_{i,j})$, $\operatorname{tr}(XX^*) = \sum_{i,j=1}^{n} |x_{i,j}|^2$. Hence, $\operatorname{tr}(AA^*) = \sum_{i,j=1}^{n} |a_{i,j}|^2$, so by the theorem above,

$$\sum_{i,j=1}^{n} |a_{i,j}|^2 = \sum_{i=1}^{n} |\lambda_i|^2 + \sum_{i \neq j} |t_{i,j}|^2,$$

proving the inequality. Equality occurs if and only if $t_{i,j} = 0$ for $i \neq j$. That is, if and only if T is diagonal. Thus, as above, if and only if A is normal. $\square$

Exercise 4.9.3. The reader should check what happens in the corollary above in the special cases when

1. A is real skew-symmetric.

2. A is complex skew-symmetric.

3. A is real symmetric.

4. A is complex symmetric.

5. A is complex skew-Hermitian.

6. A is complex Hermitian.

We now come to some consequences of Schur's inequality. These are the inequalities of Hirsh-Bendixson. The notation is as in the present section, but in addition $a = \max |a_{i,j}|$, $b = \max |b_{i,j}|$ and $c = \max |c_{i,j}|$, the matrices B and C to be defined shortly.

Corollary 4.9.4.

1. $|\lambda_i| \leq na$.

2. $|\Re(\lambda_i)| \leq nb$.

3. $|Im(\lambda_i)| \leq nc$.

Proof. By the proof of Theorem 4.9.1 we get a unitary operator U satisfying $U(A \pm A^*)U^{-1} = T \pm T^*$, that is,

$$U(\frac{A + A^*}{2})U^{-1} = \frac{T + T^*}{2}$$

and

$$U(\frac{A - A^*}{2})U^{-1} = \frac{T - T^*}{2}.$$

Letting $B = \frac{A+A^*}{2}$ and $C = \frac{A-A^*}{2}$, we get $UBU^{-1} = \frac{T+T^*}{2}$ and $UCU^{-1} = \frac{T-T^*}{2}$. Hence,

$$\sum_i^n |\frac{\lambda_i + \lambda_i^-}{2}|^2 + \sum_{i<j} |t_{i,j}|^2 = \sum_{i,j=1}^n |b_{i,j}|^2$$

and

$$\sum_i^n |\frac{\lambda_i - \lambda_i^-}{2}|^2 + \sum_{i<j} |t_{i,j}|^2 = \sum_{i,j=1}^n |c_{i,j}|^2.$$

In particular,

$$\sum_i^n |\frac{\lambda_i + \lambda_i^-}{2}|^2 \leq \sum_{i,j=1}^n |b_{i,j}|^2$$

and

$$\sum_i^n |\frac{\lambda_i - \lambda_i^-}{2}|^2 \leq \sum_{i,j=1}^n |c_{i,j}|^2.$$

Therefore,

$$\sum_i^n |\Re(\lambda_i)|^2 \leq \sum_{i,j=1}^n |b_{i,j}|^2 \leq n^2 b^2$$

and

$$\sum_i^n |\text{Im}(\lambda_i)|^2 \leq \sum_{i,j=1}^n |c_{i,j}|^2 \leq n^2 c^2.$$

Of course, from Schur's inequality we also have

$$\sum_i^n |\lambda_i|^2 \leq n^2 a^2.$$

From each of these last three inequalities the corresponding conclusion
follows readily.   □

Exercise 4.9.5. Prove if A is real, then $|\text{Im}(\lambda_i)| \leq c\sqrt{\frac{(n)(n-1)}{2}}$.

4.10   The Geometry of the Conics

4.10.1   Polarization of Symmetric Bilinear Forms

Before turning to the action of the affine group on conics we shall briefly
discuss the concept of *polarization.*

   Let V be a finite dimensional vector space over $k = \mathbb{R}$ and β :
$V \times V \longrightarrow k$ be a symmetric bilinear map on V. For the general
definition of bilinearity see Chapter 1 of Vol.II . However, the example
we have in mind is the inner product, $\langle \cdot, \cdot \rangle$ on V.

Proposition 4.10.1. *If the characteristic of k is $\neq 2$, for x, y in V we
have the following polarization identity:*

$$\beta(x,y) = \frac{1}{2}[\beta(x+y, x+y) - \beta(x,x) - \beta(y,y)].$$

*(As we will see in Chapter 5, a field of characteristic $\neq 2$ means
$1 + 1 \neq 0$.)*

Proof. Using the fact that β is bilinear and symmetric,

$$\beta(x+y, x+y) = \beta(x,x) + \beta(x,y) + \beta(y,x) + \beta(y,y)$$
$$= \beta(x,x) + \beta(y,y) + 2\beta(x,y).$$

Hence the conclusion.   □

Definition 4.10.2. Given a β as above we call the associated *quadratic
form* on V the function $q : V \longrightarrow k$ given by $q(x) = \beta(x,x)$.

   Thus if $\beta : V \times V \to k$ is a *symmetric* bilinear form, where k is a field
of characteristic $\neq 2$ and $q(x)$ is the associated quadratic form, then β
not only determines q, but by the polarization identity, is determined
by q.

   We leave the proof of the following proposition as an exercise.

Proposition 4.10.3. *If q is a quadratic form on V then*

$$q(0) = 0, \quad and \quad q(\lambda x) = \lambda^2 q(x)$$

for any $\lambda \in k$ and $x \in V$.

The reader should also keep in mind an important variant of polarization in the case of the complex field given in the section on Unitary and Orthogonal Operators of this chapter. (See 4.3.)

4.10.2 Classification of Quadric Surfaces under Aff(V)

Although one mostly deals with linear equations and systems of such. Here we shall deal with the case of a single quadratic polynomial equation of a vector variable. This is the very beginning of the subject of *affine algebraic geometry* where one studies systems of polynomial equations with coefficients in a field k on V of arbitrary degree. Here we shall also restrict our attention to the real field, $\mathbb{R}$. We note that systems of non-homogeneous linear equations do play a role in our analysis.

Recall the inner product space structure $\langle \ , \ \rangle$ on V and for $x \in V$, consider $F(x) = q(x) + l(x) + c = 0$, where $q(x)$ is the homogenous quadratic function $\langle Ax, x \rangle$, $l(x) = \langle b, x \rangle$ is a linear form on V and c is a constant. We seek to find a normal form for F (and hence for the locus of the equation $F = 0$) under Aff(V). Here A is an $n \times n$ matrix, $b \in V$ is a vector and $c \in k$.

First, since $q(x) = \langle Ax, x \rangle$, we may assume that A is symmetric. This is because each term is of the form $a_{ij}x_ix_j$. Since $x_ix_j = x_jx_i$, we can gather these terms and get $(a_{ij}+a_{ji})x_ix_j$. This is just $b_{ij}x_ix_j$, where $B = (b_{ij})$ is symmetric. Now just rename B as A. (This is polarization). We shall initially assume A is invertible which is sometimes referred to as *non-degeneracy*.

Second, we translate F by a $v \in V$. Then, we get $F(x + v) = q(x + v) + l(x + v) + c$. Since $q(x + v) = \langle A(x + v), x + v \rangle = q(x) + \langle Ax, v \rangle + \langle Av, x \rangle + q(v) = q(x) + \langle 2Ax, v \rangle + q(v)$. Now since

$$l(x + v) = \langle b, x + v \rangle = \langle x, b \rangle + \langle b, v \rangle,$$

the linear term becomes $\langle x, 2Av+b \rangle$. Choose $v \in V$ so that $2Av+b = 0$. That is, take $v = -\frac{1}{2}A^{-1}(b) \in V$. This can be done since A is invertible. Thus in this case we can translate in V to make $l(x) = 0$, if we were willing to take a different c. But, since we do not know what c is anyway, this is not much of a loss.

We have shown that in a suitable basis $q(x) = \sum_{i=1}^{n} \mu_i x_i^2$ (eigenvalues of A) using the orthogonal group only. When we bring in $GL(V)$, under the action gAg^t, properties of the field, k, enter. When if $k = \mathbb{R}$ (since $\langle \, , \, \rangle$ is positive definite), $q(x) = \sum_{i=1}^{r} x_i^2 - \sum_{i=r+1}^{p} x_i^2$, where $1 \le r \le p \le n$. But since here A is non-degenerate, $r = n$. Finally if $c \ne 0$, we can normalize c by changing scale in the range so that $c = -1$. Thus for non-degenerate F over $\mathbb{R}$ we get,

$$\sum_{i=1}^{r} x_i^2 - \sum_{i=r+1}^{n} x_i^2 = 0, \text{ or } 1.$$

So, for example, when $n = 2$ we get ellipses and degenerate ellipses (points), hyperbolas and degenerate hyperbolas (a pair of intersecting lines). But what we do not get is the parabola and degenerate parabola (a line). While when $n = 3$ and A is non-degenerate, $c = 1$ gives ellipsoids and the hyperboloid of 2 sheets while $c = 0$ gives the cone and a point. What we are missing here is the hyperbolic and elliptic paraboloid and hyperboloid of 1 sheet.

What do we need to provide the missing figures? We must have A of rank $< n$. When $n = 2$ consider $F(x, y) = x^2 - y$. Thus the A is of rank 1, $l(x, y) = y$ and $c = 0$. Then we get the parabola $y = x^2$. If $F(x, y) = y$, $A = 0$ and so has rank 0, $l(x, y) = y$ and $c = 0$. This gives the vertical line, $y = 0$, a degenerate parabola.

When $n = 3$ the situation is more complicated. If rank $A = 3$, in addition to the ellipsoid (and the point), we can take $F(x, y, z) = x^2 + y^2 - z^2$, $l = 0$ and $c = \pm 1$. If $z = \pm\sqrt{1 - (x^2 + y^2)}$, this is the hyperboloid of 2 sheets. If $z = \pm\sqrt{1 + (x^2 + y^2)}$ it is the hyperboloid of 1 sheet. Taking $c = 0$ gives the cone. If rank $A = 2$, take $F(x, y, z) = x^2 \pm y^2$, $l(x, y, z) = z$ and $c = 0$. Then the locus is either $z = x^2 + y^2$, or $z = y^2 - x^2$. These are, respectively, the elliptic paraboloid and the

hyperbolic paraboloid. (When rank $A = 1$, take $F(x, y, z) = x^2$ then
we have developable surfaces and when $A = 0$ we have a plane).

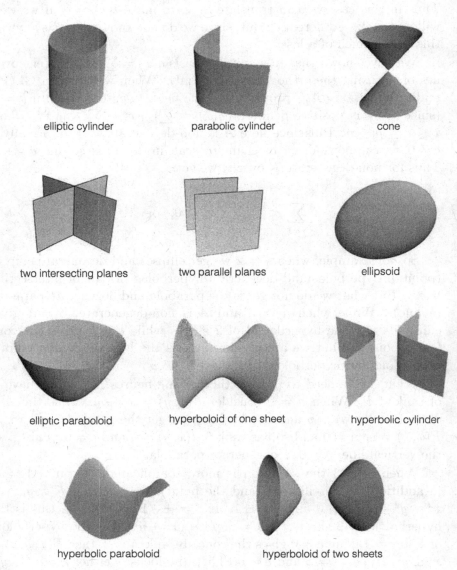

elliptic cylinder   parabolic cylinder   cone

two intersecting planes   two parallel planes   ellipsoid

elliptic paraboloid   hyperboloid of one sheet   hyperbolic cylinder

hyperbolic paraboloid   hyperboloid of two sheets

In order to get the complete list in general we go back to the effect of translation on F by a $v \in V$: $F(x+v) = q(x+v) + l(x+v) + c$, where $q(x+v) = q(x) + \langle 2Ax, v \rangle + q(v)$ and $l(x+v) = \langle x, b \rangle + \langle b, v \rangle$. Thus,

$$F(x+v) = q(x) + (l(x) + \langle x, 2Av \rangle) + (q(v) + l(v) + c). \qquad (4.1)$$

From equation 4.1 we learn when F is invariant under a translation, T_v. This occurs if and only if $Av = 0$ and $q(v) + l(v) = 0$. But $q(v) = \langle Av, v \rangle$ and since $Av = 0$, $q(v) = 0$. This leads easily to the following proposition and its corollary.

Proposition 4.10.4. *F is invariant under T_v if and only if $Av = 0$ and $\langle b, v \rangle = 0$.*

Letting $\operatorname{Ker} q = \{v \in V : Av = 0\}$, since $\operatorname{Ker} l = \{v \in V : \langle b, v \rangle = 0\}$ we get,

Corollary 4.10.5. *F is invariant under T_v if and only if v belongs to the subspace $\operatorname{Ker} l \cap \operatorname{Ker} q$.*

Our purpose here is to prove a uniqueness theorem, namely, that the normal forms that we have found so far actually comprise *all* quartics.

Definition 4.10.6. Suppose we could find a point $o \in V$ so that for all $v \in V$, $F(o+v) = F(o-v)$. We call such an o a *center* for F.

Proposition 4.10.7. *o is a center for F if and only if $b = -2A(o)$. In particular, if A is non-degenerate, then o is unique because $o = -\frac{1}{2}A^{-1}(b)$. Conversely, if A is degenerate, then o is not unique.*

Proof. Using equation 4.1 we get

$$q(o) + (l(o) + \langle o, 2Av \rangle) + (q(v) + l(v) + c)$$
$$= q(o) + (l(o) - \langle o, 2Av \rangle) + (q(v) - l(v) + c).$$

Hence, $2\langle o, 2Av \rangle + 2l(v) = 0$ for all $v \in V$. Since $l(v) = \langle b, v \rangle$ and A is symmetric $\langle 2A(o) + b, v \rangle = 0$ for all v. Hence $2A(o) + b = 0$ and $b = -2A(o)$. The converse also holds because all steps are reversible. $\square$

Exercise 4.10.8. Let $F(x) = \langle Ax, x \rangle + \langle b, x \rangle + c$, where A is symmetric, b is a vector and c is a constant. Then for an arbitrary point of $o \in V$, $\operatorname{grad} F(o) = b + 2A(o)$. Suggestion: $\frac{\partial q}{\partial x_i}(o) = 2\sum_j a_{i,j} x_j$.

Corollary 4.10.9. *o is a center for F if and only if $\operatorname{grad} F(o) = 0$.*

Proof. Since $F(o + x) = F(o - x)$ for all $x \in V$ we see

$$\frac{F(o + X_i) - F(o)}{x_i - o_i} = \frac{F(o - X_i) - F(o)}{x_i - o_i},$$

where X_i is the vector $X_i = (0, \ldots, 0, x_i, 0 \ldots 0)$. Since this last term is $-\frac{F(o) - F(o - X_i)}{x - i - o_i}$. Taking limits as $x_i \to o_i$ tells us $\frac{\partial F}{\partial x_i}(o) = 0$ for every i. Thus $\operatorname{grad} F(o) = 0$. For the converse, since $\operatorname{grad} F(o) = b + 2A(o)$. Hence if $\operatorname{grad} F(o) = 0$, $b = -2A(o)$ and F has a center at o by Proposition 4.10.7. $\qquad\square$

The set of centers for F is determined by the system of *non-homogeneous* linear equations $b + 2A(x) = 0$ the solution set of which is either a translate of a subspace, or the empty set (and as we saw earlier if A is non-degenerate there is a unique center). Let

$$\mathcal{Z}(F) = \{x \in V : F(x) = 0\}.$$

$\mathcal{Z}(F)$ is the quadric hypersurface we are interested in. It is an affine algebraic variety. A point $o \in V$ is called a *vertex* of $\mathcal{Z}(F)$ if it is a center and lies in $\mathcal{Z}(F)$. We call a *quadric central* if it has at least one center.

Lemma 4.10.10. *If a line meets a quadric in at least 3 points, then it lies wholly within it.*

Proof. Let $o + tx$ be the line, where $t \in K$. Then $F(tx) = q(tx) + l(tx) + c = t^2 q(x) + tl(x) + c = \psi(t)$. The intersection of $\mathcal{Z}(F)$ with the line is precisely the zeros of ψ. But since ψ is a polynomial in t of degree ≤ 2 it has at most 2 roots. This is impossible unless ψ is identically zero which means the line lies wholly within $\mathcal{Z}(\mathcal{F})$. $\qquad\square$

Corollary 4.10.11. *Let $\mathcal{Z}(F)$ be a quadric and o be a vertex. If $x \neq o$ lies on $\mathcal{Z}(F)$, then the whole line ox does also.*

Proof. Let $o + tx$ be the line. Then o and $o + x \in \mathcal{Z}(F)$, but then so does $o - x$ giving 3 distinct points on the variety. $\square$

In particular, given a quartic F, it is not possible that every one of the points of $\mathcal{Z}(F)$ is a vertex. For if all points x were, then $\mathcal{Z}(F)$ would contain the line through any two of its distinct points. This would mean $\mathcal{Z}(F)$ is an affine subspace, that is, $a + W$, where W is a subspace of V. But then $q = 0$, contradicting the definition of F.

Definition 4.10.12. A quadric $\mathcal{Z}(F)$ is called a *cone* if it has a vertex o and for every $x \in \mathcal{Z}(F)$ the line ox lies entirely within $\mathcal{Z}(F)$.

We now come to an important point in algebraic geometry, generally. Namely, to what extent does the variety determine the equation(s)? Or how many different equations can give the same variety? The answer usually depends on algebraic properties of the polynomial algebra.

Here we show that the variety $\mathcal{Z}(F)$ essentially depends uniquely on F. Of course, if F_1 and F_2 are proportional, they determine the same variety. As we shall see, the point here is that $K[x_1, \ldots x_n]$ has no zero divisors (see Chapter 5). We shall do more with this at the end of Chapter 9.

Theorem 4.10.13. *Suppose $\mathcal{Z}(F_1) = \mathcal{Z} = \mathcal{Z}(F_2)$. Then F_1 and F_2 are proportional.*

Proof. Let o be a point in $\mathcal{Z}$ which is not a vertex. Take o as origin. Then $F_1(x) = q_1(x) + l_1(x)$ and $F_2(x) = q_2(x) + l_2(x)$, where l_1 and l_2 are both $\neq 0$. For if, say $l_1 = 0$, then

$$\mathcal{Z} = \mathcal{Z}(q_1) = \{x \in V : x_i^2 + \ldots x_p^2 - x_{p+1}^2 - \ldots - x_r^2 = 0\},$$

which is a cone and hence contains many lines through 0, unless q_1 is positive definite. In that case $\mathcal{Z} = \mathcal{Z}(q_1) = (0)$.

Case 1. $\mathcal{Z}(F_2)$ contains lines through 0. Then since $F_2(x) = l_2(x) + q_2(x)$, on $\mathcal{Z}$ we have $tl_2(x) + t^2 q_2(x) = 0$ for all t. But then we have infinitely many roots of this equation. Therefore, the coefficients must both be zero; $l_2(x) = 0 = q_2(x)$ and so $\mathcal{Z}(F_2) = \mathcal{Z} = V$, a contradiction since we know it to be a cone.

Case 2. $\mathcal{Z} = \mathcal{Z}(q_1) = (0)$. This cannot occur because o is a point in $\mathcal{Z}$ which is not a vertex. But if $\mathcal{Z} = (0)$ there can be no such point.

Since both l_1 and l_2 are both non-zero linear functionals they each only take the value 0 at $x = 0$. Continuing the proof, we observe that points of intersection of the line $o + tx$ with $\mathcal{Z}$ are determined by either $tl_1(x) + t^2 q_1(x) = 0$, or $tl_2(x) + t^2 q_2(x) = t(l_2(x) + tq_2(x)) = 0$. Because the solutions of these equations for t must coincide, we get

$$\frac{-q_1(x)}{l_1(x)} = \frac{-q_2(x)}{l_2(x)}.$$

All this takes place in the ring of rational functions $K(x)$. Hence $l_1(x)q_2(x) = l_2(x)q_1(x)$ (in the polynomials $K[x]$), and so multiplying both sides of this equation by $l_1(x)l_2(x)$ we get the equation (in $K[x]$),

$$q_1(x)l_2(x)l_1(x)l_2(x) = q_2(x)l_1(x)l_1(x)l_2(x).$$

Since $k[x]$ has no zero divisors we can cancel terms giving,

$$l_1(x)q_2(x) = l_2(x)q_1(x), \tag{4.2}$$

this time for all $x \in V$. Now there are again two cases.

Case 1. l_1 and l_2 are proportional. Suppose $l_1 = kl_2$, where $k \neq 0 \in K$. Then from equation 4.2, canceling l_2 which is not a zero divisor we get $q_1 = kq_2$ and so $F_1 = kF_2$.

Case 2. l_1 and l_2 are not proportional. Then $\dim V^* = \dim V \geq 2$. Perform a change of basis in V^* so that $l_1(x) = x_1$ and $l_2(x) = x_2$. Equation 4.2 tells us $q_1(x)x_2 = q_2(x)x_1$. Hence there is some $l(x)$ so that $q_1(x) = l(x)x_1$ and $q_2(x) = l(x)x_2$, where $l(x)$ is linear because $l(0) = 0$ and $\deg l = 2 - 1 = 1$. This means $F_1(x) = (l(x) + 1)x_1$ and $F_2(x) = (l(x)+1)x_2$. From the first of these equations we see $\mathcal{Z}$ contains the hyperplane $x_1 = 0$, while from the second we see this is not so, a contradiction. $\square$

The following is an important corollary of the theorem.

Corollary 4.10.14. *If $\mathcal{Z}(F)$ is invariant under a parallel translation T_v, then F is also invariant under T_v.*

Proof. By assumption $\mathcal{Z}(F) = \mathcal{Z}(G)$, where $G(x) = F(x + v)$. Hence by the theorem $G = kF$, where $k \in K$ and $k \neq 0$. Comparing 2nd degree terms we see $k = 1$ and $G = F$. This means F is invariant under T_v. $\qquad\qquad\qquad\qquad\qquad\qquad\qquad\qquad\qquad\qquad\qquad\quad$ □

We can now complete our analysis of quartics. Suppose $\operatorname{Ker} l \cap \operatorname{Ker} q \neq (0)$. Denote this subspace by U. Let $u \in U$ and $x \in \mathcal{Z}(F)$. Since $u \in U$, as we know $F(x + u) = F(x)$. But because $x \in \mathcal{Z}(F)$, $F(x) = 0$ and so $F(x + u) = 0$. Hence $\mathcal{Z}(F)$ contains all of $x + U$ for every $x \in \mathcal{Z}(F)$. These are the so called cylindrical quartics. They do not involve the coordinates of U. Leaving them aside we may henceforth assume $\operatorname{Ker} l \cap \operatorname{Ker} q = (0)$. Now such things can have at most one center. For suppose $\mathcal{Z}(F)$ had two centers o and o'. Let r and r' denote the central symmetries about o and o' respectively. Then $r\mathcal{Z}(F) = \mathcal{Z}(F)$ and $r'\mathcal{Z}(F) = \mathcal{Z}(F)$. Hence $rr'\mathcal{Z}(F) = \mathcal{Z}(F)$. But by the Chain rule $d(rr') = d(r)d(r') = (-I)(-I) = I$. It follows that rr' is (a non-trivial) translation. But if an affine transformation $T = (g, v)$ has derivative I_{n+1} at every point, then $d(g) = I_n$. But since g is linear $d(g) = g$. Hence $g = I_n$ and $T = T_v$. This means $\operatorname{Ker} l \cap \operatorname{Ker} q \neq (0)$, a contradiction.

There are various ways non-cylindrical quartics can occur which we already know. Taking the origin as center we have $q(x_1, \ldots x_n) = 1$ and $q(x_1, \ldots x_n) = 0$, where q is a non-degenerate quadratic function.

We now come to the non-central quartics. Here $\operatorname{Ker} l \cap \operatorname{Ker} q = (0)$, but $\operatorname{Ker} q \neq (0)$. For if $\operatorname{Ker} q = (0)$, then A would be non-degenerate and we would be back in the case we resolved at the beginning! We may also assume $b \neq 0$ since this we have also dealt with earlier. It follows that $\dim \operatorname{Ker} l = n - 1$. Since $\operatorname{Ker} q \neq (0)$ we see that actually, $V = \operatorname{Ker} l \oplus \operatorname{Ker} q$ and hence $\dim \operatorname{Ker} q = 1$. Choose an origin lying on the quadric and a basis of V taken from $\operatorname{Ker} l$ and $\operatorname{Ker} q$. Then in these coordinates suitably normalized $q_{\operatorname{Ker} l}(x_1, \ldots, x_{n-1}) = x_n$, where $q_{\operatorname{Ker} l}$ is the restriction of q to $\operatorname{Ker} l$, giving all the others.

Chapter 5

Rings, Fields and Algebras

5.1 Preliminary Notions

In this chapter we will deal with rings with identity, both commutative and non-commutative and their important special cases, namely, fields and algebras. These ideas are fundamental to algebra and will be exposed both in elementary form in this chapter and in more advanced form in Chapter 9 in the commutative case and in Chapter 13 in the non-commutative case, while in Chapters 11 and 12 both commutative and non-commutative algebra come into view. The distinction between the commutative and the non-commutative being a fundamental one in algebra and indeed in mathematics as a whole.

Definition 5.1.1. We say that $(R, +, \cdot)$ is a *ring* if the following conditions hold.

1. $(R, +)$ is an Abelian group.

2. $(R, \cdot)$ is a binary associative operation satisfying $a(b+c) = ab + ac$ and $(b + c)a = ba + ca$. These last conditions are called the *distributive law*.

3. When the binary operation is commutative we call R commutative, otherwise R is non-commutative.

Definition 5.1.2. A *ring with identity* is a ring R where there is an element $1 \in R$ such that $a \cdot 1 = 1 \cdot a = a$, for all $a \in R$.

Example 5.1.3. The reader should check that each of the following examples are rings with any additional features that are mentioned.

1. The integers $\mathbb{Z}$, under addition and multiplication is a (commutative) ring.

2. $\mathbb{Z}_n$ is a commutative ring with identity.

3. Let R be any commutative ring with identity and define $R[x]$ as the set of all polynomials (of any degree) in one variable, with the usual operations.

4. Let R be any commutative ring with identity and define $R[x_1, \ldots, x_n]$ as the set of all polynomials in n variables, with the usual operations.

5. The set $\mathbb{Z}[i] = \{a + ib,\ a, b \in \mathbb{Z}\}$ is a commutative ring with identity, called the *Gaussian integers*.

6. The set $\mathrm{M}(n, R)$ of all $n \times n$ matrices (with $n > 1$ and entries in a commutative ring) is a *non-commutative* ring with identity I.

We ask the reader to check that $\mathbb{Z}$ and $\mathbb{Z}[i]$ are commutative rings with identity.

Exercise 5.1.4. Define the *norm* of a Gaussian integer (or of any complex number) by $N(z) = a^2 + b^2$, where $z = a + bi$. Show $N(z)N(w) = N(zw)$. Prove this implies that $(a^2 + b^2)(c^2 + d^2)$ is a sum of two squares. Thus if we wanted to express an integer as a sum of two squares it would be sufficient to do this for each of its prime factors.

We leave the proof of the following proposition as an exercise.

Proposition 5.1.5. *Let R be a ring and a, b and $c \in R$. Then:*

1. *If R has a unit element 1, then it is unique.*

2. *For each $a \in R$, $0 \cdot a = a \cdot 0 = 0$. In particular, $1 \neq 0$, unless $R = (0)$. Henceforth when we speak of a ring R with identity, we always mean $R \neq (0)$*

3. *For every a and b in R, $a(-b) = (-a)b = -(ab)$.*

4. *$a(b - c) = ab - bc$, and $(b - c)a = ba - ca$ for all a, b, c in R.*

5. *For every a, b in R and m, n in $\mathbb{Z}$, $m(na) = (mn)a$ $m(ab) = (ma)b = a(mb)$ and $(ma)(nb) = (mn)ab$.*

Definition 5.1.6. Let R be a ring with identity and S a subset of R. We say that S *generates* R if every $r \in R$ can be written as a finite sum of finite products of elements of S. We call S a *generating set* for R. If S is a finite set, we say the ring R is *finitely generated*.

So for example if R is a ring, then $R[x_1, \ldots, x_n]$ is generated by R and the x_i. However, when we want to suppress the role of R as part of the generating set sometimes we say $R[x_1, \ldots, x_n]$ is generated over R by the x_i, or $R[x_1, \ldots, x_n]$ is finitely generated over R.

Definition 5.1.7. Let R be a ring and $S \subseteq R$. We shall call S a *subring* of R is under the restricted operations S is a ring in its own right.

Definition 5.1.8. If S is a subring of R and, in addition, for all $a \in R$, $aS \subseteq$ we say S is a left ideal. Similarly, if $Sa \subseteq$ we say S is a right ideal. If $aSb \subseteq S$ for all $a, b \in R$ we say S is a two-sided ideal in R.

Definition 5.1.9. Let R and S be rings and $f : R \to S$. If f preserves both $+$ and $\cdot$, that is, $f(x + y) = f(x) + f(y)$ and $f(xy) = f(x)f(y)$ we call f a *ring homomorphism*. If f is also bijective we call it a ring isomorphism. Evidentely, isomorphic rings share all ring theoretic properties. If f is merely injective we call f an imbedding and say R is imbedded in S. When R and S are rings with identity we also require that a homomorphism satisfies $f(1) = 1$. This requirment is useful for many reasons among them it exclude the zero homomorphism.

Proposition 5.1.10. *Any ring R without identity can be imbedded as a two-sided ideal in a ring $R^\star$ with identity.*

Proof. Let $R^\star = \mathbb{Z} \oplus R$ (here we view $\mathbb{Z}$ and R as additive groups) and consider a multiplication on $R^\star$ defined by

$$(n, r) \cdot (m, s) := (nm, \; ns + mr + rs).$$

One checks easily that $R^\star$ is a ring with identity (the element $(1, 0)$), and that $\{(0, r) \mid r \in R\}$ is a subring (actually an ideal) in $R^\star$ evidently isomorphic to R as a ring. $\square$

Definition 5.1.11. Let R be a ring with 1. An element $u \in R$ is called a *unit* if there exists $v \in R$ such that $uv = 1$. We denote by G_u be the set of units of R.

Exercise 5.1.12. Let R be a ring with identity. Show G_u is a group. We call it the group of units of R

Definition 5.1.13. Let R be a ring. If $(R \setminus (0) \, , \, \cdot)$ is a group, then we say R is a *skew-field* or a *division ring*. If R is a commutative group, we call it a *field*.

Example 5.1.14. One checks easily that $\mathbb{Q}$, $\mathbb{R}$, $\mathbb{C}$ are fields. As we shall see later, the quaternions, $\mathbb{H}$, is a skew-field.

Exercise 5.1.15.   1. If $R = \mathbb{Z}$, $G_u = \{1, -1\}$.

2. If $R = \mathbb{Z}[i]$, the unit group $G_u = \{\pm 1, \pm i\}$.

3. If $R = R[x_1, \ldots, x_n]$, then G_u is $G_u(R)$

4. If k is a field, or even a skew field, G_u is $k - (0)$.

5. Let S be a commutative ring with 1 and $R = \mathrm{M}(n, S)$. Then G_u is all matrices $A \in \mathrm{M}(n, S)$ with $\det(A)$ and $\det(A)^{-1}$ units in S. The most important cases of this being $S = k$, a field, or $S = \mathbb{Z}$. Then the unit groups are $\mathrm{GL}(n, k) = \{g : \det(g) \neq 0\}$, or $\{g \in \mathrm{M}(n, \mathbb{Z}) : \det(g) = \pm 1\}$, respectively.

As we know (Proposition 5.1.5) in a ring if one of its elements x and y is 0, then $xy = 0$. However, it could happen that $xy = 0$ without either x or y being zero. We then say x and y are *zero divisors*. As we shall see if R is finite and has no zero divisors (except 0), then it is a field. This is the well known Little Wedderburn Theorem. Of course, if R is infinite this is not true as the reader can easily see by considering $\mathbb{Z}$.

Example 5.1.16. In Chapter 3 we learned matrix multiplication is associative and has I as the identity element. One checks easily that the ring $M(n, R)$ has many zero divisors. For example, if

$$N = \begin{pmatrix} 0 & 0 \\ 1 & 0 \end{pmatrix},$$

then $N \neq 0$, but $N^2 = 0$. Since $M(2, R)$ imbeds into $M(n, R)$, this shows $M(n, R)$ has many zero divisors for every n. In particular all this applies when $R = k$, a field.

Exercise 5.1.17. If k is a field, then k has no zero divisors. Even if a ring R is contained in a field it can not have zero divisors. In particular if n is not a prime number, then $\mathbb{Z}_n$ is not a field, since $\mathbb{Z}_n$ evidently has zero divisors.

Proposition 5.1.18. *The ring R has no zero divisors if and only if one can do left and right cancellations, i.e. if a, b, $c \in R$, $a \neq 0$ and $ab = ac$ (respectively $ba = ca$) then $b = c$.*

Proof. $\Longrightarrow$ Let $ab = ac$. This implies $ab - ac = 0$ and so $a(b - c) = 0$. Since R has no zero divisors and $a \neq 0$, $b = c$.
$\Longleftarrow$ Let $ab = 0$. If $a = 0$ we have nothing to prove. If $a \neq 0$, then the fact that $ab = 0 = a0$ tells us $b = 0$. $\qquad\square$

Definition 5.1.19. A commutative ring R with 1 and with no zero divisors is called an *integral domain*. This is where commutative algebra is done.

Proposition 5.1.20. *A finite integral domain R is a field.*

Proof. Let $a \neq 0$ in R. We have to prove that a has an inverse. Consider the map $\varphi : R - \{0\} \longrightarrow R - \{0\}$ defined by $x \mapsto \varphi(x) = ax$. Since $x \neq 0$, $\varphi(x) \neq 0$. Since R is an integral domain, by cancellation the map φ is injective. Because of this and the fact that R is finite it follows that $R - \{0\}$ and $\varphi(R - \{0\})$ have the same number of elements. Therefore φ is surjective. Since $1 \in R - \{0\}$ there must be an $x \in R - \{0\}$ such that $\varphi(x) = 1$, i.e. $ax = 1$ and so $x = a^{-1}$. $\square$

This gives us some more important fields.

Corollary 5.1.21. *For every prime p, $\mathbb{Z}_p$ is a field.*

Proof. If p divides the product of integers ab, then it must divide a or b (or both). This follows immediately from the Fundamental Theorem of Arithmetic 0.5.12. This means $\mathbb{Z}_p$ is an integral domain. Since it is finite it is a field by Proposition 5.1.20. $\square$

Later we shall see non-commutative versions of Proposition 5.1.20, for example, in the Little Wedderburn Theorem.

We conclude our discussion of integral domains with some important definitions. Note the similarity to the case when the ring is $\mathbb{Z}$.

Definition 5.1.22. Let R be an integral domain.

1. Two elements a and b of R such that $a = ub$ for some unit $u \in R$ are said to be *associate* in R.

2. If $r \in R$ is non-zero and is not a unit we say that r is *irreducible* in R if $r = ab$ implies a or b is a unit. Otherwise r is said to be *reducible*.

3. A non-zero element p is said to be *prime* if it is not a unit and $p \mid ab$ implies $p \mid a$ or $p \mid b$ for all $a, b \in R$.

4. We say that R is a *unique factorization domain* (UFD), or simply that R is *factorial*, if every non-zero element $m \in R - (0)$ which is not a unit can be written as a (finite) product of irreducible elements and moreover, this factorization is unique in the sense

that if $t_1 t_2 ... t_k$ and $s_1 s_2 ... s_l$ are two factorizations of m into irreducibles, then $k = l$ and there is a permutation σ of $\{1, ..., k\}$ such that t_i and $s_{\sigma(i)}$ are associates for $1 \leq i \leq k$.

We now come to the concept of a quotient field. It is a direct and straightforward generalization of making fractions from integers.

Proposition 5.1.23. *Any integral domain R can be imbedded in a unique field called its quotient field. It is clearly the smallest field containing R.*

Proof. Consider the set

$$A = \{(a, b) | a, b \in R \text{ with } b \neq 0\}$$

and define

$$(a, b) \sim (c, d) \text{ if and only if } ad = bc.$$

As the reader can easily verify, this is an equivalence relation. Let $\frac{a}{b}$ denote the equivalence class of (a, b) in $k = A/\sim$, and define the operations

$$\frac{a}{b} + \frac{c}{d} = \frac{ad + bc}{bd}$$

and

$$\frac{a}{b} \cdot \frac{c}{d} = \frac{ac}{bd}.$$

One checks that these operations are well defined, and equipped with these operations k becomes a field. Check this!

In addition, we can identify R with the subset

$$R = \left\{ \frac{a}{1} \;\middle|\; a \in R \right\}$$

and $1 \equiv \frac{1}{1}$. The last identification is possible because if $a \neq 0$, then $\frac{a}{b} \times \frac{b}{a} = 1$ (b is $\neq 0$ by definition). $\qquad\square$

We note that commutativity was heavily used in this proof.

Example 5.1.24. When R is the integers $\mathbb{Z}$, then, of course, the quotient field is $\mathbb{Q}$. Whereas if R is the ring of polynomials in one or severable variables over a given field, then the quotient field is the *rational function field* in those same variables and we write $R(x_1, \ldots, x_n)$.

5.2   Subrings and Ideals

Let R be a ring. We defined a subring S of R in 5.1.7. Equivalently, a non-empty subset S of a ring R is a subring of R if whenever a, $b \in S$, $a - b \in S$ and $ab \in S$. We also defined an ideal in 5.1.8. Along the same lines a subset I of R is a two-sided ideal if $a, b \in I$, then $(a - b) \in I$, and if $a \in I$ and $r \in R$, $ar \in I$ and $ra \in I$. When R is commutative we just say I is an ideal.

The significance of a two-sided ideal, or in the case of a commutative ring an ideal, is that these ideals I have the same relationship to rings as normal subgroups have to groups. Namely, they are the kernels of homomorphisms. Or put another way, they make the quotient R/I a ring. Of course in the ring R, $\{0\}$ and R are (trivial) two-sided ideals. An ideal I is called a *proper ideal* if I is not trivial.

Definition 5.2.1. We call the ring R a *simple* ring, if R has only the trivial two-sided ideals.

An important ideal in a ring R is its *center*, $\mathcal{Z}(R)$.

Definition 5.2.2. $\mathcal{Z}(R) = \{z \in R \mid zx = xz \text{ for all } x \in R\}$.

We ask the reader to verify this is a two-sided ideal.

Let $S \subset R$, where R is a ring. Then the subring generated by S, denoted by $\langle S \rangle$, is the set of all finite sums of finite products of elements of S. Similarly, and the ideal $I(S)$ generated by S is

$$I(S) = \left\{ \sum_{i=1}^{n} r_i s_i \mid \text{where } r_i \in R, \ s_i \in S \text{ and } n \text{ any positive integer} \right\}.$$

Exercise 5.2.3.   1. Show that $\langle S \rangle$ is the intersection of all subrings of R containing S. Notice there is such a subring. Thus, $\langle S \rangle$ is the smallest subring of R containing S.

2. Show that $I(S)$ is the intersection of all ideals of R containing S. Notice there is such an ideal. Thus, $I(S)$ is the smallest ideal of R containing S.

Exercise 5.2.4. Let R be a ring with 1. An ideal I is proper if and only if $1 \notin I$.

5.3   Homomorphisms

Slightly earlier we defined a ring homomorphism $f : R \longrightarrow S$ between rings (see 5.1.9).

Definition 5.3.1. Let $f : R \longrightarrow S$ be a ring homomorphism. We define the *kernel* of f as

$$\text{Ker} f = \{x \in R : f(x) = 0\}.$$

Proposition 5.3.2. $\text{Ker} f$ *is a two-sided ideal of* R.

Proof. If $x, y \in \text{Ker} f$, then $f(x + y) = f(x) + f(y) = 0 + 0 = 0$, and so $x + y \in \text{Ker} f$. Now let $x \in \text{Ker} f$ and $y \in R$. In addition, $f(xy) = f(x)f(y) = 0f(y) = 0$. Hence $xy \in \text{Ker} f$ and similarly for yx.   □

Conversely, given a two-sided ideal I in R, there is a homomorphism that has I as its kernel. To prove this we need to define what we mean by a *quotient ring*.

Definition 5.3.3. Let R be a ring and I a two-sided ideal. We define the quotient ring R/I to be

$$R/I = \{x + I, \text{ with } x \in R\}.$$

Multiplication in R/I is given by

$$(x + I)(y + I) = xy + I.$$

This multiplication is well defined. To see that, we have to show that if $(x_1 - x) \in I$ and $(y_1 - y) \in I$, then $(xy - x_1y_1) \in I$.

Indeed, if $(x_1 - x) \in I$, then $x_1 = x + i_1$ for some $i_1 \in I$. Similarly, $y_1 = y + i_2$, $i_2 \in I$. Then, $x_1y_1 = xy + i_1y + xi_2 + i_1i_2$. Therefore, since $(i_1y + xi_2 + i_1i_2) \in I$, $(xy - x_1y_1) \in I$.

With this multiplication and the obvious addition, R/I becomes a ring. The reader should verify that this is so. We remark that if I is merely a left or right ideal, multiplication in R/I is not well defined; only two-sided ideals I can make R/I into a ring. Let $\pi(x) = x + I$, for $x \in R$. Since $\pi : R \to R/I$ is a surjective map called the *canonical homomorphism*.

Evidently we get.

Proposition 5.3.4. *If I is a two-sided ideal in R, then R/I is a ring and π is a surjective ring homomorphism $\pi : R \to R/I$ with $\operatorname{Ker} \pi = I$.*

5.3.1   The Three Isomorphism Theorems

We now come to the three isomorphism theorems for rings. These are completely analogous to the corresponding theorems for groups, with ideals playing the role of normal subgroups and the maps being exactly the same. In view of this and since the group theoretic results were proved in considerable detail, we will leave it to the reader to check these things here.

Theorem 5.3.5. *(First Isomorphism Theorem)* If $f : R \longrightarrow R/I$ is a ring homomorphism, then f induces a ring isomorphism

$$R/\operatorname{Ker} f \cong \operatorname{Im} f.$$

This is essentially the proposition just above.

Theorem 5.3.6. *(Second Isomorphism Theorem)* Let I and J be two-sided ideals of a ring R. Then, there is a ring isomorphism

$$I/(I \cap J) \cong (I + J)/J.$$

Theorem 5.3.7. (*Third Isomorphism Theorem*) *If* $I \subset J$ *are two-sided ideal in* R, *then* J/I *is an ideal in* R/I *and there is a ring isomorphism*

$$(R/I)/(J/I) \cong R/J.$$

We now come to the correspondence theorem which lies behind the third isomorphism theorem.

Theorem 5.3.8. *Let* I *be a two sided ideal of* R. *There is a one-to-one correspondence between the set of all these ideals of* R *which contain* I *and the set of all two sided ideals of* R/I, *given by* $J \to J/I$. *So, every ideal in* R/I *is of the form* J/I *for* $I \subset J \subset R$.

5.3.2   The Characteristic of a Field

We now define the *characteristic of a ring with* 1 R, which we denote by $\text{char}(R)$.

Consider the subring of R generated by 1 and define

$$\text{char}(R) := \min\{m \in \mathbb{Z}^+ \mid m1 = 0\},$$

if this minimum exists. Otherwise, the subring generated by 1 is isomorphic to the ring $\mathbb{Z}$, and we say that the characteristic of R is 0.

Now, suppose that $\text{char}(R) = n > 0$. Then, evidently the subring generated by 1 is isomorphic to the ring $\mathbb{Z}_n$. However, that means that $\mathbb{Z}_n$ is a subring of R. If R is an integral domain, Exercise 5.1.17 tells us n is prime. Thus the characteristic of an integral domain is either 0 or a prime, p. When the integral domain itself is finite, obviously it has characteristic p. In particular, all this applies to fields.

As an example, an infinite field of characteristic p is given by rational functions in one variable, with coefficients from $\mathbb{Z}_p$.

Definition 5.3.9. A field which contains no subfield except itself, is called a *prime field*.

Example 5.3.10. The field $\mathbb{Q}$ of rational numbers is a prime field. Indeed, suppose that $A \subset \mathbb{Q}$ is a subfield. Then, there must exist an $a \in A$, with $a \neq 0$. Since A is a subfield, a^{-1} must be in A, hence

$1 = aa^{-1}$ is in A. Therefore, A must contain all the integers, and so the inverses of all the non-zero integers, which implies that A must contain all fractions $\frac{n}{m} = nm^{-1}$. Hence $A = \mathbb{Q}$.

Obviously any field containing the ring $\mathbb{Z}$ has characteristic zero. In particular, the fields $\mathbb{Q}$, $\mathbb{R}$ and $\mathbb{C}$ have characteristic zero.

Proposition 5.3.11. *For every prime p, the field $\mathbb{Z}_p$ is a prime field.*

Proof. This follows since the additive group of $\mathbb{Z}_p$ contains no proper subgroups.   □

Proposition 5.3.12. *Every field k contains a prime field.*

Proof. Assume that k is not a prime field. Then it must have a nontrivial proper subfield F. Since $0, 1$ are in all subfields, $F \supseteq Z$ the additive group generated by 1. If $p1 = 0$ for some prime p then F is isomorphic with $\mathbb{Z}_p$. Otherwise, $Z \cong \mathbb{Z}$. But then, $F \cong \mathbb{Q}$, the quotient field of $\mathbb{Z}$.   □

Proposition 5.3.13. *A homomorphism of a field k to any ring R is either the zero map or is injective.*

In commutative algebra in Chapter 9 we shall encounter significant generalizations of this.

Proof. Let $f : k \longrightarrow R$ be a homomorphism. We will show that if f is not injective, then f must be the zero homomorphism. For there must be an a and b in k with $a \neq b$ and $f(a) = f(b)$. Let $c = a - b$. Then $f(c) = f(a - b) = f(a) - f(b) = 0$. Now, since $a \neq b$, $c \neq 0$, so it has an inverse c^{-1}. For any $x \in k$, $f(x) = f(c(c^{-1}x)) = f(c)f(c^{-1}x) = 0 \cdot f(c^{-1}x) = 0$. Thus f is the zero homomorphism.   □

Exercise 5.3.14. Let k be a field of characteristic $p \neq 0$. Then the map $f : k \longrightarrow k$ defined by $x \mapsto f(x) = x^p$ is an injective homomorphism (see [3]).

Proof. Since k is a field, so commutative, using the binomial theorem 0.4.2 for any two elements a, b of k, we obtain

$$f(a+b) = (a+b)^p = a^p + \sum_{i=0}^{p-1} \binom{p}{i} a^{p-i} b^i + b^p.$$

Now, we observe for any $i = 1, ..., p-1$

$$\binom{p}{i} = \frac{p!}{i!(p-i)!}$$

is divisible by p. Therefore, since k has characteristic p, $\binom{p}{i} a^{p-i} b^i = 0$. Hence, $f(a+b) = a^p + b^p = f(a) + f(b)$. In addition, since k is commutative $f(ab) = (ab)^p = a^p b^p$. Also, $f(1) = 1^p = 1 \neq 0$. So f is a non-zero homomorphism, hence by 5.3.13, f is an injective homomorphism.   □

5.4   Maximal Ideals

Let R be a commutative ring with identity.

Definition 5.4.1. We say that I is a *maximal ideal* in R if it is a proper ideal and it is not contained in another proper ideal.

The following is important both in the commutative and non-commutative cases.

Proposition 5.4.2. *Let R be a ring with 1. Any proper ideal two sided I is contained in a maximal two sided ideal.*

Proof. Consider the set

$$\mathfrak{A} = \{I' \mid I' \text{ is a two sided ideal}, 1 \notin I', I \subseteq I'\}.$$

We observe that $\mathfrak{A} \neq \varnothing$ because $I \in \mathfrak{A}$. Now, we define a partial ordering in $\mathfrak{A}$ by

$$I' \preceq I'' \text{ iff } I' \subseteq I''.$$

We must show that every chain in $\mathfrak{A}$ has a cap.

Let $\{I_\lambda,\ \lambda \in \Lambda\}$ be a chain in $\mathfrak{A}$. Then

$$\mathfrak{J} = \bigcup_{\lambda \in \Lambda} I_\lambda \in \mathfrak{A}.$$

Indeed, $\mathfrak{J}$ is a two sided ideal because if $a, b \in \mathfrak{J}$ then there are $\lambda_1, \lambda_2 \in \Lambda$ such that $a \in I_{\lambda_1}$, $b \in I_{\lambda_2}$ and so $a - b \in I_{\lambda_3} \subset \mathfrak{J}$. The same is true for ra and ar where $a \in \mathfrak{J}$ and $r \in R$. Also $\mathfrak{J}$ does not contain 1, since none of the I_λ contains 1. By Zorn's lemma there is a maximal element. $\square$

We now give an example showing that the existence of an identity is necessary.

Example 5.4.3. Take any divisible group G written additively (e.g. the group $\mathbb{Q}^+$, or $\mathbb{R}^+$), and define a trivial multiplication on it by setting $g \cdot h = 0$ for all g and h in G. Then G becomes a ring without identity and any ideal I of this ring corresponds exactly to a subgroup of G. As we know from 1.14.29 G has no maximal subgroups. So as a ring, G has no maximal ideals.

Theorem 5.4.4. *Let R be a commutative ring with identity.*

1. *R is a field if and only if R has no non-trivial ideals.*

2. *If I is an ideal in R, then R/I is a field if and only if I is a maximal ideal.*

3. *If R is an arbitrary (not necessarily commutative) ring with identity, then R/I is a simple ring if and only if and I is a maximal two-sided ideal[1].*

Proof. 1) $\Longrightarrow$ Suppose $R = k$ is a field and I be an ideal $\neq \{0\}$. Then, there is an $x \in I$ with $x \neq 0$. Take xx^{-1}. Then $xx^{-1} = 1 \in I$, and so $I = k$.

---

[1]The corresponding statement to 1, namely R is a simple ring if and only if R is isomorphic with a matrix ring, will be dealt with in Wedderburn's Big Theorem in Volume II, Chapter 13.

$\Longleftarrow$ Let R be a commutative ring with identity and without ideals, except $\{0\}$, and R. Assuming 2, since $\{0\}$ is a maximal ideal, $R/\{0\} \cong R$ is a field.

2) $\Longrightarrow$ If R/I is a field and I is an ideal. Then I is contained in a maximal ideal, say J. The third isomorphism theorem 5.3.7 tells us

$$(R/I)/(J/I) \cong R/J.$$

But R/I is, by hypothesis, a field. Since J/I is an ideal in it, J/I $\{0\}$ or the field itself so that $J/I = \{0\}$ or $J/I = R/I$. Thus $J = I$ or $J = R$, i.e. I is maximal.

$\Longleftarrow$ Suppose I is maximal. If $\bar{x} = x + I$ is a non-zero element of R/I. Then $x \notin I$. Consider the ideal (x, I) generated by x and I,

$$(x, I) = \{ax + i \mid a \in R, \ i \in I\}.$$

$(x, I) \supset I$ and is larger than I. Since I is a maximal ideal, $(x, I) = R$ and since $1 \in R$, $1 = ax + i$ for some $a \in R$ and $i \in I$. Passing to the quotient by applying the projection map, $\bar{1} = \bar{a}\bar{x} + \bar{i}$ and since $\bar{i} = \bar{0}$ we see $\bar{1} = \bar{a}\bar{x}$ and so $\bar{a}$ is the inverse of $\bar{x}$ and R/I is a field.

We leave the proof of 3, which is very similar to 1 and 2, to the reader.   □

Exercise 5.4.5. Is there a commutative ring R, an ideal $I \subset R$ and an ideal $J \subset I \subset R$ such that we have the following two ring isomorphisms:

$$R/I \cong \mathbb{Q}(\sqrt{2}) \quad \text{and} \quad R/J \cong \mathbb{Q}?$$

Proof. The answer is no. Indeed, since $R/J \cong \mathbb{Q}$ is a field, J is a maximal ideal and since $J \subset I$, $I = R$ or $I = J$. If $I = J$, then $R/I \cong \mathbb{Q}$. If $I = R$, then $R/I = \{0\}$.   □

5.4.1 Prime Ideals

Here R is a commutative ring with 1.

Definition 5.4.6. The ideal I of R ($I \neq R$) is called a *prime ideal* if R/I is an integral domain.

Proposition 5.4.7. *I is a prime ideal if and only if $a \cdot b \in I$ implies $a \in I$ or $b \in I$.*

Since I being maximal is equivalent to R/I being a field, and an integral domain is more general than a field, a prime ideal is more general than a maximal ideal. However, in certain cases these coincide. As we shall see shortly, an ideal in $\mathbb{Z}$, or $\mathbb{Z}[i]$ or $k[x]$ where k is a field is prime if and only if it is maximal.

Exercise 5.4.8. Prove the only ideals in $\mathbb{Z}$ are of the form (n). (As we shall see shortly (n) is a prime ideal if and only if n is a prime number).

Proposition 5.4.9. *Every non-zero commutative ring with 1 has a minimal prime ideal.*

Proof. Since R is a non-zero ring it must contain a maximal ideal, which is a prime ideal. Denote by $\mathfrak{U}$ the set of all prime ideals of R ordered by inclusion. If $\mathfrak{V}$ is a totally ordered subset of $\mathfrak{U}$ and

$$I = \bigcap_{P \in \mathfrak{V}} P,$$

I s an ideal. We show it is actually a prime ideal. If $ab \in I$, then $ab \in P$ for all $P \in \mathfrak{V}$. Now, let $\mathfrak{J} = \{P \in \mathfrak{V} \mid b \in P\}$, and $K = \bigcap_{P \in \mathfrak{J}} P$. Since $\mathfrak{V}$ is a totally ordered set, either $K = I$, so $b \in I$ which means that I is prime, or $I \subset K$ and for all $P \in \mathfrak{V}$ such that P is properly contained in K. Then $b \notin P$. This means for all these P, $a \in P$ since they are prime. Therefore, $a \in I$, i.e. in either case I is prime. By Zorn's lemma we have the existence of a maximal element, which in our case is a minimal prime ideal. $\qquad\qquad\square$

Definition 5.4.10. Let R be a commutative ring with 1. An element $p \in R$, which is not zero or a unit, is called *prime* if when $p = ab \in R$ either a or b is a unit.

5.5   Euclidean Rings

In this section R is again a commutative ring with 1. The content of this section is inspired by the most important property of the ring $\mathbb{Z}$, namely the Euclidean algorithm.

Definition 5.5.1. We call R a *Euclidean ring* if it possesses a map

$$g : R \longrightarrow \mathbb{Z}^+,$$

called the *grade*, satisfying

1. $g(xy) \geq g(x)$ for $y \neq 0$.

2. If $a, b \in R$ and $a \neq 0$, then, $b = qa + r$, where $q, r \in R$ and $g(r) < g(a)$ or $r = 0$.

Actually, it is not necessary to assume R has a unit element, since its existence is guaranteed by the following proposition:

Proposition 5.5.2. *If a commutative ring $R \neq \{0\}$ satisfies the two conditions of a Euclidean ring, then it must have a unit element.*

Proof. Consider the set

$$\{g(a) \mid a \in R, \ a \neq 0\}.$$

This is a non-empty subset of $\mathbb{Z}^+$ and therefore it has a minimal element, say $g(m)$. So, for each $a \in R$ with $a \neq 0$, $g(a) \geq g(m)$. $\square$

Definition 5.5.3. When R is a commutative ring with identity we can sharpen things. An ideal I is said to be *generated* by $S \subseteq R$, and denoted by $I = (S)$, if every $x \in I$ is a finite sum $x = \sum_{i=1}^{n} s_i r_i$, where $s_i \in S$ and $r_i \in R$. We call the elements of S *ideal generators* of I. If S is finite we then say the ideal I is *finitely generated*.

The complexity of the ring theory of R is reflected in how extensive the ideals of that ring happen to be. For example, are all the ideals finitely generated? Can they ascend or descend infinitely? As we saw when k is a field there are only trivial ideals. The next best thing would be to have ideals involving only a single generator (as in $\mathbb{Z}$). This is the role of Euclidean rings and more generally principal ideal domains.

Theorem 5.5.4. *In a Euclidean ring every ideal has a single generator.*

If an ideal in a ring has a single generator it is called a principal ideal. A ring all of whose ideals are principal ideals is called a principal ideal domain or PID. Hence another way to state theorem 5.5.4 is: A Euclidean ring is a PID.

Proof. Let I be an ideal of R. Consider the grades of all the elements of I

$$\{g(a) \ : \ a \in I, \ a \neq 0\}.$$

Let a be an element of I of the lowest possible grade. Then $I = (a)$. Evidently, $(a) \subseteq I$. If $b \in I$, then since $a \neq 0$, $b = qa + r$, where $g(r) < g(a)$. Because b and a are in I, $b - qa = r$ is also in I. Therefore $g(r) = 0$ and so $r = 0$ and $b = qa$. □

Here are the most important Euclidean rings. Our first example below was already dealt with in Proposition 1.5.5, but for continuity we also include it here.

Example 5.5.5. Let $R = \mathbb{Z}$. For i) take $g(x) = |x|$. Then,

$$|xy| = |x|.|y| \geq |x| \text{ if } y \neq 0$$

and for ii) we can assume $a > 0$. Then we consider

$$\frac{b}{a} = \lambda \in \mathbb{Q} \text{ and } |\lambda - q| \leq \frac{1}{2} < 1.$$

Hence,

$$\left|\frac{b}{a} - q\right| < 1 \text{ which implies } -1 < \frac{b}{a} - q < 1.$$

Multiplying by a $(a \neq 0)$, $-a < b - qa < a$. Therefore, $-a < r < a$, i.e. $|r| < |a|$.

Example 5.5.6. The Ring of Gaussian Integers. This is the ring $\mathbb{Z}[i]$ defined as

$$\mathbb{Z}[i] = \{a + bi \mid a, b \in \mathbb{Z}, \ i = \sqrt{-1}\}.$$

We ask the reader to check that $\mathbb{Z}[i]$ is a commutative ring with identity, $1 + 0i$. We now prove $\mathbb{Z}[i]$ is a Euclidean ring. We take as grade, g, the norm, i.e.

$$g(z) = N(z) = |z|^2 = N(a + bi) = a^2 + b^2.$$

To prove property i) we have $|zw| = |z| \cdot |w|$ for every $z, w \in \mathbb{C}$. So $|zw|^2 = |z|^2|w|^2$ and thus $N(zw) = N(z)N(w) \geq N(z)$.

For property ii), since $a \neq 0$, we divide b by a in $\mathbb{C}$ (which is a field) letting $\lambda = \frac{b}{a}$. This complex number λ lies somewhere on the plane. If it hits one of the points of the lattice $\mathbb{Z}^2$, take $r = 0$. If not, the worst case for λ is for it to lie at the midpoint of one of the boxes determined by the lattice. Consider such a q. Its distance from λ is given by $N(\lambda - q) \leq \frac{1}{2}$. Hence $N(\lambda - q) < 1$. Since $N(a) > 0$, $N(a)N(\lambda - q) = N(a(\lambda - q)) < N(a)$. Thus $N(b - aq) < N(a)$. Calling $b - aq = r$, we see that $N(r) < N(a)$.

Exercise 5.5.7. Consider the following sets under complex number addition and multiplication

$$\{a + b\sqrt{-2},\ a,\ b \in \mathbb{Z}\}$$

and

$$\{a + b\sqrt{-3},\ a, b \in \mathbb{Z}\}.$$

Are they rings? Are they Euclidean rings? Prove or disprove.

After the integers, the next example is the most important example of a Euclidean ring in this book.

Proposition 5.5.8. *Let k be a field, and $k[x]$ be the ring of polynomials in one variable. Then $k[x]$ is a Euclidean ring.*

We will prove that given $f(x)$ and $g(x) \neq 0$, one can find $q(x)$ and $r(x)$ so that $f(x) = q(x)g(x) + r(x)$, where $0 \leq \deg r < \deg g$. Here the "grade" is deg. The reader should notice if k were not a field, this argument would fail.

Proof. Let $f(x) = \sum_{i=0}^{n} a_i x^i$, where $a_n \neq 0$, $n = \deg(f)$ and $g(x) = \sum_{i=0}^{m} b_i x^i$, where $b_m \neq 0$, $m = \deg(g)$. We can assume $n \geq m$; otherwise we just write $f = 0g + f$.

Since k is a field,

$$f(x) - \frac{a_n}{b_m}x^{n-m}g(x) = f_1(x) \in k[x]$$

and has lower degree than $f(x)$. Therefore, by induction on the $\deg(f)$, $f_1 = q_1 g + r$, where $\deg(r) < \deg(g)$. Hence

$$f(x) = (\frac{a_n}{b_m}x^{n-m} + q_1)g(x) + r(x),$$

where $\deg(r) < \deg(g)$. $\square$

Here is a deliberately misplaced exercise.

Exercise 5.5.9. Are $\mathbb{Z}[x]$ and $k[x,y]$, where k is a field, Euclidean rings? Why or why not?

Theorem 5.5.10. *Let R be an integral domain and Euclidean ring, and $I \neq 0$ be an ideal in R. Then, the following conditions are equivalent:*

1. *I is a maximal ideal.*

2. *I is a prime ideal.*

3. *$I = (p)$, where p is a prime element in R.*

Proof. $1 \implies 2$ is trivial.
Regarding $2 \implies 3$, since R is a principal ideal domain, $I = (q)$. Suppose q factors as $q = ab$, where neither a nor b are units. Now, $ab \in (q)$ implies that $a \in (q)$ or $b \in (q)$. Assume that $a \in (q)$. Then $a = kq$, and so $q|a$. Also, $ab = kqb$; hence since $ab = q$, $q = kqb$. Therefore, since R is an integral domain, $1 = kb$, which means a is a unit and therefore I is prime.
To prove $3 \implies 1$, Suppose $I = (p)$ with p a prime. Take an ideal $\mathcal{J}$ bigger than I, i.e. $I \subset \mathcal{J} \subset R$. Because R is a Euclidean ring, $\mathcal{J} = (a)$. Now, $p \in I$, so $p \in \mathcal{J}$, and therefore $p = ka$ for some $k \in R$. But p is prime, hence either k or a is a unit. Therefore $I = (a) = \mathcal{J}$. $\square$

5.5.1   Vieta's Formula and the Discriminant

Here we study some important details concerning polynomials in 1 variable, namely, the relationship of the coefficients of a polynomial to its roots. For this it will be necessary to assume the field K is algebraically closed. That is, any polynomial in $K[x]$ of degree ≥ 1 has all its roots in K. For example, $\mathbb{C}$ is algebraically closed. Also, any field can be (essentially) uniquely imbedded in an algebraically closed field (see Appendix A of Volume II).

Theorem 5.5.11. *(Vieta's Formulas) Let*

$$f(x) = a_n x^n + a_{n-1} x^{n-1} + \cdots + a_1 x + a_0$$

be a polynomial with coefficients in an algebraically closed field K. Here $a_n \neq 0$. Vieta's formulas relate the roots, $x_1,...,x_n$ (counting multiplicities) to the coefficients a_i, $i = 0,...,n$ as follows:

$$\sum_{1 \leq j_1 \leq \cdots \leq j_k \leq n} x_{j_1} x_{j_2} \cdots x_{j_k} = (-1)^k \frac{a_{n-k}}{a_k}. \tag{1}$$

Proof. Since $x_1,...,x_n$ are the roots of $f(x)$.

$$a_n x^n + a_{n-1} x^{n-1} + \cdots + a_1 x + a_0 = a_n (x - x_1)(x - x_2) \cdots (x - x_n).$$

Expanding the right side of the above equation, each of the 2^n terms will be of the form

$$(-1)^{n-k} x^k x_1^{\alpha_1} x_2^{\alpha_2} \cdots x_n^{\alpha_n},$$

where $\alpha_i = 0$ or 1 depending on the appearance, or not, of x_i in the product term. By grouping the terms of the same degree together we get the equivalent formulas,

$$\begin{cases} x_1 + x_2 + \cdots + x_n = -\frac{a_{n-1}}{a_n}. \\ (x_1 x_2 + x_1 x_3 + \cdots + x_1 x_n) + \cdots + (x_{n-1} x_1 + \cdots + x_{n-1} x_n) = \frac{a_{n-2}}{a_n}. \\ \vdots \\ x_1 x_2 \cdots x_n = (-1)^n \frac{a_0}{a_n}. \end{cases}$$

$\square$

We now turn to the discriminant Δ of a polynomial. Again, let f be a polynomial of degree $n \geq 1$ with coefficients in the algebraically closed field, K, and $x_1,...,x_n$ are the roots of f in K (counting multiplicities).

Definition 5.5.12. The *discriminant* of f is defined as:

$$\Delta(f) = a_n^{2n-2} \prod_{1 \leq i < j \leq n} (x_i - x_j)^2.$$

By Vieta's formula, $\Delta(f)$ is a polynomial in the coefficients of f. For example, when $K = \mathbb{C}$ and f is a quadratic polynomial, $ax^2 + bx + c$, the quadratic formula gives the roots as $\frac{-b \pm \sqrt{b^2 - 4ac}}{2a}$. Therefore, $\Delta(f) = b^2 - 4ac$.

The purpose of the discriminant, which is an immediate consequence of the definition, is:

Proposition 5.5.13. *A polynomial $f \in K[x]$ of degree $n \geq 1$ has n distinct roots if and only $\Delta(f) \neq 0$.*

5.5.2   The Chinese Remainder Theorem

Let R be a Euclidean ring and $\{a_1, ..., a_k\}$ be pairwise relatively prime elements of R. Let (a_i) be the principal ideal generated by a_i, and

$$\pi_i : R \longrightarrow R/(a_i)$$

be homomorphisms, and let

$$\pi : R \longrightarrow \prod_{i=1}^{k} R/(a_i), \text{ such that } \pi(r) = \big(\pi_1(r), ...\pi_k(r)\big).$$

The following result is called the Chinese remainder theorem for Euclidean rings.

Theorem 5.5.14. *(Chinese Remainder Theorem) The homomorphism π is surjective.*

Before proving the theorem, we make some remarks and state a proposition. Saying that π is surjective is equivalent to saying that we can find an $x \in R$ such that

$$x = r_i \quad (\mathrm{mod}\ (a_i)) \quad i = 1, ..., k.$$

Now,

$$\mathrm{Ker}\ \pi = \{x \in R \ : \ \pi_i(x) = \bar{0}, \ \forall\ i\}$$
$$= \bigcap_{i=1}^{k}(a_i) = (a_1 \cdots a_k).$$

Therefore,

$$R \big/ (a_1 \cdots a_k) \cong \prod_{i=1}^{k} R(a_i).$$

For example, let $x \in R$ and $x = p_1^{l_1} \cdots p_k^{l_k}$. Taking $a_i = p_i^{l_i}$, the set $a_1, ..., a_k$ satisfies the hypothesis. Therefore,

$$R/(x) \cong \prod_{i=1}^{k} R/(p_i^{l_i}).$$

We leave the proof of the following proposition to the reader.

Proposition 5.5.15.   *Let R be a commutative ring with 1 and $(I_i)_{i=1,...,k}$ be a finite family of proper ideals. Then,*

1. *The sum $I_1 + ... + I_k = \{a_1 + a_2 + ... + a_k \mid a_i \in I_i\}$ is an ideal of R.*

2. *The product $I_1 \cdots I_k = \{\sum_{j=1}^{n} a_{1j} \cdots a_{kj} \mid a_{ij} \in I_i,\ n \in \mathbb{Z}^+\}$ is a proper ideal of R.*

3. *The direct product $(R/I_1) \times \cdots \times (R/I_k)$ is a commutative ring with $1 \neq 0$.*

4. The map
$$\pi : R \longrightarrow (R/I_1) \times \cdots \times (R/I_k)$$

defined by
$$x \mapsto (x + I_1, \ldots, x + I_k)$$

is a ring homomorphism with kernel $\mathrm{Ker}(\pi) = I_1 \cap \ldots \cap I_k$.

Notice that if I_1 and I_2 are two ideals then the following diagram commutes:

$$I_1I_2 \subset I_1 \cap I_2 \quad\quad\quad\quad\quad I_1 \cup I_2 \subset I_1 + I_2,$$

where all the arrows are inclusion maps.

We now return to the proof of the theorem.

Proof. For each i let
$$b_i = \prod_{j \neq i} a_j \in R.$$

Therefore, a_i and b_i are relatively prime for any i. Indeed, suppose that p is a prime in R which divides both a_i and b_i for some i. Then p must divide a_j for $j \neq i$, which contradicts our assumption. Therefore, there is an α_i and β_i in R such that $\alpha_i a_i + \beta_i b_i = 1$. Let

$$\gamma_i = 1 - \alpha_i a_i = \beta_i b_i \in R.$$

This means that $\gamma_i \in (a_j)$ for each $j \neq i$. In other words, $\pi_j(\gamma_i) = \overline{0}$, if $j \neq i$.
On the other hand, since $\pi_i(a_i) = 0$,

$$\pi_i(\gamma_i) = \pi_i(1) - \pi_i(\alpha_i a_i) = \overline{1} - \pi_i(\alpha_i)\pi_i(a_i) = \overline{1}.$$

But
$$\pi(\gamma_i) = (\bar{0}, ..., \bar{0}, \bar{1}, \bar{0}, ..., \bar{0}) \quad \text{(with } \bar{1} \text{ in the } i\text{-th spot)}.$$

Now, let $\bar{r}_1, ..., \bar{r}_k$ be in $\prod_{i=1}^{k} R/(a_i)$ and calculating we get

$$\pi\left(\sum_{i=1}^{k} r_i \gamma_i\right) = \sum_{i=1}^{k} \pi(r_i \gamma_i = \sum_{i=1}^{k} \pi(r_i)\pi(\gamma_i)$$

$$= \sum_{i=1}^{k} \bar{r}_i (\bar{0}, ..., \bar{0}, \bar{1}, \bar{0}, ..., \bar{0})$$

$$= \sum_{i=1}^{k} (\bar{0}, ..., \bar{0}, \bar{r}_i, \bar{0}, ..., \bar{0})$$

$$= (\bar{r}_1, ..., \bar{r}_k),$$

which means that π is surjective. $\qquad\qquad\qquad\qquad\qquad\square$

We can generalize these results to any commutative ring R with identity as follows.

Definition 5.5.16. Two ideals I and J are called *co-prime* if $I + J = R$.

Define the product of the I and J as

$$IJ = \{ab \mid a \in I, \ b \in J\}.$$

This is the smallest ideal containing all the elements ab with $a \in I$, $b \in J$.

Proposition 5.5.17. *If I and J are co-prime ideals, then $I \cap J \leq IJ$.*

Proof. If the ideals I and J are co-prime, $I + J = R$. Then, since $1 \in R = I + J$, there exist $a \in I$ and $b \in J$ with $1 = a + b$. Therefore, for any $r \in I \cap J$, $ar \in IJ$ and $rb \in IJ$. Hence

$$r = r1 = r(a + b) = ra + rb = ar + rb \in IJ,$$

which implies that $I \cap J \leq IJ$. $\qquad\qquad\qquad\qquad\qquad\square$

We now formulate a more general Chinese Remainder Theorem. For simplicity we work with only two ideals.

Theorem 5.5.18. *(Chinese Remainder Theorem) Given ideals I and J of R, then*

1. *$IJ \le I \cap J$.*

2. *If I and J are co-prime ideals, then $IJ = I \cap J$.*

3. *If I and J are co-prime ideals, then the map*

$$\varphi : R/(IJ) \longrightarrow R/I \times R/J, \quad such\ that\ \varphi(x+IJ) = (x+I,\ x+J),$$

   defines a ring isomorphism

$$R/(IJ) \cong R/I \times R/J.$$

Proof. 1) Since I and J are ideals $ab \in I$ and $ab \in J$. Hence $ab \in I \cap J$. Thus, $I \cap J$ is an ideal containing all the elements ab with $a \in I$ and $b \in J$ and so $I \cap J$ contains the smallest such ideal, i.e., $IJ \le I \cap J$.

2) From 1) and Proposition 5.5.17, we get $IJ = I \cap J$.

3) φ is well defined. Indeed, suppose that $x + IJ = y + IJ$, i.e. $x - y \in IJ$. Then, by 1), $x - y \in I \cap J$, which means that $x - y \in I$ and $x - y \in J$. From this follows

$$\varphi(x + IJ) = (x+I,\ x+J) = (y+I,\ y+J) = \varphi(y+IJ).$$

Now, by definition, φ is a ring homomorphism.

To show that φ is injective, suppose that $(x+I,\ x+J) = (y+I,\ y+J)$, i.e., that $x + I = y + I$ and $x + J = y + J$. Then, we get $x - y \in I$ and $x - y \in J$. Hence $x - y \in I \cap J$. By 2) this implies that $x - y \in IJ$, hence $x + IJ = y + IJ$.

Finally, to prove that φ is onto, consider any $(x+I,\ y+J) \in (R/I) \times (R/J)$. We want to find $s \in R$ such that $\varphi(s + IJ) = (x + I,\ y + J)$. Since I and J are co-prime, write $1 = a + b$ with $a \in I$ and $b \in J$ and let $s := ay + bx$. Then,

$$\begin{aligned}
\varphi(s + IJ) &= (s + I \ , \ s + J) = (ay + bx + I, \ ay + bx + J) \\
&= (bx + I \ , \ ay + J) = ((1-a)x + I, \ (1-b)y + J) \\
&= (x - ax + I \ , \ y - by + J) = (x + I, \ y + J).
\end{aligned}$$

Hence, φ is surjective. $\square$

Exercise 5.5.19. Prove that in $\mathbb{Z}$ two principal ideals (a) and (b) are co-prime if and only if $(a, b) = 1$.

In this case, the Chinese Remainder Theorem says that

$$\mathbb{Z}/(mn) \cong \mathbb{Z}/(m) \times \mathbb{Z}/(n).$$

5.6   Unique Factorization

Since any integer x has positive and negative divisors, a product ab can also be written $(-a)(-b)$. In particular, a prime $p = (-p) \times (-1)$. We consider these to be trivial factorizations. We now generalize this to an integral domain R. Each $a \in R$ is divisible by any unit u and $a \in R$, $a = u \cdot u^{-1} \cdot a$. For this reason we make the following definition.

Definition 5.6.1. We say that the factorization $x = ab$ is a *trivial factorization* if either a or b is a unit of R. Consistent with our earlier definition of primes of R, an element of R that has only a trivial factorization is called prime.

We emphasize that since we are in a commutative ring, when we speak of unique factorization we also allow permutation of the elements in addition to multiplying by units. Moreover, we say that $a, b \in R - \{0\}$ are *associates*, and write $a \sim b$, if $a = ub$, where u is a unit. We invite the reader to check that $\sim$ is an equivalence relation.

Definition 5.6.2. Let a and b, with $b \neq 0$, be elements of R. We say that b *divides* a (or equivalently, b is a *divisor* of a), and write $b \mid a$, if there is a $c \in R$ such that $a = bc$. If b does not divide a, we write $b \nmid a$.

Proposition 5.6.3. *Let R be a Euclidean ring. Then every $x \in R$ can be uniquely factored into primes. Moreover, if a prime p divides a product ab, then p divides either a or b (or both).*

We call such a commutative ring with unit, a unique factorization domain (UFD).

In particular, this proposition applies to the Gaussian integers and to the polynomial ring, $k[x]$, where k is a field. The proofs of these statements are completely analogous to the one given for the ring $\mathbb{Z}$ (see Chapter 0).

In general an integral domain may not have unique factorization. An example of this is provided by the ring $\mathbb{Z}[\sqrt{-5}]$. Here

$$6 = 2 \cdot 3 = (1 + \sqrt{-5})(1 - \sqrt{-5}).$$

If $a + b\sqrt{-5}$ divides 2, then the norm $N(a + b\sqrt{-5}) = a^2 + 5b^2$ must divide $N(2) = 4$. But this can happen only if $a = \pm 2$ and $b = 0$, or if $a = \pm 1$ and $b = 0$, showing that 2 is prime of this ring. Similarly 3 is prime. Moreover, evidently $1 + \sqrt{-5}$ and $1 - \sqrt{-5}$ are also primes. So the factorization of 6 is not unique.

Exercise 5.6.4. Consider the ring $R = \{a + b\sqrt{-3}, \ a, \ b \in \mathbb{Z}\}$. Show unique factorization fails here.

Since knowing what the units are in an integral domain is necessary to deal with unique factorization questions, we ask the reader to calculate the group of units in the following cases.

Exercise 5.6.5. Let m not a perfect square be in $\mathbb{Z}$. Consider

$$\mathbb{Z}[\sqrt{m}] = \{a + b\sqrt{m} \ \mid \ a, \ b \in \mathbb{Z}\}.$$

Prove $\mathbb{Z}[\sqrt{m}]$ is an integral domain and find its units.

Exercise 5.6.6. Let $n \geq 2$. If d is not a perfect n^{th} power, then the nth root of d is irrational.

5.6.1   Fermat's Two-Square Thm. & Gaussian Primes

Here we will answer the question: *What are the primes of the Gaussian integers?* In general, when one enlarges the ring one increases the possibilities for factorization. Thus a prime in the smaller ring may not remain prime in a larger one. In particular, a prime in $\mathbb{Z}$ is not always a prime in $\mathbb{Z}[i]$.

Let us consider the prime 5. In the ring $\mathbb{Z}[i] = \{a + ib, \ a, b \in \mathbb{Z}\}$ the integer 5 can be factorized as

$$(1 + 2i)(1 - 2i) = 5,$$

which shows that 5 is not a prime in $\mathbb{Z}[i]$ (as we will see this is due to the fact that $5 = 1^2 + 2^2$). Notice that also

$$(2 + i)(2 - i) = 5.$$

This does not violate unique factorization since i is a unit and $i(2 - i) = 2i + 1 = 1 + 2i$, which means that $2 - i$ and $1 + 2i$ are associates.

Proposition 5.6.7. *If the norm of a Gaussian integer is prime in $\mathbb{Z}$, then the Gaussian integer is prime in $\mathbb{Z}[i]$.*

Proof. Let $z \in \mathbb{Z}[i]$ and have a prime norm, i.e. $N(z)$ is prime and suppose z can be factored in $\mathbb{Z}[i]$, as $z = wu$. Then taking norms,

$$N(z) = N(w)N(u).$$

This is an equation in positive integers, and $N(z)$ is prime in $\mathbb{Z}$, so either $N(w)$ or $N(u)$ is ± 1. Therefore w or u is a unit, so $N(z)$ is prime. $\square$

The converse of Proposition 5.6.7 is not valid; a Gaussian prime may not have a prime norm. For instance, 3 has norm 9, and as we just saw 3 is prime in $\mathbb{Z}[i]$.

We now turn to Fermat's 2 square theorem[2] which will categorize those primes of $\mathbb{Z}$ which remain primes in $\mathbb{Z}[i]$. However, before doing

---

[2] More than 50 different proofs of this theorem have been published, most of them using either the fact that $\mathbb{Z}[i]$ is a UFD., or Dirichlet's Approximation Theorem.

so we sort through the primes p of $\mathbb{Z}$ as follows: 2 is prime in $\mathbb{Z}$, but $2 = (1+i)(1-i)$ in $\mathbb{Z}[i]$ and neither of these factors is a unit. Now the other (odd) primes can be put in residue classes mod 4 and they will all either have residue 1 such as 5 or 13 or have residue 3 such as 3 or 7.

Fermat's 2 square theorem states the primes of $\mathbb{Z}$ of the form $p = 4n + 1$ for some n (or the prime 2) are composite in $\mathbb{Z}[i]$, while the primes of the form $p = 4n + 3$ remain prime in $\mathbb{Z}[i]$. In the former case $p = a^2 + b^2$ where a and b are integers ≥ 1.[3]

Theorem 5.6.8. *(Fermat's Two-Square Theorem)* *If p is an odd prime, then there are integers a and b such that $p = a^2 + b^2$ if and only if $p \equiv 1 \pmod 4$. The numbers a and b are uniquely determined apart from order and sign. This occurs if and only if $p = (a+bi)(a-bi)$ is a non-trivial factorization of p in $\mathbb{Z}[i]$.*

In particular, a prime p in $\mathbb{Z}$ is composite in $\mathbb{Z}[i]$ if and only if it is a sum of two squares of integers.

Proof. ($\Rightarrow$). First suppose that $p = a^2 + b^2$. Then a and b cannot both be even or both odd, for otherwise p would be even. Assume that a is even and b is odd, so $a^2 \equiv 0 \pmod 4$, $b^2 \equiv 1 \pmod 4$. Therefore $p \equiv 1 \pmod 4$.

($\Leftarrow$). Since p is odd we can write

$$x^{p-1} - 1 = \left(x^{\frac{p-1}{2}} - 1\right)\left(x^{\frac{p-1}{2}} + 1\right).$$

This polynomial has exactly $p - 1$ zeros in the multiplicative group of $\mathbb{Z}_p$. Hence the right-hand side has $p - 1$ zeros too. But the factors have at most $(p-1)/2$ zeros each, which means they have exactly $(p-1)/2$ distinct zeros each. In particular, there is a $\bar{g}$ such that $\bar{g}^{\frac{p-1}{2}} = -\bar{1}$. This can be written

$$\left(g^{\frac{p-1}{4}}\right)^2 \equiv -1 \pmod p$$

---

[3]The Two-Square Theorem, as it is known today, was stated without proof by Fermat in 1640, though he claimed to have a proof. The first known proof of the theorem was published by Euler (1755) which took several years of effort.

because here we assume $p \equiv 1 \pmod 4$. Let $m = g^{\frac{p-1}{4}}$. Then

$$p \mid m^2 + 1 = (m+i)(m-i).$$

If p were a prime in $\mathbb{Z}[i]$, it would divide at least one of the factors on the right, say $p \mid m + i$. But if we conjugate, we get $p \mid m - i$ and subtracting we get $p \mid 2i$. Since this is impossible, we conclude p is not a prime in $\mathbb{Z}[i]$. Therefore it can be factored

$$p = (a + ib)(c + id),$$

where neither factor is a unit. Taking norms

$$p^2 = N(p) = N(a+ib)N(c+id) = (a^2 + b^2)(c^2 + d^2).$$

Since both factors are > 1 and p is prime in $\mathbb{Z}$, $p = a^2 + b^2$. $\qquad\square$

We leave as an exercise the proof of the following corollary.

Corollary 5.6.9. *If a prime $p \in \mathbb{Z}$ is composite in $\mathbb{Z}[i]$, then p has exactly two Gaussian prime factors (up to a unit multiple), which are conjugate and have norm p.*

An Application to Diophantine Geometry

Here we consider an "algebraic curve" and deal with a question of "Diophantine geometry" due to Mordell. The term Diophantine refers to finding integer solutions.

In 1920 Mordell proved (in [82]) that there are only finitely many Diophantine solutions of the equation[4] of the form $y^3 = x^2 + n$, where $n \in \mathbb{Z}$. That is, the graph of this equation passes through at most finitely many integer points (points in $\mathbb{Z}^2$). He proved this, as often happens in mathematics, by going beyond the smaller domain $\mathbb{Z}$ to the larger Gaussian integers and exploiting the more powerful fact of unique factorization in this larger ring. The question of the number of rational solutions of this equation is still an open problem and is connected

---

[4]This equation is called *Mordell's equation*.

with one of the most outstanding open problems in Number Theory, namely, the Birch and Swinnerton-Dyer conjecture. We shall now solve the special case $n = 1$. Namely, $y^3 = x^2 + 1$. The reader will notice that the same line of argument works if 1 is replaced by any perfect square.

Proposition 5.6.10. *The only integer point on the curve $y^3 = x^2 + 1$ is $(0, 1)$.*

Proof. If x is odd, then $x^2 + 1$ is never a cube as it is congruent to 2 (mod 4) (check!). We can therefore assume x is even. Let (x, y) be a point on the curve. In the ring $\mathbb{Z}[i]$ our equation can be written as

$$y^3 = (x + i)(x - i).$$

A prime element of $\mathbb{Z}[i]$ that divides both $x + i$ and $x - i$ also divides their difference $(x + i) - (x - i) = 2i$, so it is equal up to units to $1 + i$. But since x is even $1 + i$ does not divide $x + i$, which implies that $x + i$ and $x - i$ are co-prime in $\mathbb{Z}[i]$. Their product is a cube, and so since $\mathbb{Z}[i]$ has unique factorization, each of them must be a product of a unit and a cube in $\mathbb{Z}[i]$. As all units in $\mathbb{Z}[i]$ are cubes $-1 = (-1)^3$, $1 = 1^3$, $i = i^3$ and $-i = (-i)^3$, there must exist *integers* a, b in $\mathbb{Z}[i]$ such that

$$x + i = (a + bi)^3.$$

Hence,

$$x = a(a^2 - 3b^2) \quad \text{and} \quad 1 = (3a^2 - b^2)b.$$

From this we obtain $b = -1$ and $a = 0$. Therefore the only solution of our equation is $x = 0$, $y = 1$ □

Exercise 5.6.11. Prove that the following equations have no integral solution.

1. $x^3 = y^2 + 5$.

2. $x^3 = y^2 - 6$.

5.7 The Polynomial Ring

Let R be an integral domain and consider

$$R[x] = \left\{ f(x) = \sum_{i=0}^{n} a_i x^i, \; a_i \in R, \; n < \infty \right\}.$$

Definition 5.7.1. $R[x]$ is called a *polynomial ring with coefficients in R*. We shall assume $a_n \neq 0$, and call n its *degree*, writing $\deg(f) = n$.

To add (or substract) polynomials, $f(x) = a_n x^n + \ldots + a_0$ and $g(x) = b_m x^m + \ldots + b_0$ we just add (or substract) the coefficients of the respective monomials of the same degree. The degree of $(f + g)$ will then be given by $\deg(f + g) = \max\{n, m\}$.

For multiplication,

$$f(x)g(x) = \sum_{i=0}^{n} \sum_{j=0}^{m} a_i b_j x^{i+j}.$$

Obviously, $\deg(f \cdot g) = n + m$.

We ask the reader to check that under these operations $R[x]$ is a commutative ring with identity.

Similarly, we can consider the polynomial ring in several variables $x_1, \ldots, x_n$:

$$R[x_1, \ldots, x_n] = \left\{ f(x_1, \ldots, x_n) = \sum_{i_1, \ldots, i_n} a_{i_1} \ldots a_{i_n} x_1^{i_1} \ldots x_n^{i_n}, \; i_1 + \cdots + i_n = m \right\}.$$

Here the degree of a monomial $a_{i_1} \ldots a_{i_n} x_1^{i_1} \ldots x_n^{i_n}$ is $i_1 + \ldots + i_n = m$ and the *degree* of f is the maximum of the degrees of the monomials with non-zero coefficients.

Because the constants are polynomials of degree zero, $R \subseteq R[x]$ is a subring and we can consider

$$(R[x])[y] := R[x, y]$$

and repeat this as many times as needed to obtain $R[x_1, ...x_n] = R[x_1][x_2, ..., x_n] = R[x_1, ..., x_{n-1}][x_n]$.

The reader will recall the group of units of $R[x_1, ..., x_n]$ is just the unit group of R.

Proposition 5.7.2. *If R has no zero divisors, then $R[x]$ has no zero divisors, and by induction the same holds for $R[x_1, ..., x_n]$.*

Proof. Let $f(x) = a_0 + a_1 x + \cdots + a_n x^n$ and $g(x) = b_0 + b_1 x + \cdots + b_m x^m$. Suppose $f(x), g(x) \neq 0$. Now, if $f(x)g(x) \equiv 0$, then $a_n b_m = 0$ and since R has no zero divisors, $a_n = 0$ or $b_m = 0$, a contradiction.   □

Therefore, since R is an integral domain so is $R[x]$ and we can construct its quotient field, $R(x)$, of *rational functions* even if its elements are not functions, but rational expressions over R.

As we have learned, when R is a field and $n \geq 2$ this is not a Euclidean ring (because it is not a PID). Nevertheless $R[x_1, ..., x_n]$ has unique factorization if R does.

Proposition 5.7.3. *For $f(x) \in R[x]$, $a \in R$ is a root of $f(x) = 0$ if and only if $x - a$ is a factor of $f(x)$, i.e. $f(a) = 0$ if and only if $f(x) = (x - a)g(x)$ for some $g(x) \in R[x]$). In addition, $f(x)$ has at most $\deg f$ distinct roots in R.*

Proof. Examining the proof that $k[x]$ is a Euclidean ring we see that since $x - a$ is monic, the argument there works to prove

$$f(x) = (x - a)q(x) + r, \quad r \in R.$$

Therefore, $f(a) = 0$ if and only if $r = 0$. The second statement follows from the first by induction of $\deg(f)$.   □

5.7.1   Gauss' Lemma and Eisenstein's Criterion

As we know, an element a of a ring R is *irreducible* if it is not a unit and whenever $a = bc$, either b or c is a unit.

Let $f(x) \in R[x]$ be the polynomial

$$f(x) = a_n x^n + a_{n-1} x^{n-1} + \cdots + a_1 x + a_0.$$

Definition 5.7.4. The *content* of f is defined as

$$\text{cont}(f) = \gcd(a_0, a_1, ..., a_n)$$

(in R). The polynomial $f(x)$ is said to be *primitive* if its content is 1.

Thus $\text{cont}(f)$ is well-defined up to a unit in R.

Theorem 5.7.5. *If R has unique factorization, so does $R[x]$ (and, by induction, so does $R[x_1, ..., x_n]$). In particular, $k[x_1, ..., x_n]$, where k is a field, has unique factorization.*

Proof. First we prove the existence of a factorization. If $a \in R$ is irreducible then a is also irreducible in $R[x]$. For if $a(x) \in R[x]$, write $a(x) = dF(x)$, where $d = \text{cont}(a)$. Then $\text{cont}(F) = 1$. We can certainly factor d into a product of irreducibles in R. Either $F(x)$ is irreducible in $R[x]$, or it factors properly as a product of lower degree polynomials (since $\text{cont}(F) = 1$). All the factors will also have content 1 (since a divisor of any factor would divide F). We repeat this process (a finite number of times), and so get a factorization of $F(x)$, and hence of $a(x)$, as a product of irreducibles in $R[x]$.

For the uniqueness of the factorization, it suffices to prove each irreducible element of $R[x]$ generates a prime ideal in $R[x]$. For irreducibles $p \in R$ this is clear, since $R[x]/(p) = (R/(p))[x]$, which is an integral domain. The general case will follow from the lemma below.   □

Lemma 5.7.6. *Let $f(x)$, $g(x) \in R[x]$ with $\text{cont}(f) = \text{cont}(g) = 1$. Then $\text{cont}(fg) = 1$. More generally, for $f(x)$, $g(x) \in R[x]$, $\text{cont}(fg) = \text{cont}(f)\text{cont}(g)$.*

Proof. Suppose a prime $p \in R$ divides each of the coefficients of $f(x)g(x)$. Then $f(x)g(x) = 0$ in $(R/(p))[x]$, which is an integral domain. Thus p either divides all coefficients of $f(x)$ or p divides all coefficients of $g(x)$. Hence either $f(x)$ or $g(x)$ must be congruent to 0 in $(R/(p))[x]$. But this contradicts our assumption $\text{cont}(f) = \text{cont}(g) = 1$.

In the general case, write $f = dF$, $g = d_1 G$, where $\text{cont}(F) = \text{cont}(G) = 1$. Then $f(x)g(x) = dd_1 F(x)G(x)$, so, by the first part of the lemma, $\text{cont}(fg) = dd_1 = \text{cont}(f)\text{cont}(g)$.   □

Theorem 5.7.7. (Gauss' Lemma) *Let k be the field of fractions of R. If $f(x) \in R[x]$ factors in $k[x]$, then $f(x)$ factors in $R[x]$ with factors of the same degrees as the $k[x]$ factors. In particular, if $f(x) \in R[x]$ is irreducible, then $f(x)$ is also irreducible in $k[x]$.*

Proof. Every element of $k[x]$ can be written as $\alpha(x)/a$, where $\alpha(x) \in R[x]$ and $a \in R$. Suppose in $k[x]$ we have $f(x) = (\alpha(x)/a)(\beta(x)/b)$, with $a, b \in R$ and $\alpha(x), \beta(x) \in R[x]$. Then $abf(x) = \alpha(x)\beta(x) \in R[x]$. Consider an irreducible factor p of ab in R. Then $\alpha(x)\beta(x) = 0$ in $(R/p)[x]$. Thus p either divides all coefficients of $\alpha(x)$ or p divides all coefficients of $\beta(x)$. We can then cancel a factor p in the $R[x]$-equation $abf(x) = \alpha(x)\beta(x)$, without leaving $R[x]$. By induction on the number of prime factors of ab in R, we conclude that $f(x) = \alpha_k(x)\beta_k(x) \in R[x]$, where $\deg \alpha_k = \deg \alpha$ and $\deg \beta_k = \deg \beta$. $\square$

Theorem 5.7.8. *(Eisenstein's Criterion) Let p be a prime and suppose that*
$$f(x) = a_n x^n + \cdots + a_0 \in \mathbb{Z}[x].$$
If p divides a_i for any $i = 0, 1, ..., n-1$, but $p \nmid a_n$ and $p^2 \nmid a_0$, then $f(x)$ is irreducible over $\mathbb{Q}$.

Proof. By Gauss's Lemma, we need only show that $f(x)$ does not factor into polynomials of lower degree in $\mathbb{Z}[x]$. Suppose the contrary and let
$$f(x) = (b_k x^k + \cdots + b_0)(c_l x^l + \cdots + c_0)$$
be a factorization in $\mathbb{Z}[x]$, with b_k and c_l not equal to zero and $k, l < n$. Since p^2 does not divide $a_0 = b_0 c_0$, either b^0 or c_0 is not divisible by p. Suppose that $p \nmid b_0$. Then $p \nmid c_0$ since p does divide a_0. Now, since $p \nmid a_n$ and $a_n = b_k c_l$, neither b_k nor c_l is divisible by p. Let m be the smallest value of j such that $p \nmid c_j$. Then
$$a_m = b_0 c_m + b_1 c_{m-1} + \cdots + b_m c_0,$$
is not divisible by p, since each term on the right-hand side of the equation is divisible by p except for $b_0 c_m$. Therefore, $m = n$ since a_i is divisible by p for $m < n$. Hence, $f(x)$ cannot be factored into polynomials of lower degree and therefore must be irreducible. $\square$

The following is a useful corollary. An exercise further down the line in this chapter gives a significant generalization.

Corollary 5.7.9. *Let a be a positive integer that is not a perfect square. Then $x^2 - a$ is irreducible over $\mathbb{Q}[x]$. Thus the $\sqrt{a}$ is irrational. Similarly if p is prime, $x^n - p$ is irreducible over $\mathbb{Q}[x]$ so any n-th root of any prime is irrational.*

Proof. The prime factorization of a is $p_1^{e_1} \cdots p_k^{e_k}$, where at least one of the primes involved has exponent not a multiple of 2. Call it p. Since the exponent of p is odd, p^2 does not divide a and Eisenstein's criterion applies. The second statement works the same way. This proves irreducibility in $\mathbb{Z}[x]$. But then, by Gauss' lemma we get irreducibility over $\mathbb{Q}[x]$. □

Here are two further examples.

1. $5x^3 + 18x + 12$ is irreducible over $\mathbb{Q}[x]$, using the prime $p = 3$.

2. $2x^6 + 25x^4 - 15x^3 + 20x - 5 \in \mathbb{Z}[x]$ has content 1, and is irreducible in $\mathbb{Q}[x]$ by the Eisenstein criterion using the prime 5. By the Gauss lemma, it is irreducible in $\mathbb{Z}[x]$.

5.7.2 Cyclotomic Polynomials

Before defining cyclotomic polynomials we must first introduce some auxiliary concepts.

Definition 5.7.10. (*n-th Root of Unity*) Let n be a positive integer. A complex number ω is an n-th *root of unity* if $\omega^n = 1$.

As is well known for each $n \geq 2$ there are n distinct n-th roots of unity. These are

$$\{e^{\frac{2\pi k}{n}} \mid 1 \leq k \leq n\}.$$

Exercise 5.7.11. Show that under multiplication the n-th roots of unity form a group isomorphic to $(\mathbb{Z}_n, +)$.

Definition 5.7.12. (**Primitive n-th Root of Unity**) A *primitive n-th root of unity* is an n-th root of unity whose order is n.

That is, ω is a root of the polynomial $x^n - 1$ but not of the polynomial $x^m - 1$ for any $m < n$.

Exercise 5.7.13. Let p be a prime. Prove that any non-trivial p-th root of unity is primitive. If p is not a prime, there must be non-trivial p-th roots of unity which are not primitive.

Notice that if ω is a primitive n-th root of unity, then $\langle \omega \rangle$ contains n distinct elements, and so ω is a generator of the group of the n-th roots of unity.

It is easy to see that the primitive n-th roots of unity are

$$\{e^{\frac{2\pi i}{n} k} \ : \ 1 \le k \le n, \ \text{and} \ \gcd(k, n) = 1\}.$$

Definition 5.7.14. (n-th Cyclotomic Polynomial) For any positive integer n the n-th *cyclotomic polynomial*[5] is the (monic) polynomial $\Phi_n(x)$, defined by

$$\Phi_n(x) = (x - \omega_1)(x - \omega_2) \cdots (x - \omega_m),$$

where $\omega_1, \ldots, \omega_m$ are the primitive n-th roots of unity.

It follows that

$$\Phi_n(x) = \prod_{\substack{1 \le k \le n \\ \gcd(n,k)=1}} \left(x - e^{\frac{2\pi i}{n} k} \right).$$

That $\Phi_n(x)$ is a monic polynomial is obvious since when written as a product of linear factors every x term has coefficient 1. In addition, its degree is $\phi(n)$, where ϕ is the Euler ϕ-function, since the number of integers k, such that $1 \le k \le n$ and g.c.d$(k, n) = 1$ is by definition $\phi(n)$.

Example 5.7.15. The first cyclotomic polynomials are

$$\Phi_1(x) = x - 1, \ \Phi_2(x) = x + 1, \ \Phi_3(x) = x^2 + x + 1, \ \Phi_4(x) = x^2 + 1,$$

$$\Phi_5(x) = x^4 + x^3 + x^2 + x + 1.$$

---

[5]The word cyclotomic comes from the Greek and means "cut the circle", which has to do with the position of the roots of unity ζ on the unit circle in the complex plane.

Proposition 5.7.16. *For any positive integer n,*

$$x^n - 1 = \prod_{d \mid n} \Phi_d(x).$$

Proof. Let ω be a root of $\Phi_d(x)$, with $d \mid n$. Then, ω is also a d-th root of unity. If $q \in \mathbb{Z}$ and $n = dq$ we get

$$\omega^n = (\omega^d)^q = 1^q = 1,$$

which shows that ω is a root of the polynomial $x^n - 1$.

Now, if ω is a root of $x^n - 1$ then ω is an n-th root of unity and if its order is d, ω will be a primitive d-th root of unity. Therefore, ω is also a root of $\Phi_d(x)$. Since the n-th roots of unity form a group of n elements, $d \mid n$; hence ω is a root of $\Phi_d(x)$ for some d that divides n. We just proved that the polynomials $x^n - 1$ and $\prod_{d \mid n} \Phi_d(x)$ share all their roots. In addition, both are monic polynomials and hence, they are equal. $\square$

Theorem 5.7.17. *For any positive integer n we have $\Phi_n(x) \in \mathbb{Z}[x]$.*

Proof. We proceed by induction. As $\Phi_1(x) = x - 1$, the base case of $n = 1$ is clearly true.

Now, suppose for all $k < n$ we have that $\Phi_k(x)$ is a polynomial with integer coefficients. We will show that so is $\Phi_n(x)$. Define

$$p_n(x) = \prod_{\substack{d \mid n \\ d \neq n}} \Phi_d(x).$$

By inductive hypothesis $p_n(x) \in \mathbb{Z}[x]$. Clearly $p_n(x)$ is monic. Thus, by the division algorithm there exist polynomials $q(x)$ and $r(x)$ such that

$$x^n - 1 = p_n(x)q(x) + r(x),$$

where $\deg r(x) < \deg p_n(x)$ or $r(x) = 0$. Now take the $n - \phi(n)$ roots of $p_n(x)$ and substitute them into the above equation. It is easy to see $r(x)$ has at least $n - \phi(n)$ roots. But then if r is non-zero, it has degree at least $n - \phi(n)$. As $\deg p_n(x) = n - \phi(n)$, we get a contradiction since $\deg r(x) < \deg p_n(x)$. Thus $r(x) = 0$ everywhere and $x^n - 1 = p_n(x)q(x)$. Because $x^n - 1 = p_n(x)\Phi_n(x)$, it immediately follows that $q(x) = \Phi_n(x)$, and thus $\Phi_n(x) \in \mathbb{Z}[x]$. $\qquad\square$

Theorem 5.7.18. *For p a prime, $\Phi_p(x)$ is irreducible.*

Proof. To prove this it suffices to show that $\Phi_p(x + 1)$ is irreducible. Note that

$$\Phi_p(x + 1) = \frac{(x + 1)^p - 1}{x}$$

$$= x^{p-1} + \binom{p}{1}x^{p-2} + \dots + \binom{p}{p-1},$$

which is obviously irreducible by Eisenstein's criterion (see Theorem 5.7.8). $\qquad\square$

5.7.3   The Formal Derivative

To detect multiple and simple roots of a polynomial in $k[x]$, we now turn to the formal, or algebraic derivative.

Definition 5.7.19. Let k be a field. The *algebraic* or *formal derivative* $f'(x)$, or $Df(x)$ of the polynomial $f(x) = \sum_{i=0}^{n} c_i x^i \in k[x]$, is the polynomial, $f'(x) = \sum_{i=1}^{n} i c_i x^{i-1}$, where the integer i is considered to be in the prime field of k.

These are its important features.

1. $(cf)'(x) = c \cdot f'(x)$.

2. $(f + g)'(x) = f'(x) + g'(x)$.

3. $(fg)'(x) = f(x) \cdot g'(x) + f'(x) \cdot g(x)$.

The first two of these are easy to check and we leave this to the reader. Using the first two, the third reduces to taking $f(x) = x^i$ and $g(x) = x^j$. Then $(fg)' = (x^{i+j})' = (i + j)x^{i+j-1}$, while $f(x)g'(x) + f'(x)g(x) = jx^{j-1}x^i + ix^{i-1}x^j$ which also is $(i + j)x^{i+j-1}$.

Here is the connection to simple roots. It will be employed in Chapter 11 on Galois Theory.

Lemma 5.7.20. *If $f \in k[x]$ and $\gcd(f, f') = 1$, then f has only simple roots.*

Proof. Assume the contrary, i.e. $f = g^2 h$ with g, h in $k[x]$. Then:

$$D(f) = D(g^2 h) = D(g^2)h + g^2 D(h) = 2gD(g)h + g^2 D(h)$$
$$= g(2D(g)h + gD(h)).$$

Therefore, g divides $D(f)$, and as it divides also f it must divide $\gcd(f, D(f))$, so $\gcd(f, D(f)) \neq 1$, a contradiction. $\qquad\square$

From this we will see that when k has characteristic zero this criterion yields only simple roots.

Corollary 5.7.21. *Over a field k of characteristic zero, all polynomials have distinct roots in any extension field.*

Proof. Without loss of generality, we can assume f is irreducible. The derivative of x^i is ix^{i-1} and since k has characteristic 0, $i \neq 0$. Thus f' is a non-zero polynomial of degree $< \deg f$. Since f is irreducible, $\gcd(f, f')$ is either 1 or f, and the latter must be excluded because f cannot possibly divide f'. $\qquad\square$

Here is another consequence.

Proposition 5.7.22. *The polynomial $x^n - 1$ has no repeated factors in $\mathbb{Z}[x]$ for any $n \geq 1$.*

Proof. Instead of $\mathbb{Z}[x]$ we will work in the larger ring $\mathbb{Q}[x]$. Then:

$$x^n - 1 - \frac{1}{n}x(x^n - 1)' = x^n - 1 - \frac{1}{n}x \, n \, x^{n-1} = -1,$$

hence $\gcd(x^n - 1, (x^n - 1)') = 1$, and by Lemma 5.7.20 $x^n - 1$ has no repeated factors in $\mathbb{Q}[x]$ and a fortiori in $\mathbb{Z}[x]$. $\qquad\square$

Proposition 5.7.23. *Let p be a prime number and $f(x)$ be a polynomial with coefficients from $\mathbb{Z}_p$. Then,*

$$f(x^p) \equiv f(x)^p \pmod{p}.$$

Proof. This identity is a simple corollary of the fact that $\big(a(x) + b(x)\big)^p \equiv a(x)^p + b(x)^p \pmod{p}$ for polynomials $a(x)$, $b(x)$, which itself follows from the binomial theorem.   □

Theorem 5.7.24. $\Phi_n(x)$ *is irreducible over* $\mathbb{Z}[x]$.

Proof. The idea for this proof is simple. We first show that if ω is a primitive n-th root of unity and p is a prime such that $\gcd(p, n) = 1$, then ω^p is a root of the minimum polynomial of ω. Let the minimum polynomial for ω be $f(x)$. Since $\Phi_n(\omega) = 0$, $f(x) \mid \Phi_n(x)$. Hence there exists some polynomial $g(x)$ with integer coefficients such that $\Phi_n(x) = f(x) \cdot g(x)$. Suppose ω^p is not a root of $f(x)$. Then $g(\omega^p) = 0$. Let the minimum polynomial for ω^p be $h(x)$ and write $g(x) = h(x) \cdot k(x)$. Then, $\Phi_n(x^p) = f(x^p) \cdot h(x^p) \cdot k(x^p)$. We observe $h(x^p)$ has ω as a root. It follows that $f(x) \mid h(x^p)$. Let $h(x^p) = f(x) \cdot \ell(x)$ and so $\Phi_n(x^p) = \ell(x) \cdot k(x^p) \cdot f(x) \cdot f(x^p)$. Reducing modulo p we have

$$\Phi_n(x)^p \equiv f(x)^{p+1} \cdot k(x)^p \cdot \ell(x) \pmod{p},$$

where here we have used the identity $f(x^p) \equiv f(x)^p \pmod{p}$ for a polynomial $f(x)$, proved just above. Now, take an irreducible divisor $\pi(x)$ of $f(x)$ in $\mathbb{Z}_p[x]$ and note that $\pi(x)$ must divide $\Phi_n(x)$ more than once; otherwise, the right hand side would only be divisible by $\pi(x)$, p times. This is a contradiction to proposition 5.7.22, so our original assumption was incorrect and ω^p is a root of f.

Observe that we had no restrictions on ω and almost no restrictions on p. Now, we are ready to prove the irreducibility of $\Phi_n(x)$. By using the above result we know if p is prime and relatively prime to n, then the minimum polynomial of ω_n has the root ω_n^p. Evidently, for any pq where p, q are not necessarily distinct primes, and do not divide n, we have ω_n^{pq} is a root. By applying induction, we can show that all numbers

k relatively prime to n have ω_n^k is a root. But this polynomial is $\Phi_n(x)$. It follows that $\Phi_n(x)$ is the minimum polynomial of ω_n. This implies it is irreducible. $\qquad\square$

As an application of cyclotomic polynomials we will prove Dirichlet's theorem. To do this we need the following lemma:

Lemma 5.7.25. *For every integer n, there exists an integer $A > 0$ such that all prime divisors $p > A$ of the values of $\Phi_n(c)$ at integer points c are congruent to 1 modulo n. In other words, prime divisors of values of the n-th cyclotomic polynomial are either* small *or congruent to 1 modulo n.*

Proof. Consider the polynomial $f(x) = (x-1)(x^2-1)\cdots(x^{n-1}-1)$. The polynomials $f(x)$ and $\Phi_n(x)$ have no common roots, so their gcd in $\mathbb{Q}[x]$ is 1. Hence $a(x)f(x) + b(x)\Phi_n(x) = 1$ for some $a(x), b(x) \in \mathbb{Q}[x]$. Let A denote the common denominator of all coefficients of $a(x)$ and $b(x)$. Then for $p(x) = Aa(x)$, $q(x) = Ab(x)$ we have $p(x)f(x) + q(x)\Phi_n(x) = A$, and $p(x), q(x) \in \mathbb{Z}[x]$. Assume that a prime number $p > A$ divides $\Phi_n(c)$ for some c. Then c is a root of $\Phi_n(x)$ modulo p, which implies, $c^n \equiv 1 \pmod{p}$. Notice that n is the order of c modulo p. Indeed, if $c^k \equiv 1 \pmod{p}$ for some $k < n$, then c is a root of $f(x)$ modulo p. But the equality $p(x)f(x) + q(x)\Phi_n(x) = A$ shows that $f(x)$ and $\Phi_n(x)$ are relatively prime modulo p. Recall that $c^{p-1} \equiv 1 \pmod{p}$ by Fermat's little theorem, and so $p-1$ is divisible by n, the order of c. That is, $p \equiv 1 \pmod{n}$. $\qquad\square$

Theorem 5.7.26. *(Dirichlet's Theorem) For every integer n, there exist infinitely many primes $p \equiv 1 \pmod{n}$.*

Proof. Assume that there are only finitely many primes congruent to 1 modulo n, and let $p_1, ..., p_m$ be those primes and let A be as above. Consider

$$c = A! p_1 p_2 \cdots p_m.$$

The number $k = \Phi_n(c)$ is relatively prime to c (since the constant term of Φ_n is ± 1), and so is not divisible by any of the primes $p_1, ..., p_m$, and has no divisors $d \leq A$ either. Take any prime divisor p of k: then Lemma 5.7.25 shows that $p \equiv 1 \pmod{n}$. If $k = \pm 1$, it has no prime divisors, in which case we just replace k by Nk where N is a number large enough so that Nk is greater than all the roots of the equation $\Phi_n(x) = \pm 1$.   □

Corollary 5.7.27. *Let p be a prime number and x be an integer. Then, every prime divisor q of $1 + x + \dots + x^{p-1}$ satisfies either $q \equiv 1 \pmod{p}$, or $q = p$.*

Proof. Since $1 + x + \dots + x^{p-1} = \frac{x^p-1}{x-1} = \frac{x^p-1}{\Phi_1(x)}$, we get $1 + x + \dots + x^{p-1} = \Phi_p(x)$. From Theorem 5.7.26 we know that $q \equiv 1 \pmod{p}$, or $q \mid p$, that is, $q = p$ (since p is prime).   □

Exercise 5.7.28. Find all integer solutions of the equation

$$\frac{x^7 - 1}{x - 1} = y^5 - 1.$$

Proof. The above equation is equivalent to

$$1 + x + \dots + x^6 = (y - 1)(1 + y + \dots + y^4).$$

From Corollary 5.7.27 it follows that every prime divisor p of $1 + x + \dots + x^6$ satisfies either $p = 7$, or $p \equiv 1 \pmod{7}$. This implies that $(y - 1) \equiv 0 \pmod{7}$ or $(y - 1) \equiv 1 \pmod{7}$, which means that $y \equiv 1 \pmod{7}$, or $y \equiv 2 \pmod{7}$. If $y \equiv 1 \pmod{7}$, then $1 + y + \dots + y^4 \equiv 5 \neq 0, 1 \pmod{7}$, which is a contradiction. On the other side, if $y \equiv 2 \pmod{7}$, then $1 + y + \dots + y^4 \equiv 31 \equiv 3 \neq 0, 1 \pmod{7}$, also a contradiction. Therefore, the given equation has no integer solutions.   □

5.7.4   The Fundamental Thm. of Symmetric Polynomials

Let $k[x_1, \dots, x_n]$ be the polynomial ring in n variables with coefficients from the field k and let the symmetric group, S_n, act on $k[x_1, \dots, x_n]$ by permuting the variables:

$$\sigma \cdot p(x_1, \dots, x_n) = p(x_{\sigma(1)}, \dots, x_{\sigma(n)}), \ \sigma \in S_n.$$

Definition 5.7.29. A polynomial fixed under this action is called a *symmetric* polynomial.

The set of all symmetric polynomials of $k[x_1, \ldots, x_n]$ is a subalgebra of $k[x_1, \ldots, x_n]$ usually denoted by $k[x_1, \ldots, x_n]^{S_n}$.

The following polynomials of $k[x_1, \ldots, x_n]$ are evidently symmetric. They are the *elementary symmetric polynomials*.

$$s_1 = \sum_{1 \leq i \leq n} x_i = x_1 + x_2 + \cdots + x_n$$

$$s_2 = \sum_{1 \leq i < j \leq n} x_i x_j = x_1 x_2 + x_1 x_3 + \cdots + x_1 x_n + \cdots + x_{n-1} x_n$$

and, more generally, for every $k = 1, \ldots, n$, we define the k-th *elementary symmetric* polynomial s_k as

$$s_k = \sum_{i_1 < \ldots < i_k} x_{i_1} \cdots x_{i_k},$$

and finally $s_n = x_1 \cdots x_n$.

Since these n polynomials are symmetric, so is any polynomial, $f(s_1, \ldots, s_n)$, in n variables in them. The fundamental theorem of symmetric polynomials[6] states that the converse is also true. Namely, any symmetric polynomial is a polynomial in $s_1, \ldots, s_n$.

For example, the polynomial $x_1^2 + \cdots + x_n^2$, which is obviously symmetric, can be expressed as $s_1^2 - 2s_2$.

To prove the fundamental theorem of symmetric polynomials we need the notion of the *lexicographic ordering* of monomials in the variables $\underline{x} = (x_1, \ldots, x_n)$.

Definition 5.7.30. We say that the monomial $\underline{x}^I = x^{i_1} \cdots x^{i_n}$ is *lexicographically* smaller than the monomial $\underline{x}^J$, and we write

$$\underline{x}^I <_{\text{lex}} \underline{x}^J$$

---

[6]Edward Waring (1736-1798) gave formulas expressing any symmetric polynomial in terms of the elementary symmetric functions. The algorithmic proof we present here is due to Carl Friedrich Gauss (1777-1855).

if the first non-zero entry in the difference of the exponent sequences

$$J - I = (j_1 - i_1, \ldots, j_n - i_n)$$

is positive.

As example is the following ordering,

$$x_1^2 > x_1 x_2^4 > x_1 x_2^3 > x_1 x_2 x_3^5.$$

Definition 5.7.31. The *leading monomial*, denoted by $\mathrm{lm}(p)$, of a non-zero polynomial $p \in k[x_1, \ldots, x_n]$ is the greatest monomial with respect to this ordering in p.

One can easily check that $\mathrm{lm}(pq) = \mathrm{lm}(p)\,\mathrm{lm}(q)$ and $\mathrm{lm}(s_m) = x_1 \cdots x_m$.

Theorem 5.7.32. *(Fundamental Theorem of Symmetric Polynomials)* Let $p \in k[x_1, \ldots, x_n]$ be a symmetric polynomial. Then, p can be expressed as a polynomial in $s_1, \ldots, s_n$. That is, $k[x_1, \ldots, x_n]^{S_n} = k[s_1, \ldots, s_n]$.

Proof. Let $p \in k[x_1, \ldots, x_n]$ be any symmetric polynomial. Let $\underline{x}^I$ be the leading monomial of p. Then, $i_1 \geq i_2 \geq \ldots \geq i_n$. This is because, by the symmetry of p, one can find a permutation which applied to $\underline{x}^I$ would give a lexicographically larger monomial (with the same non-zero coefficient) in p as $\underline{x}^I$. Now, consider the expression

$$s_n^{i_n} s_{n-1}^{i_{n-1}-i_n} \cdots s_1^{i_1-i_2}. \tag{1}$$

The leading monomial of this polynomial is

$$(x_1 \cdots x_n)^{i_n} (x_1 \cdots x_{n-1})^{i_{n-1}-i_n} \cdots x_1^{i_1-i_2},$$

i.e. the monomial $\underline{x}^I$. Hence, subtracting a scalar multiple of the expression (1) from p, we can cancel the term with the monomial $\underline{x}^I$, obtaining a symmetric polynomial with a strictly smaller leading monomial. Repeating this step a finite number of times, one can express p as a polynomial in the s_k. □

This important fact constitutes just the tip of the iceberg as it leads to the following general question: Suppose a finite group (or a more general group such as a compact group, or even a reductive Lie group) G acts on a finite dimensional vector space V over a field k and V has a basis $x_1, \ldots, x_n$, so that G acts on $k[x_1, \ldots, x_n]$. Consider the *algebra of invariants*, $k[x_1, \ldots, x_n]^G$, that is, those polynomials which are G-fixed. Is this a *finitely generated subalgebra* of the polynomial algebra? When $k = \mathbb{R}$, or $\mathbb{C}$ this is true (see e.g. Theorem (2.6.4) of [1]) for any compact group acting continuously on V. In the case G is finite, it certainly holds for any action and any field k. These ideas lead us to *invariant theory* and Hilbert's 14-th problem. Here we just take the opportunity to point this avenue out to the reader and will leave it at that.

Definition 5.7.33. A set of polynomials is called *algebraically independent* if there are no algebraic relations between them.

Proposition 5.7.34. *The elementary symmetric polynomials are algebraically independent.*

Proof. Suppose the contrary and let $q \in k[x_1, \ldots, x_n]$ be a polynomial such that

$$q(s_1, \ldots, s_n) = 0.$$

We will show that q is the zero polynomial. To do this, we rewrite $q(s_1, \ldots, s_n)$ as a polynomial in the x_i's. Choose the term $a\underline{x}^I$ of q with the highest lexicographic order. Therefore, $q(s_1, \ldots, s_n)$ as a polynomial in the x_i's, has as leading term

$$ax_1^{i_1+\cdots+i_n} x_2^{i_2+\cdots+i_n} \cdots x_n^{i_n},$$

which, by assumption, is zero. Hence, $a = 0$. $\square$

This brings us to the following.

Corollary 5.7.35. *Every symmetric polynomial can be written* uniquely *as a polynomial in the elementary symmetric polynomials.*

Proof. Let p be a symmetric polynomial, and assume that p can be written in two different ways as a polynomial in $s_1,...,s_n$, i.e.

$$p = q_1(s_1, \ldots, s_n) = q_2(s_1, \ldots, s_n),$$

for some $q_1, q_2 \in k[x_1, \ldots, x_n]$. Then,

$$0 = q_1(s_1, \ldots, s_n) - q_2(s_1, \ldots, s_n)$$

and by the Proposition 5.7.34, $q_1 = q_2$.   □

Exercise 5.7.36. Let k be a positive integer. An interesting example of a symmetric polynomial is:

$$p(x_1,\ldots,x_n) = \sum_{i=1}^{n} x_i^k.$$

Write p as a polynomial in the s_i.

Exercise 5.7.37. Show that the elementary symmetric functions express the coefficients of a monic polynomial of one variable in terms of its roots.

$$\prod_{i=1}^{n}(X - x_i) = X^n + \sum_{k=1}^{n}(-1)^k s_k X^{n-k}.$$

5.7.5   There is No Such Thing as a Pattern

Suppose one is given a finite sequence of numbers $\{a_0, a_1, \ldots, a_n\}$, coming from a field k of characteristic zero, such as, for example, the first few squares of integers, or the first elements of the Fibonacci sequence $\{1, 1, 2, 3, 5, 8, 13 \ldots\}$. Is there any uniqueness to continuing this "pattern"? That is, is there a *polynomial* rule which tells one what a_{n+1} must be? In this section we will show that there is not! Indeed, we can choose $a_{n+1} = \alpha$ *arbitrarily* and there will be a polynomial rule, f, for which $f(i) = a_i$, $i = 0, \ldots n$, but $f(n + 1) = \alpha$. Actually, as we shall see, this can be done for any finite sequence $a_{n+j} = \alpha_j$, $1 \le j \le k$. So what we seek is a polynomial $f \in K[x]$ (the algebraic closure) so that $f(i) = a_i$ for $i = 0, \ldots n$ and $f(n + j) = \alpha_j$ for $j = 1 \ldots k$.

Let us first simplify the notation and just call $\alpha_j = a_{n+j}$ and $n+k = m$. Then we want to find a polynomial $f \in K[x]$ so that $f(i) = a_i$ for $0 \le i \le m$. For each $i = 0, \ldots n, \ldots, m$, let $\phi_i(x) = \prod_{j \ne i}(x - j)$. Then each ϕ_i is a polynomial of degree m, $\phi_i(i) \ne 0$ since k has characteristic 0, and if $i \ne j$, $\phi_i(j) = 0$. Since $\phi_i(i) \ne 0$, let $f_i = \frac{\phi_i}{\phi_i(i)}$. Then f_i is also a polynomial of degree m and $f_i(j) = \delta_{i,j}$ the Kronecker delta, where $i, j = 0 \ldots m$. Take $f = \sum_{i=0}^{m} a_i f_i$. Then f is a polynomial of degree m and evidently, $f(i) = a_i$ for all $i = 0, \ldots m$.

5.7.6 Non-Negative Polynomials

Here we ask the question, which was conjectured by Minkowski[7], as to whether a real polynomial p in n variables that is everywhere non-negative must be a sum of squares, $q_1^2 + \ldots q_k^2$, of such polynomials? As we shall see, the answer is affirmative for polynomials where $n = 1$, but negative otherwise.

First let $n = 1$. Of course if the degree of p is odd, then the hypothesis of positivity can never be realized. Therefore as a test case we should take p to have degree 2, i.e.,

$$p(x) = ax^2 + bx + c.$$

Evidently, the positivity forces $a > 0$. Then the minimum value of p occurs at $x = \frac{-b}{2a}$ and at that point $p(x) = -\frac{b^2}{4a} + c$. Hence here our hypothesis amounts to $a > 0$ and $c \ge \frac{b^2}{4a}$. Then, completing the square, we see that

$$ax^2 + bx + c = a(x + \frac{b}{2a})^2 + (c - \frac{b^2}{4a})$$

and since, as we just observed, this last term is a non-negative constant, we can take its square root in $\mathbb{R}$. Similarly, since $a > 0$ the first term is also a square.

---

[7]Hermann Minkowski, (1864-1909) Polish/Russian-Jewish mathematician who was a close friend of Hilbert, created the subject, *The Geometry of Numbers* and used it to solve problems in number theory (particularly quadratic forms in n variables with integer coefficients), mathematical physics. In 1907 he showed that his former student Albert Einstein's special theory of relativity (1905), could, indeed should, be understood geometrically as a theory of four-dimensional space-time, since known as "Minkowski space-time".

Theorem 5.7.38. *A real polynomial p in one variable which takes only non-negative values is a sum of 2 squares of real polynomials in one variable.*

Using the Lagrange identity in two variables

$$(a_1^2 + a_2^2)(b_1^2 + b_2^2) = (a_1 b_1 + a_2 b_2)^2 + (a_1 b_2 - a_2 b_1)^2,$$

we see that the set of these polynomials which can be written as the sum of 2 squares is closed under multiplication. That is, if $p(x) = q_1(x)^2 + q_2(x)^2$ and $r(x) = s_1(x)^2 + s_2(x)^2$, then pr is itself a sum of 2 squares. This identity is just $N(zw) = N(z)N(w)$ in disguise. Later we shall see a more elaborate Lagrange identity in four variables, which bears the same relationship to a norm identity, but this time with regard to the quaternions rather than the complex numbers.

We now prove this result on polynomials in one variable of degree $2k$ by induction, the case $k = 1$ having just been dealt with.

Now, let p be a real polynomial in x of even degree $k > 2$. Suppose p has no real root. Then all its roots occur as conjugate pairs of complex roots and p factors as a product of things of the form $a > 0$ times $x - (a_j \pm i b_j)$. Since this latter term yields $x^2 - 2a_j x + (a_j^2 + b_j^2)$, we see that p is a product of $a > 0$ and $\frac{k}{2}$ *real* quadratic polynomials. Since each of these has a positive leading coefficient and also $a_j^2 + b_j^2 \geq a_j^2$ we see by the case $k = 1$ above that each of these real quadratic polynomials takes only non-negative values. Peeling off all but one and using induction on k we get the product of $k - 1$ of them is a sum of 2 squares. Hence the product of all of them is also a sum of 2 squares. Tossing the $\sqrt{a}$ on the first one completes the proof in this case.

Now, suppose p has a real root α of multiplicity m. Then $p(x) = (x - \alpha)^m \cdot r(x)$, where $r(\alpha) \neq 0$. Since p takes on only non-negative values, m must be even. Otherwise, when x is on the left (respectively the right side of α) we will get both positive and negative values and since $r(\alpha) \neq 0$ and therefore by continuity $r(x) \neq 0$ for x near α, this will make p somewhere take on negative values. Hence $p(x) = q(x)^2 \cdot r(x)$, where $q = (x - \alpha)^{\frac{m}{2}}$ is a real polynomial and r takes only non-negative values because p does. Again by induction r is a sum of 2 squares and hence so is p.

This leaves us to consider the case of a polynomial in two or more variables, a very different proposition from the one variable case, particularly since the proof above made strong use of the fact that $\mathbb{R}[x]$ is a Euclidean ring and $\mathbb{R}[x, y]$ is not. For this reason alone one should be skeptical of a positive result here. However, one reason to be slightly optimistic is that (as opposed to the one variable case where actually p was a sum of 2 squares) we have the flexibility of writing it as a sum of any finite number of squares.

To refute the positive statement, following the argument of Motzkin, we must find a $p \in \mathbb{R}[x, y]$ which is everywhere non-negative, but is not the sum of any finite number of squares of elements in $\mathbb{R}[x, y]$. We shall do this by use of the inequality which says that the geometric mean of a finite number $a_1, \ldots, a_n$ of positive numbers is bounded by the arithmetic mean (for a proof of this see e.g. [84] p. 23). Thus,

$$(a_1 \cdots a_n)^{\frac{1}{n}} \le \frac{a_1 + \cdots + a_n}{n}.$$

We shall be interested in the case $n = 3$. We take $a_1 = x^2$ and $a_2 = y^2$ and to make $a_1 \cdots a_n$ as simple as possible take $a_3 = \frac{1}{x^2 y^2}$. Then the geometric mean is 1 and the inequality says that

$$1 \le \frac{1}{3}(x^2 + y^2 + \frac{1}{x^2 y^2}).$$

Let $p(x, y) = x^4 y^2 + x^2 y^4 - 3x^2 y^2 + 1$. Then, p takes on only non-negative values. Certainly, if $x = 0$ or $y = 0$ this is so. Suppose for some x and y, both $\neq 0$, $p(x, y) < 0$. Then, the same would be true of $p(x, y)/x^2 y^2 = x^2 + y^2 - 3 + 1/x^2 y^2 < 0$. Therefore, $\frac{1}{3}(x^2 + y^2 + \frac{1}{x^2 y^2}) < 1$, a contradiction. This proves that p, a polynomial of degree 6, is non-negative for all choices of x, y. Assume it can be written as a sum of squares, $q_1^2 + \cdots + q_s^2$. Then each of the q_i would have to be of degree ≤ 3 and hence of the form,

$$q_i = a_i + b_i xy + c_i x^2 y + d_i xy^2 + e_i x + f_i y + g_i x^2 + h_i y^2.$$

However, taking x or $y = 0$ we see that

$$q_1^2(x, 0) + \cdots + q_s^2(x, 0) = 1 = q_1^2(0, y) + \cdots + q_s^2(0, y),$$

which means each of the functions of 1 variable, $q_i(x,0)$ and $q_i(0,y)$ are bounded. From this we see $e_i = f_i = g_i = h_i = 0$, so that actually,

$$q_i = a_i + b_i xy + c_i x^2 y + d_i xy^2.$$

It follows that the coefficient of $x^2 y^2$ in p (which is -3) is $b_1^2 + \cdots b_s^2$, which being non-negative cannot be -3, a contradiction.

We remark that this entire section actually holds for "formally real fields[8]". Hilbert, in his famous address before the International Congress of Mathematicians (Hilbert's 17-th Problem), slightly reformulated this incorrect statement, making it a *correct conjecture* that was proved 27 years later by E. Artin. Here is the statement of Artin's result.

Theorem 5.7.39. *Let p be a polynomial in $\mathbb{R}[x_1,\ldots,x_n]$ which takes on only non-negative values. Then $p = q_1^2 + \cdots + q_s^2$, where each $q_i \in \mathbb{R}(x_1,\ldots,x_n)$.*

5.7.6.1   Exercises

Exercise 5.7.40. Let k be a field and $f(x)$ and $g(x)$ be polynomials with $\deg(g) = m \geq 1$. Then $f(x)$ can be written uniquely as

$$f(x) = a_0(x) + a_1(x)g(x) + \ldots a_r(x)g(x)^r,$$

where $\deg(a_i) < m$.

Exercise 5.7.41. Let k be a field and $f(x) \in k[x]$.

1. If $a \in k$, show that the evaluation map $f \mapsto f(a)$ is a ring epimorphism from $k[x]$ to k.

---

[8]A *formally real field* is a field k that satisfies one of the following equivalent statements:

1. -1 is not a sum of squares in k.

2. There is an element in k that is not a sum of squares, and $\operatorname{char}(k) \neq 2$.

3. If any sum of squares in k equals 0, then each of the elements must be 0.

One can prove that in this case k can be equipped with a (not necessarily unique) ordering that makes it an *ordered field*.

2. Show this map is injective if and only if k is infinite.

3. Now suppose k is infinite and $f(x_1, \ldots, x_n) \in k[x_1, \ldots, x_n]$. Show that if $f \neq 0$, there exist $a_1, \ldots, a_n \in k$ so that $f(a_1, \ldots, a_n) \neq 0$. Hint: the case $n = 1$ has just been proved. Use induction to reduce to that case.

4. Show that if k is infinite, then the evaluation map is a ring isomorphism.

5. Suppose k is finite. Let $f(x_1, \ldots, x_r)$ be a polynomial of degree $n < r$ with constant term 0. Show there exist $(a_1, \ldots, a_r) \neq 0$ with $f(a_1, \ldots, a_r) = 0$.

Exercise 5.7.42. Prove that the rings $\mathbb{Z}[x]$ and $\mathbb{Q}[x]$ are not isomorphic.

Exercise 5.7.43. Decide which of the following are ideals of $\mathbb{Z}[x]$:

1. The set I of all polynomials whose constant term is a multiple of 3.

2. The set I of all polynomials whose coefficient of x^2 is a multiple of 3.

3. The set I of all polynomials whose constant term, coefficient of x, and coefficient of x^2 are zero.

4. $\mathbb{Z}[x^2]$ (i.e. the polynomials in which only even powers of x appear).

5. The set I of polynomials whose coefficients sum to zero.

6. The set I of polynomials $p(x)$ such that $p'(0) = 0$.

Exercise 5.7.44. Let $\mathbb{Q}[\pi]$ be the ring of all real numbers of the form

$$r_0 + r_1\pi + \cdots + r_n\pi^n, \quad \text{with } r_i \in \mathbb{Q}, \, n \geq 0.$$

Is $\mathbb{Q}[\pi]$ isomorphic to the polynomial ring $\mathbb{Q}[x]$?

Exercise 5.7.45. Let $R = \mathbb{Z}[\sqrt{5}] = \{a + b\sqrt{5} \mid a, b \in \mathbb{Z}\}$.
   Let $I = \{a + b\sqrt{5} \in R \mid a \equiv b \pmod{4}\}$. Is I an ideal in R, and if it is, write down the quotient ring R/I.

Exercise 5.7.46. Show that $p(x) = 6x^5 + 14x^3 - 21x + 35$ and $q(x) = 18x^5 - 30x^2 + 120x + 360$ are irreducible in $\mathbb{Q}[x]$.

5.8   Finite Fields and Wedderburn's Little Theorem

Here we show any finite field has order p^n some prime p and also that a finite division ring must be commutative.

Lemma 5.8.1. *Let F be a finite field of characteristic p. Then, the prime subfield of F is isomorphic to $\mathbb{Z}_p$.*

Proof. Write $\mathbb{Z}_p = \{0, 1, 2, \ldots, p-1\}$, and define a map

$$\phi : \mathbb{Z}_p \longrightarrow F \text{ such that } m \mapsto \phi(m) := m \cdot 1 = \underbrace{1 + 1 + \cdots + 1}_{m},$$

where 1 is the multiplicative unit in F. One checks easily that ϕ is a field homomorphism. We claim that ϕ is injective. Assume the contrary, i.e. for some $p > a > b \geq 0$ we get $\phi(a) = \phi(b)$. Then, $c = a - b > 0$, so $c \in \mathbb{Z}_p - \{0\}$, which implies that there exists $c^{-1} \in \mathbb{Z}_p$. Now,

$$\phi(1) = \phi(c \cdot c^{-1}) = \phi(c) \cdot \phi(c^{-1}) = \big(\phi(a) - \phi(b)\big)\phi(c^{-1}) = 0.$$

But since φ is a field homomorphism, $\phi(1) = 1 \neq 0$. This is a contradiction. Therefore, since ϕ is injective $\text{Im}(\phi)$ is a prime field of F. $\square$

Proposition 5.8.2. *Let F be a finite field with p elements and suppose K is a finite field with $F \subset K$. If $[K : F] = n$, then K has p^n elements.*

Proof. K is a vector space over F and since K is finite, it must be of finite dimension as a vector space over F. Since $[K : F] = n$, K must have a basis of n elements over F. Let $e_1, \ldots, e_n$ be such a basis. Then, each element $v \in K$ is written as $v = \lambda_1 e_1 + \lambda_2 e_2 + \cdots + \lambda_n e_n$, with $\lambda_1, \lambda_2, \ldots, \lambda_n$ in F.

Now, the number of elements in K is equal to the number of such linear combinations, which in turn is equal to the number of values of $\lambda_1, \lambda_2, \ldots, \lambda_n$ in F. Each λ_i can take p values as F has p elements. So the number of different combinations is p^n, which implies $|K| = p^n$. $\square$

Corollary 5.8.3. *A finite field F has p^n elements, where the prime number p is the characteristic of F.*

Proof. Because of Lemma 5.8.1, F as a vector space over $\mathbb{Z}_p$ of dimension $\dim_{\mathbb{Z}_p}(F) = n$. Let $\xi_1,...,\xi_n$ be a basis. Then, if $a \in F$,

$$a = \lambda_1 \xi_1 + \cdots + \lambda_n \xi_n$$

for some $\lambda_1,...,\lambda_n$ in $\mathbb{Z}_p$. Since each λ_i can take p different values as $\mathbb{Z}_p$ has only p elements, the number of different combinations is $= p^n$, which implies $|F| = p^n$.   □

Corollary 5.8.4. *If the finite field F has p^m elements, then every $x \in F$ satisfies $x^{p^m} = x$.*

Proof. If $x = 0$, then $x^{p^m} = x$. Now, the non-zero elements of F form a group under multiplication of order $p^m - 1$. Since for any element x of a finite group, G, $x^{|G|} = 1_G$, we see $x^{p^m-1} = 1$ and so $x^{p^m} = x$.   □

Exercise 5.8.5. Let F be a finite field and $\alpha \neq 0$ and $\beta \neq 0$ be elements of F. Then there exist a and b in F such that $1 + \alpha a^2 + \beta b^2 = 0$.

We now turn to the proof of Wedderburn's Little Theorem[9] which states that any finite integral domain is a division ring and therefore a field. In other words, for finite rings, there is no distinction between skew-fields and fields.

Our proof requires that we know that the theory of vector spaces (basis and dimension) over a division ring works just as well as over a field.

---

[9]The original proof was given in 1905 by J. H. M. Wedderburn (1882-1948) in [114] but it had a gap. Leonard Eugene Dickson, shortly after Wedderburn, gave another proof ([29]), based on which Wedderburn then produced a correct one. In 1927, E. Artin in [9], after remarking that Wedderburn's first proof from [114] was not valid, gave a new proof close to Wedderburn's main ideas, using some group theory, divisibility properties of polynomials, and proving that all maximal subfields of the algebra are conjugate. In 1931, Witt in a one-page paper [120], used only cyclotomic polynomials and elementary group theory to prove Wedderburn's theorem. In 1952, Zassenhaus, in his rather long group theoretic proof, [124], derives Wedderburn's theorem from the theorem (proved in the same paper) that a finite group is Abelian if, for each Abelian subgroup, the normalizer coincides with the centralizer.

Theorem 5.8.6. (Wedderburn Little Theorem) *A finite division ring F is a field.*

Proof. Since F is a finite division ring its characteristic is p, a prime. Let $\mathcal{Z}(F)$ be its center. Then by lemma 5.8.1 $\mathcal{Z}(F)$ is a finite field containing the prime field P. Therefore $\mathcal{Z}(F)$ is a vector space over P. Since $|P| = p$, $|\mathcal{Z}(F)| = p^m = q$, where $m = \dim_P \mathcal{Z}(F)$. Moreover, F is itself a vector space over $\mathcal{Z}(F)$. Therefore $|F| = q^n$, where $n = \dim_{\mathcal{Z}(F)} F$. For all this see proposition 5.8.2.

Then, for $x \in F^\times$, let $Z_F(x)$ denote the centralizer of x in F and $Z_{F^\times}(x)$ the centralizer of x in $F^\times$. We note that $Z_F(x)$ is a sub-division ring of F containing $\mathcal{Z}(F)$. Therefore, $|Z_F(x)| = q^{\delta(x)}$, where $\delta(x) = \dim_{\mathcal{Z}(F)} Z_F(x)$. Then, n is a multiple of each $\delta(x)$ and $\delta(x) < n$ unless $x \in \mathcal{Z}(F)$. The order of

$$|Z_{F^\times}(x)| = \frac{q^n - 1}{q^{\delta(x)} - 1}.$$

Now apply the class equation to the multiplicative group $F^\times$.

$$q^n - 1 = q - 1 + \sum_{x \in (F^\times - \mathcal{Z}(F^\times))} \frac{q^n - 1}{q^{\delta(x)} - 1}$$

where we sum over representatives.

Now suppose $n > 1$.

Consider the polynomials $u^n - 1 = \Phi_n(u) = \prod_{i=1}^{n}(u - \alpha_i)$ and $u^{\delta(x)} - 1 = \prod_{i=1}^{\delta(x)}(u - \beta_i)$, where the α_i range over the primitive n-th roots of 1 and the β_i range over the $\delta(x)$-th roots of 1. Since $\delta(x)$ is a proper divisor of n for all x outside of $\mathcal{Z}(F)$, none of the β_i is a primitive n-th root of 1. Hence $\Phi_n(u)$ is relatively prime to $u^{\delta(x)} - 1$. As we know both these polynomials have their coefficients in $\mathbb{Z}$. Hence, $\Phi_n(u)$ divides $\frac{u^n - 1}{u^{\delta(x)} - 1}$ in $\mathbb{Z}[u]$. Now evaluate at the integer q. It follows that $\Phi_n(q)$ divides $\frac{q^n - 1}{q^{\delta(x)} - 1}$. By the class equation this forces $\Phi_n(q)$ to divide $q - 1$. But $\Phi_n(q) = \prod_{i=1}^{n}(q - \alpha_i)$, where the α_i range over the primitive n-th roots of 1. Since the integer $q > 1$, $|q - \alpha_i| > q - 1$ for each i. This means $\Phi_n(q)$ is strictly larger than $q - 1$, a contradiction. Therefore $n = 1$ and F is commutative. $\square$

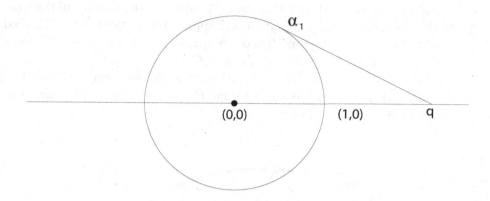

5.8.1   Application to Projective Geometry

One of the most amazing theorems in the geometry of the real projective plane was discovered by Girard Desargues (1591-1661).

Desargues' Theorem states that given two triangles ABC and $A'B'C'$ such that the lines AA', BB' and CC' pass through a single point O, then the pair of lines AB and $A'B'$, AC and $A'C'$ and BC and $B'C'$ meet in points R, P, and Q which are colinear!

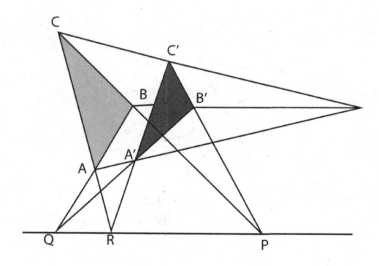

Closely associated with Desargues' theorem is the wonderful theorem of the Alexandrian mathematician, Pappus (born about 290 C.E, died about 350 C.E.). Pappus' Theorem (projective plane version) states that given the three points A, B, and C on line l and A', B', and C' on line m, if segments AB' and $A'B$ meet at R, BC' and $B'C$ meet at P, and CA' and $C'A$ meet at Q then P, Q, and R are collinear. The theorem is illustrated below.

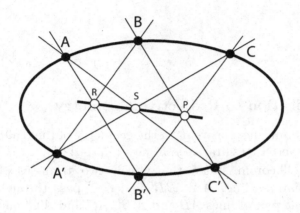

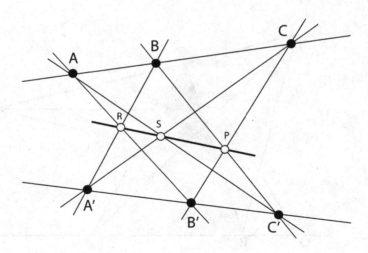

It turns out that Desargues' theorem holds in a projective geometry if and only if that geometry takes its coordinates from a division ring. Moreover, Pappus' Theorem holds in a projective geometry if and only if that geometry can be coordinatized using numbers from a field. Thus, it is possible to construct infinite projective planes where Pappus' Theorem holds but Desargues' does not. However, since every finite division ring is actually a field by Wedderburn's Little Theorem, every finite Desarguean projective geometry is also Pappian. *There is no known completely geometrical proof of this fact.*

5.9 k-Algebras

Definition 5.9.1. An *algebra* over k, or a *k-algebra* $\mathcal{A}$ is a *finite* dimensional vector space $\mathcal{A}$ over k together with a bilinear *multiplication*

$$\mathcal{A} \times \mathcal{A} \longrightarrow \mathcal{A}, \quad (x, y) \mapsto xy$$

such that, for any a, $b \in k$ and x, y, $z \in \mathcal{A}$: $a(xy) = (ax)y = x(ay)$.

Thus, here *neither associativity, nor existence of a unit* is assumed. However in most cases these algebras are associative and have a unit, as follows.

Definition 5.9.2. A k-algebra $\mathcal{A}$ is *associative* if for all x, y, $z \in \mathcal{A}$, $(xy)z = x(yz)$. $\mathcal{A}$ has a unit 1 if $x1 = x = 1x$ for all $x \in \mathcal{A}$. A k-algebra $\mathcal{A}$ is said to be *commutative* if $xy = yx$ for all x, $y \in \mathcal{A}$. The *dimension* of a k-algebra $\mathcal{A}$ is its dimension as a vector space.

Many concepts in associative k-algebras with identity are analogous to those of ring theory, and play a similar role. However, they often have sharper conclusions with easier proofs because they are much less general. For example, a two-sided ideal in an associative k-algebra, in addition to being invariant under multiplication from the left and right, is a *subspace* as well. Similarly, the center, $\mathcal{Z}(\mathcal{A})$, of $\mathcal{A}$ is automatically a *subspace*.

A subalgebra of a k-algebra $\mathcal{A}$ is both a subspace and a subalgebra of $\mathcal{A}$ as a ring. If I is a two-sided ideal in a k-algebra $\mathcal{A}$ with unit, then the quotient $\mathcal{A}/I$ becomes a k-algebra with the appropriate definition of multiplication and unit element.

Definition 5.9.3. A *homomorphism of k-algebras* is a linear map $f : \mathcal{A} \to \mathcal{B}$ such that

1. $f(xy) = f(x)f(y)$ for all $x, y \in A$, and

2. $f(1_A) = 1_B$.

Now, if f happens to be bijective, then it is a k-algebra *isomorphism* and $\mathcal{A}$ and $\mathcal{B}$ share all k-algebra theoretic properties. If $\mathcal{A} = \mathcal{B}$, then we call f an *automorphism*.

As an exercise we leave the verification of the following proposition to the reader.

Proposition 5.9.4. $\mathcal{Z}(A)$ *is a subalgebra of $\mathcal{A}$. If $f : \mathcal{A} \to \mathcal{B}$ is an associative algebra homomorphism, then $f(\mathcal{Z}(\mathcal{A})) \subseteq \mathcal{Z}(\mathcal{B})$. In particular, if f is an isomorphism $f(\mathcal{Z}(\mathcal{A})) = \mathcal{Z}(\mathcal{A})$.*

We now define the *structure constants* of an associative algebra.

Let $x_1, \ldots, x_n$ be a basis of $\mathcal{A}$. Then, $x_i x_j = \sum_l c_{i,j}^l x^l$, for uniquely determined constants $c_{i,j}^l \in k$, called the *structure constants* (with respect to this basis). These completely determine the multiplication on $\mathcal{A}$. The question is: what is the use of structure constants? If one wants to check whether a given k-algebra is associative, for example, one only needs to check the associativity on x_i, x_j and x_l, which leads to a finite verification.

Exercise 5.9.5. Formulate explicitly the equation for associativity, commutativity and the existence of a unit, in terms of the structure constants.

Example 5.9.6. Let $\mathcal{A} = \mathrm{M}(n, k) \simeq \mathrm{End}_k(V)$ (with matrix multiplication), where V is a vector space of dimension n over a field k. As we know, $\mathrm{End}_k(V)$ and so also $\mathrm{M}(n, k)$ is associative and has a unit. A basis for it is given by the matrices $E_{i,j}$, where $i, j = 1, \ldots, n$, and all entries are 0 except the ij-th, which is 1. Moreover,

$$E_{ij} E_{lm} = \begin{cases} 0 & \text{if } j \neq l, \\ E_{im} & \text{if } j = l. \end{cases}$$

Proposition 5.9.7. *One can characterize the center of a matrix algebra and its two-sided ideals as follows:*

1. *The center $\mathcal{Z}(\mathrm{End}_k(V))$ is the scalar matrices.*

2. *$\mathrm{End}_k(V)$ is a simple k-algebra, that is, it has only the trivial two-sided ideals.*

Proof.

1. Evidently, scalar matrices are in $\mathcal{Z}(\mathrm{End}_k(V))$. Let

$$A = \sum_{i,j} a_{ij} E_{ij} \in \mathcal{Z}(\mathrm{End}_k(V)).$$

Then $AX = XA$ for all $X \in \mathrm{End}_k(V)$. Let $X = E_{i_0 j_0}$. It follows that,

$$\sum_i a_{i i_0} E_{i j_0} = \sum_j a_{i_0 j} E_{i_0 j}.$$

Since the set of E_{ij} is linearly independent in $\mathrm{M}(n, k)$,

$$a_{i i_0} = 0 = a_{i_0 j}.$$

Therefore, $i_0 = j_0$ and in this case, $a_{i_0 i_0} = $ constant λ. In particular, $a_{ii} = a_{jj}$ for all i and j. So, $A = \lambda I$.

2. Let I be a two-sided ideal in the k-algebra $\mathrm{M}(n, k)$ and $X = \sum_{i,j} x_{ij} E_{ij}$ be a non-zero element of I. Suppose $x_{i_0 j_0} \neq 0$. Then

$$E_{i_0 i_0} X E_{j_0 j_0} \in I.$$

But this is

$$\sum_{i,j} x_{ij} E_{i_0 i_0} E_{ij} E_{j_0 j_0}.$$

If $i \neq i_0$ or $j \neq j_0$, the corresponding term is 0. So, we get

$$E_{i_0 i_0} X E_{j_0 j_0} = x_{i_0 j_0} E_{i_0 j_0}.$$

Since we are over a field, dividing we get $E_{i_0 j_0} \in I$. Therefore,

$$E_{s i_0} E_{i_0 j_0} E_{j_0 t} = E_{st} \in I, \quad \text{for all } s \text{ and } t.$$

Hence, $I = \mathrm{M}(n, k)$.

$\square$

We summarize the relations involving the matrix units as follows:

1. $E_{i,i}^2 = E_{i,i}$.

2. $E_{i,i} E_{j,j} = 0$, if $i \neq j$.

3. $\sum_{i=1...n} E_{i,i} = I$.

These are called the *fundamental matrix unit relations*. The reason for this terminology is the following lemma.

Lemma 5.9.8. *For $i = 1 \ldots n$, let A_i be non-zero operators on V satisfying the following conditions:*

1. $A_i^2 = A_i$.

2. $A_i A_j = 0$ *if $i \neq j$.*

3. $\sum_{i=1...n} A_i = I$.

Then V is the direct sum of the $A_i(V)$ and each A_i has rank 1.

Proof. Equation 3 tells us the $A_i(V)$ generate V while 1 and 2 tell us the sum is direct. Since each $A_i(V)$ is non-zero it must have dimension 1. $\square$

We can now see the effect of an automorphism on the matrix units. Let f be an automorphism of $\text{End}_k(V)$. Then clearly $f(E_{i,i}) = A_i$ satisfy the hypothesis of Lemma 5.9.8. Hence V is the direct sum of the $A_i(V)$.

We leave the following as an exercise to the reader.

Exercise 5.9.9. Show T, the triangular matrices, is a subalgebra of $\text{End}_k(V)$ and N, the strictly triangular matrices, is an ideal in T. Show D, the diagonal matrices, is also an (Abelian) subalgebra of T (which is not an ideal) and $T/N = D$. Find $\mathcal{Z}(T)$ and $\mathcal{Z}(N)$.

Exercise 5.9.10. Let T commute with D, where D is diagonal with *distinct* entries. Show T is diagonal.

We now show that every automorphism of $\text{End}_k(V)$ is inner.

Theorem 5.9.11. *Every automorphism α of the k-algebra $\text{End}_k(V)$ is inner. That is, there is some $g \in \text{GL}(V)$ such that $\alpha(T) = gTg^{-1}$ for all $T \in \text{End}_k(V)$.*

Proof. Let $A_{i,j} = \alpha(E_{i,j})$, where the $E_{i,j}$ are the matrix units. Then each $A_{i,i} \neq 0$, $A_{i,i}^2 = A_{i,i}$, $A_{i,i}A_{j,j} = \delta_{i,j}A_{j,j}$ and $\sum_{i=1}^{n} A_{i,i} = I$. For each $i = 1 \ldots n$ choose a non-zero $w_i \in A_{i,i}(V)$. If $\sum c_i w_i = 0$, then

$$0 = A_{j,j}(0) = A_{j,j}\left(\sum c_i w_i\right) = \sum A_{j,j}(c_i w_i) = c_j w_j,$$

so that each $c_j = 0$. Thus the w_i are linearly independent and therefore are a basis of V. Hence there is an invertible operator g_0 satisfying $g_0(e_i) = w_i$, where the e_i is the standard basis of V. For $T \in \text{End}_k(V)$ let $\beta(T) = g_0^{-1}\alpha(T)g_0$. Then the reader can easily check that β is also an automorphism of the algebra $\text{End}_k(V)$. Moreover, $\beta(E_{i,i}) = E_{i,i}$ so the effect of replacing α by β is that we can assume α fixes each $E_{i,i}$. Hence $\alpha(E_{i,j}) = \alpha(E_{i,i}E_{i,j}E_{j,j}) = E_{i,i}\alpha(E_{i,j})E_{j,j}$. It follows that $\alpha(E_{i,j})(e_k) = 0$ if $j \neq k$ and $\alpha(E_{i,j})(e_j) \in ke_i$. Hence $\alpha(E_{i,j}) = a_{i,j}E_{i,j}$ for some non-zero scalar $a_{i,j}$. But also $\alpha(E_{i,j}E_{j,k}) = \alpha(E_{i,k})$. Therefore the scalars $a_{i,j}$ must satisfy the relations $a_{i,j}a_{j,k} = a_{i,k}$ and since each all the $a_{i,i} = 1$, $a_{i,j}^{-1} = a_{j,i}$ for all i, j. Letting $b_i = a_{i,1}$ we see $a_{i,j} = a_{i,1}a_{1,i} = b_ib_j^{-1}$. Then $g = \text{diag}\, b_1, \ldots b_n$ is invertible and $ge_{i,j}g^{-1} =$

$b_i b_j^{-1} E_{i,j} = a_{i,j} E_{i,j}$. By the above, for all i, j, $g^{-1} E_{i,j} g = E_{i,j}$. Thus $\alpha(T) = gTg^{-1}$ for all T. $\square$

A family of linear operators (A) on a vector space V acts irreducibly if there is no non-trivial subspace W of V invariant under all of (A). As a prelude of things to come we have the following proposition.

Proposition 5.9.12. $\mathrm{End}_k(V)$ *operates irreducibly on* V.

Proof. Suppose $(0) < W < V$ is a subspace invariant under all of $\mathrm{End}_k(V)$. Write $V = W \oplus U$ as a direct sum of k-spaces. Let $m = \dim W$ and $n - m = \dim U$. Thus each matrix of $\mathrm{End}_k(V)$ has the form

$$\begin{pmatrix} A & B \\ 0 & C \end{pmatrix}.$$

Therefore, $n^2 = \dim(\mathrm{End}_k(V)) = m^2 + (n-m)^2 + m(n-m)$. Hence $n^2 - m^2 = (n-m)n$ and so $n + m = n$ and $m = 0$, a contradiction. $\square$

Another important example of a k-algebra is the *group algebra*.

Example 5.9.13. Let G be a group, which we take to be finite, since we want the group algebra to be finite dimensional. Let k be a field and define the *group algebra* $k(G)$ to be the set of all finite k-linear combinations of elements of G, written in some order, i.e.,

$$k(G) = \Big\{ \sum_{i \in I} \lambda_i g_i, \ \lambda_i \in k, \ g_i \in G \Big\}.$$

We define addition coordinate-wise and multiplication as follows:

$$\Big(\sum_I \lambda_i g_i \Big) \Big(\sum_J \mu_j g_j \Big) = \sum_{i,j} \lambda_i \mu_j g_i g_j.$$

In other words, the structure constants are given by $g_i \times g_j = g_i g_j$. Since the generators, namely G, are associative so is $k(G)$. Also, the identity of G serves as the identity of $k(G)$. In this way, $k(G)$ becomes a k-associative algebra, called the *group algebra* of G. Its dimension is $|G|$.

5.9.1 Division Algebras

Definition 5.9.14. A non-zero algebra $\mathcal{A}$ is called a *division algebra* if each $a \neq 0$ in $\mathcal{A}$ has a multiplicative inverse in $\mathcal{A}$.

Here we shall give a number of examples of division algebras, both commutative and non-commutative. We first consider division algebras over $\mathbb{R}$. Of course, $\mathbb{R}$ is itself such a division algebra.

Next comes the complex numbers, $\mathbb{C}$. This is a two dimensional vector space over $\mathbb{R}$ which is generated over $\mathbb{R}$ by 1 and i. We take as structure constants $1 \cdot 1 = 1$, $1 \cdot i = i \cdot 1 = i$ and $i \cdot i = -1$. We leave to the reader as an exercise the proof that this is a commutative division algebra over $\mathbb{R}$, isomorphic as an $\mathbb{R}$ algebra to the complex numbers.

Now, let G be the quaternion group, take $k = \mathbb{R}$ and define,

$$\mathcal{A} = \text{group algebra of } G \text{ over } \mathbb{R} \text{ is an 8-dim. vector space over } \mathbb{R}.$$

As we saw just above, this is an associative $\mathbb{R}$ algebra with a unit. However since we can get $-1, -i, -j$ and $-k$ by taking appropriate coefficients we can just restrict ourselves to the linear span of $1, i, j$ and k resulting in a 4 dimensional algebra,

$$\mathbb{H} = \{a.1 + bi + cj + dk \mid a, b, c, d \in \mathbb{R}\}.$$

$\mathbb{H}$ is an associative algebra over $\mathbb{R}$ with identity, namely 1 and, of course, is non-commutative.

Let $q = a.1 + bi + cj + dk$ be any element of $\mathbb{H}$. We shall call $bi + cj + dk$ the imaginary part of q and define $\bar{q}$ the *conjugate* of q by

$$\bar{q} = a.1 - bi - cj - dk.$$

Then

$$q\bar{q} = (a^2 + b^2 + c^2 + d^2).1 \equiv a^2 + b^2 + c^2 + d^2 = N(q),$$

where $N(q) = a^2 + b^2 + c^2 + d^2$ is called the norm of q. One checks that $N(q) = N(\bar{q})$ and that $N(q) > 0$ unless $q = 0$. Since, as we know, $\mathbb{H}$ is an $\mathbb{R}$-algebra,

$$q\left(\frac{\bar{q}}{N(q)}\right) = 1.$$

Therefore, $\frac{\bar{q}}{N(q)}$ is the multiplicative inverse of q and so $\mathbb{H}$ is a *division* algebra.

Notice that the construction of $\mathbb{H}$ is rather similar to that of $\mathbb{C}$. Evidently, conjugation is a bijective map of $\mathbb{H}$ of order 2. However, notice in the following proposition that conjugation is an anti-automorphism, not an automorphism. Thus the only reason that in $\mathbb{C}$ (and $\mathbb{R}$) it is an automorphism is that $\mathbb{C}$ happens to be commutative!

Proposition 5.9.15. *Let q and r be quaternions. Then $\overline{qr} = \bar{r}\bar{q}$ and $N(qr) = N(q)N(r)$.*

Proof. To check the first statement, since the conjugation map is $\mathbb{R}$-linear and we are in an $\mathbb{R}$-algebra, it is sufficient to check this for q and r basis elements, which we leave to the reader. $N(qr) = qr(\overline{qr}) = q(r\bar{r})\bar{q} = r\bar{r}(q\bar{q}) = N(q)N(r)$. Here we used the fact that $r\bar{r}$ is a real number, and therefore it commutes with the quaternions. $\square$

The following proposition shows that it is not easy to get to be a finite dimensional $\mathbb{R}$-division algebra.

Proposition 5.9.16. *Any division algebra, $\mathcal{A}$, over $\mathbb{R}$ with identity, of odd dimension is isomorphic to $\mathbb{R}$, and therefore has dimension 1.*

Proof. Let $a \neq 0$ be any non-zero element of $\mathcal{A}$. Consider the linear map $L_a : \mathcal{A} \to \mathcal{A}$, given by $L_a(x) = ax$. Since the dimension of $\mathcal{A}$ is odd, the Bolzano-Cauchy intermediate value theorem of Calculus tells us that L_a must have a real eigenvalue, λ. Let $v \neq 0$ be the corresponding eigenvector. Then $L_a(v) = av = \lambda v$ which implies $(a - \lambda 1_{\mathcal{A}})v = 0$, where $1_{\mathcal{A}}$ is the identity element of $\mathcal{A}$. Since $\mathcal{A}$ is a division algebra and $a \neq 0$, $a = \lambda 1_{\mathcal{A}}$, i.e. $a \in \mathbb{R}1_{\mathcal{A}}$. Hence $\mathcal{A} = \mathbb{R}1_{\mathcal{A}}$. $\square$

As we shall see, this proposition is a special case of the Frobenius theorem, which says the only finite dimensional division algebras over $\mathbb{R}$ are the reals, the complex numbers and the quaternions. (Thus the only commutative ones are the first two.) This will be proven in Chapter 13 with this proposition playing a role.

We now construct finite quaternion rings $\mathbb{H}/\mathbb{Z}_p$, where p is a prime (that is, the quaternions over $\mathbb{Z}_p$) and ask the following question:

Is $\mathbb{H}/\mathbb{Z}_p$ a skew field?

Proposition 5.9.17. *If p is a prime, then, $\mathbb{H}/\mathbb{Z}_p$ is not (even) a skew field.*

Proof. Let $p > 2$ and assume that $\mathbb{H}/\mathbb{Z}_p$ is a skew field. Since it is finite, Theorem 5.8.6 above tells us that $\mathbb{H}/\mathbb{Z}_p$ is actually a field. Hence, this gives the following relations:

$$ij = ji, \quad ij = k, \quad ji = (p-1)k. \tag{1}$$

But then, from (1), one also gets

$$0 = ij - ji = (p-1)k - k = (p-2)k,$$

a contradiction, since $p > 2$.

The case $p = 2$ is trivial, since $\mathbb{H}/\mathbb{Z}_p$ is clearly not a skew field. Just take $q = 1 + i$, and note that $q \cdot q = 0$, even though $q \neq 0$. $\square$

When $p = 3$, an alternative proof of the above proposition is the following. In $\mathbb{H}/\mathbb{Z}_3$, $(1 + i - k)(1 + i + j) = 3i = 0$ and so there are zero divisors.

For more on the structure of $\mathbb{H}/\mathbb{Z}_p$ see Aristidou [8].

An interesting "division algebra" over $\mathbb{R}$ which, however, is *not associative*, is the *Cayley numbers* or *Octonions*, **O**. **O** is an 8-dimensional algebra. (For references, see [13], [30] or [62].)

The octonions are defined as follows:

$$\mathbf{O} = \{a + bu : a, b \in \mathbb{H}\} = \mathbb{H} \oplus \mathbb{H}u,$$

where u is a fixed non-zero element of **H**. The vector space operations are defined as usual. The multiplication goes as follows:

$$(a + bu)(c + du) = (ac - \bar{d}b) + (da + b\bar{c})u,$$

where $\bar{d}$ is the usual quaternionic conjugate. This multiplication is $\mathbb{R}$-linear. We extend the conjugation to $\mathbf{O}$ by setting $\overline{(a+bu)} = \bar{a} - bu$ and we define $\operatorname{Im}(x) = (x - \bar{x})/2$. If $x = a + bu$, then

$$(x - \bar{x})/2 = (a + bu - \bar{a} + bu)/2 = \operatorname{Im}(a) + bu,$$

so a is a pure imaginary quaternion and b is arbitrary. Thus the space $\mathbf{P} = \operatorname{Im}(\mathbf{O})$ of *purely imaginary* Cayley numbers is 7-dimensional.

We define a basis in $\mathbf{O}$ by

$$4e_0 = 1, \qquad e_1 = i, \qquad e_2 = j, \qquad e_3 = k,$$
$$e_4 = u, \qquad e_5 = iu, \qquad e_6 = ju, \qquad e_7 = ku.$$

Exercise 5.9.18. Check that $\mathbf{O}$ is an $\mathbb{R}$-algebra with identity which is *non-associative*, and *every non-zero element has a multiplicative inverse*.

5.9.1.1   Exercises

Exercise 5.9.19. Let R be an integral domain with 1 (not necessarily commutative) and prove the following statements.

1. If R is finite, then it is a division ring.

2. Let R be an integral domain with identity. Show R has no idempotents $(x^2 = x)$ except $x = 0, 1$.

3. Let R be an integral domain with identity. Show R has no nilpotents $(x^n = 0)$ except $x = 0$.

Exercise 5.9.20. Let R be a ring (not necessarily commutative), $f(x) = a_0 + a_1 x + \ldots a_n x^n$, where $x, a_i \in R$ and show the following:

1. For every positive integer i and ring elements x and a that $x^i - a_i^i = (x^{i-1} + ax^{i-2} + a^2 x^{i-3} ldotsa^{i-1})$.

2. Hence left multiplication by a_i and summation over i yield

$$\sum_{i=0}^{n} a_i x^i - \sum_{i=0}^{n} a_i a^i = \sum_{i=1}^{n} a_i (x^{i-1} + ax^{i-2} + a^2 x^{i-3} + \ldots + a^{i-1})(x-a).$$

3. Hence $f(x) = q(x)(x - a) + f(a)$, where $f(a) = \sum_{i=0}^{n} a_i a^i$ and $q(x) = \sum_{j=1}^{n} q_j x^{j-1}$. Here $q_j = a_j + a_{j+1}a + \ldots + a_n a^{n-1}$.

4. Thus the remainder theorem can be made to give explicit q and remainder and this works even if R is non-commutative!

Chapter 6

R-Modules

6.1 Generalities on R-Modules

In this chapter R will be a ring with identity. In the earlier sections R will be commutative, while in the last section we will be especially interested in non-commutative rings.

The notion of an R-module is a generalization of the idea of a vector space (as well as a ring). A module[1] is formally just like a vector space, except we replace the field of scalars by R. In the last four sections of this chapter we will show why modules can be used to do things that can not be done using vector spaces.

Definition 6.1.1. Let R be a ring with 1 and $(M, +)$ an Abelian group. A *left R-module* (or simply an *R-module*) means there is a map $R \times M \longrightarrow M$, $(r, m) \mapsto rm$ such that for all $r, s \in R$ and $m, n \in M$ the following axioms hold:

1. $r(m + n) = rm + rn$.

2. $(r + s)m = rm + sm$.

3. $(rs)m = r(sm)$.

4. $1m = m$ for all $m \in M$.

---

[1]R. Dedekind was the first to use modules in his work on number theory in 1870.

This last condition is sometimes expressed by calling the R-module M *unital*. In a similar way, one can define *right R-modules*.

The obvious first example of an R-module M is the case when $R = k$, a field. Then M is a vector space, so a module is a (substantial) generalization of a vector space.

Any Abelian group, A, can be regarded as a $\mathbb{Z}$-module. Here, if $n \in \mathbb{Z}$ and $a \in A$, we take na to be

$$
na = \begin{cases} \underbrace{a + a + \cdots + a}_{n} & \text{if } n > 0, \\ 0 & \text{if } n = 0, \\ \underbrace{-a - a - \cdots - a}_{n} & \text{if } n < 0. \end{cases}
$$

We leave to the reader to check that this is indeed a module and conversely any $\mathbb{Z}$ module is just an Abelian group.

A further example of an R-module is when the ring R acts on itself by left multiplication.

Given an R-module M we have the usual notions of a *submodule* N. Namely, N is a subgroup of the additive group of M and is stable under multiplication by R. Similarly, a *module homomorphism f* is just a homomorphism of the additive group structure satisfying $f(rm) = rf(m)$ for $r \in R$ and $m \in M$.

When R is a field k, and M is a vector space, an R-module homomorphism is just a linear transformation.

As a final example of an R-module, consider two R-modules, M and M'. Let $\operatorname{Hom}_R(M, M')$ denote the set of all R-module homomorphisms from M to M'. We add such homomorphisms pointwise and let R operate on a homomorphism f by $rf(x) = f(rx)$, $x \in M$. Then, an easy check shows that with these operations $\operatorname{Hom}_R(M, M')$ is itself an R-module. It is called the set of *intertwining operators from M to M'*. This module is tailored to study non-commutative phenomena and will play an important role in Representation Theory in Chapter 12.

We leave the proofs of the next few rather routine facts to the reader.

Lemma 6.1.2. *Let M be a module over a ring R with identity, and N be a non-empty subset of M. Then*

1. $r0_M = 0_M$ *for all* $r \in R$.

2. $0_R m = 0_M$ *for all* $m \in M$.

3. $r(-m)r = -rm = -r(m)$ *for all* $m \in M$, $r \in R$.

Proposition 6.1.3. *Let* $N \subset M$. *Then* N *is a submodule of* M *if and only if for all* x, $y \in N$ *and* $r \in R$ *the following hold.*

1. $x - y \in N$.

2. $rx \in N$.

We ask the reader to check that the following are submodules.

1. If V is a vector space over the field k, then the submodules are just the subspaces.

2. If G is an Abelian group (i.e. a $\mathbb{Z}$-module), the submodules of G are simply its subgroups.

3. If R is a commutative ring with identity regarded as a module over itself by left translation, its submodules are just its ideals.

Proposition 6.1.4. *Now suppose* $f : M \to M'$ *is an* R-module ho-momorphism. *Then* $\operatorname{Ker} f$ *is a submodule of* M, $f(M)$ *is a submodule of* M' *and* f *induces an* R-module homomorphism $\tilde{f} : M/\operatorname{Ker} f \to M'$ *such that* $f = \pi \circ \tilde{f}$.

As usual, since it is only $f(M)$ that is involved here, we may as well assume f is surjective.

More generally, if $f : M \to M'$ is a surjective R-module homomor-phism, if N' is a submodule of M', then $f^{-1}(M')$ is a submodule of M and $M/f^{-1}(M') \cong M'/N'$. Moreover, $f \circ \pi' = \pi \circ \tilde{f}$.

Let M be an R-module, and N a submodule. The factor group M/N (as an additive Abelian group) has as elements cosets of the form $m + N$ for $m \in M$. We can make M/N into an R-module by defining, for $m \in M$, $r \in R$,

$$r(m + N) := rm + N.$$

The reader can check this action is well defined and the module axioms are satisfied.

Definition 6.1.5. The R-module M/N with this action is called the *quotient module* of M by N.

The canonical epimorphism $\pi : M \longrightarrow M/N$ is then a *module homomorphism*.

Example 6.1.6. Let $n \in \mathbb{Z}$, $n \geq 2$. Then $\mathbb{Z}/n\mathbb{Z}$ is a natural $\mathbb{Z}$-module.

The following concepts are completely analogous to the corresponding ones for $\mathbb{Z}$-modules (Abelian groups), or indeed for any group, given in Chapter 1.

Let M be an R-module over a commutative ring with identity. M is called *torsion free* if for any $x \neq 0 \in M$ there is no $r \neq 0 \in R$ such that $rx = 0$. We shall say M is a *torsion module* if for $x \in M$ there is some $r(x) \neq 0 \in R$ such that $rx = 0$. These elements are called torsion elements.

Exercise 6.1.7. Show that M_0 the set of torsion elements is a submodule of M. Show that M/M_0 is torsion free.

Definition 6.1.8. When S is a subset of the R-module M, we consider lin.sp.$_R S$, that is all finite linear combinations of elements of S with coefficients from R. This is evidently a submodule N of M. It is called *the submodule generated* by S and S is called a *generating set*. In particular, we shall say M is finitely generated if S is finite. That is, there are a finite number of elements $\{x_1, \ldots x_k\}$ in M so that the submodule $\{\sum_{i=1}^{k} r_i x_i : r_i \in R\}$ is M. We call k the *length* of this generating set.

6.1.1   The Three Isomorphism Theorems for Modules

Let M and N be two R-modules, and $\phi : M \rightarrow N$ be an R-homomorphism. As the following three isomorphism theorems are analogous to those for rings (or groups) and with similar proofs, we leave the details of these to the reader.

Theorem 6.1.9. *(First Isomorphism Theorem) If the module ho-momorphism ϕ is surjective, it induces an isomorphism*

$$M/\operatorname{Ker}\phi^{-} \cong N.$$

Theorem 6.1.10. *(Second Isomorphism Theorem) If L and N are submodules of the R-module M, then*

$$(L+N)/N \cong L/(L \cap N).$$

Theorem 6.1.11. *(Third Isomorphism Theorem) If L, and N are submodules of the R-module M and $N \subset L$, then L/N is a submodule of M/N and*

$$(M/N)/(L/N) \cong M/L.$$

6.1.2   Direct Products and Direct Sums

Let $(M_i)_{i \in I}$ be a family of R-modules and let $P = \prod_{i \in I} M_i$ be their cartesian product. Then, by definition, an element of P is a function

$$f : I \longrightarrow \bigcup_{i \in I} M_i, \quad \text{such that } f(i) \in M_i.$$

In addition P is defined by taking two elements f and g of P, and declaring their sum $f + g$ to be

$$f + g : I \longrightarrow \bigcup_{i \in I} M_i, \; : \; (f+g)(i) := f(i) + g(i) \in M_i.$$

$$\mu : R \times P \longrightarrow P \; : (r, f) \mapsto \mu(r, f) := rf : I \longrightarrow \bigcup_{i \in I} M_i$$

given by

$$(rf)(i) = r[f(i)] \in M_i, \quad \forall \, i \in I.$$

The reader can easily verify that this addition and scalar multiplication give a module structure P which becomes an R-module, known as the *direct product* of the family $(M_i)_{i \in I}$.

Now, let S be the subset of P containing all $f \in P$ such that $f(i) = 0$ for all except a finite number of indices $i \in I$. Clearly, S is a submodule of P. Therefore, S is a module over R, known as the *direct sum* of the family $(M_i)_{i \in I}$, and denoted by

$$S = \sum_{i \in I} M_i.$$

If I is finite, of course these coincide and we write

$$S = M_1 \oplus M_2 \oplus \cdots \oplus M_n = \bigoplus_{i=1}^{n} M_i.$$

In this case $R = S$ and the module is called either *direct product*, or *direct sum*, with the latter terminology the most popular. With the direct sum S and the direct product P of a family $(M_i)_{i \in I}$ of R-modules we have two families of homomorphisms.

First, for each $j \in I$ we define a map

$$s_j : M_j \longrightarrow S \quad \text{such that} \quad [s_j(x)](k) = \begin{cases} x & \text{if } j = k \\ 0 & \text{if } j \neq k \end{cases}$$

for each $x \in M_j$. It is easy to see that s_j is an injective homomorphism of the module M_j into the module S, and it is called the *natural projection* of M_j into $\bigoplus M_i$.

On the other hand we define the maps

$$p_k : P = \prod_{i \in I} M_i \longrightarrow M_k \quad \text{such that} \quad p_k(f) = f(k)$$

for any $f : I \longrightarrow \bigcup_{i \in I} M_i$ in P.

Once again one checks that for each k, p_k is a surjective homomorphism of the module P onto the module M_k called the *natural projection* map of the direct product P onto the module M_k.

6.2   Homological Algebra

According to C. A. Weibel, [115], Homological algebra is a tool used
to prove non-constructive existence theorems in algebra and algebraic
topology by providing obstructions to carrying out various kinds of constructions. When there is no obstruction the construction is possible.
Homological algebra had its origins in the 19th century, in the works of
Riemann (1857) and Betti (1871) on Betti numbers, and later (1895)
by Poincaré ([98], [99]) who invented the word *homology* and used it to
study manifolds which were the boundaries of higher-dimensional ones.
A 1925 observation of Emmy Noether, [92], shifted the attention from
these numbers to the *homology groups* and developed algebraic techniques such as the idea of a module over a ring which we are studying
now. These are both absolutely crucial ingredients in the modern theory
of homological algebra as well as algebra itself, yet for the next twenty
years homology theory was to remain confined to the realm of topology.
Algebraic techniques were developed for computational purposes in the
1930s, but homology remained a part of topology until the 1940s. In
1942 came the paper by Samuel Eilenberg and Saunders MacLane [32].
In it we find Hom and Ext defined for the first time, and along with them
the notions of a *functor* and *natural isomorphism*. These were needed to
provide a precise language for discussing the properties of $\mathrm{Hom}(A, B)$.
In particular, the latter varies naturally, *contravariantly* in A and *co-
variantly* in B. Only three years later this language was expanded to
include *category* and *natural equivalence*. However, this terminology
was not widely accepted by the mathematical community until the appearance of H.Cartan and Eilenberg's book in 1956. This book [22]
crystallized and redirected the field completely. Their systematic use
of *derived functors*, defined via *projective* and *injective resolutions* of
modules, unified all the previously disparate homology theories. It was
a true revolution in mathematics, and as such it was also a new beginning. The search for a general setting for derived functors led to the
notion of Abelian categories, and the search for non-trivial examples of
projective modules led to the rise of algebraic K-theory. As a result, homological algebra was here to stay. Several new fields of study emerged

from the Cartan-Eilenberg revolution. The importance of regular local rings in algebra grew out of results obtained by homological methods in the late 1950s. These methods now play a role in group theory, Lie algebras and algebraic geometry. The sheer list of terms that were first defined in their book may give the reader an idea of how much of this is due to the existence of one book that made homological algebra an important tool in mathematics.

6.2.1   Exact Sequences

A finite or infinite sequence of modules and homomorphisms

$$\cdots \longrightarrow M_{k-1} \xrightarrow{\varphi} M_k \xrightarrow{\chi} M_{k+1} \longrightarrow \cdots$$

is called an *exact sequence* if at every module other than the ends (if any) of the sequence, the image of the input homomorphism is equal with the kernel of the output homomorphism.

$$\mathrm{Im}(\varphi) = \mathrm{Ker}(\chi).$$

An exact sequence of the form

$$0 \longrightarrow M_1 \xrightarrow{\varphi} M \xrightarrow{\chi} M_2 \longrightarrow 0$$

is called a *short exact sequence*. When we have a short exact sequence, then (just as for groups)

$$M_2 \cong M/\mathrm{Ker}(\chi) = M/\mathrm{Im}(\phi), \quad \text{and} \quad \mathrm{Im}(\phi) \cong M_1.$$

Thus a short sequence generalizes the case where M_1 is a submodule of M and $M_2 = M/M_1$. A useful example is the case where $M^{\cdot} = \mathbb{Z}$:

$$0 \longrightarrow \mathbb{Z} \xrightarrow{\times n} \mathbb{Z} \xrightarrow{\pi} \mathbb{Z}/n\mathbb{Z} \longrightarrow 0$$

where $\times n$ is the multiplication by the integer n.

Definition 6.2.1. If $f : M \longrightarrow N$ is an R-module homomorphism, we define

$$\mathrm{Coker}(f) = N/\mathrm{Im}(f), \quad \text{and} \quad \mathrm{Coim}(f) = M/\mathrm{Ker}(f) \cong \mathrm{Im}(f).$$

In this way, associated with the R-module homomorphism $f : M \to N$, there are two short exact sequences

$$0 \longrightarrow \mathrm{Ker}(f) \longrightarrow M \longrightarrow \mathrm{Coim}(f) \longrightarrow 0$$

and

$$0 \longrightarrow \mathrm{Im}(f) \longrightarrow N \longrightarrow \mathrm{Coker}(f) \longrightarrow 0,$$

as well as the following exact sequence

$$0 \longrightarrow \mathrm{Ker}(f) \longrightarrow M \xrightarrow{f} N \longrightarrow \mathrm{Coker}(f) \longrightarrow 0,$$

obtained by piecing together the two previous short exact sequences.

Theorem 6.2.2. *Let*

$$0 \longrightarrow M_1 \xrightarrow{\alpha_1} M \xrightarrow{\alpha_2} M_2 \longrightarrow 0$$

be a short exact sequence of R-modules. The following statements are equivalent.

1. *$M \cong M_1 \oplus M_2$.*

2. *There is an R-homomorphism $\beta_1 : M \to M_1$ such that $\beta_1 \circ \alpha_1 = Id_{M_1}$.*

3. *There is an R-homomorphism $\beta_2 : M_2 \to M$ such that $\alpha_2 \circ \beta_2 = Id_{M_2}$.*

Proof. $1 \Longrightarrow 2$ Since the sequence is exact, α_1 is $1 : 1$ and therefore

$$M = M_1 \oplus M_2 = \alpha_1(M_1) \oplus M_2.$$

Hence there is the projection map $\pi_1 : M \to \alpha_1(M_1)$. Define $\beta_1 : M \to M_1$ by $\beta_1 = \alpha_1^{-1} \circ \pi_1$. Clearly, $\beta_1 \circ \alpha_1 = Id_{M_1}$.

$2 \Longrightarrow 3$ From exactness of the sequence it follows that α_2 is surjective, so for any $m_2 \in M_2$, there is an $m \in M$ such that $\alpha_2(m) = m_2$. We define

$$\beta_2 : M_2 \longrightarrow M \quad : \quad \beta_2(m_2) = m - \alpha_1\beta_1(m)$$

for all $m_2 \in M_2$. The map β_2 is well defined. Indeed, if there is also an $\bar{m} \in M$ such that $\alpha_2(\bar{m}) = m_2$, then, $\alpha_2(m - m_2) = 0$, and so $m - \bar{m} \in \mathrm{Ker}(\alpha_2) = \mathrm{Im}(\alpha_1)$, which implies that $m - \bar{m} = \alpha_1(m_1)$ for some $m_1 \in M_1$. Then,

$$
\begin{aligned}
m - \alpha_1\beta_1(m) &= \bar{m} + \alpha_1(m_1) - \alpha_1\beta_1(\bar{m} + \alpha_1(m_1)) \\
&= \bar{m} + \alpha_1(m_1) - \alpha_1\beta_1(\bar{m}) - \alpha_1\beta_1(m_1) \\
&= \bar{m} + \alpha_1(m_1) - \alpha_1\beta_1(\bar{m}) - \alpha_1(m_1) \\
&= \bar{m} - \alpha_1\beta_1(\bar{m}).
\end{aligned}
$$

Finally, $\alpha_2\beta_2(m_2) = \alpha_2(m) - \alpha_2\alpha_1(m) = \alpha_2(m) = m_2$, since $\alpha_2 \circ \alpha_1 = 0$.

$3 \Longrightarrow 1$ Let $m \in M$. Then $m = \alpha_2\beta_2(m) + [m - \alpha_2\beta_2(m)]$. Since

$$\alpha_2(m - \beta_2\alpha_2(m)) = \alpha_2(m) - \alpha_2(m) = 0,$$

$$m - \beta_2\alpha_2(m) \in \mathrm{Ker}(\alpha_2 = \mathrm{Im}(\alpha_1).$$

As $\alpha_2(m) \in M_2$,

$$M = \alpha_1(M_1) + \beta_2(M_2).$$

To show that this is a direct sum, we prove

$$\alpha_1(M_1) \cap \beta_2(M_2) = \{0\}.$$

Let $x \in \alpha_1(M_1) \cap \beta_2(M_2)$. Then $x = \alpha_1(x_1) = \beta_2(x_2)$ for some $x_1 \in M_1$ and $x_2 \in M_2$. Hence,

$$0 = \alpha_2\alpha_1(x_1) = \alpha_2(x) = \alpha_2\beta_2(x_2) = x_2.$$

So $x_2 = 0$. Therefore, $x = \beta_2(x_2) = \beta_2(0) = 0$, and $M = \alpha_1(M_1) \oplus \beta_2(M_2)$. But $M_1 \cong \alpha_1(M_1)$ and $M_2 \cong \beta_2(M_2)$, and so $M \cong M_1 \oplus M_2$. $\square$

Definition 6.2.3. An exact sequence which satisfies one of the statements of theorem 6.2.2 is called a *split* short exact sequence. The map β_1 is called a *retraction* and the map β_2 a *cross section* or simply a *section*.

There are certain tools concerning R-modules which came into Algebra from Algebraic Topology and which are very important in the development of homology and cohomology groups. They also turn out to be very useful in other areas. The method we use is commonly called *diagram chasing*.

Lemma 6.2.4. *(The Four Lemma) Suppose that in the diagram*

$$
\begin{array}{ccccccc}
M_1 & \xrightarrow{\alpha_1} & M_2 & \xrightarrow{\alpha_2} & M_3 & \xrightarrow{\alpha_3} & M_4 \\
f_1 \downarrow & & f_2 \downarrow & & f_3 \downarrow & & f_4 \downarrow \\
N_1 & \xrightarrow{\beta_1} & N_2 & \xrightarrow{\beta_2} & N_3 & \xrightarrow{\beta_3} & N_4
\end{array}
$$

the rows are exact and each square commutes. If f_1 is surjective and f_4 is injective, then, the following equations hold.

1. $Im(f_2) = \beta_2^{-1}(Im(f_3))$.

2. $\mathrm{Ker}(f_3) = \alpha_2(\mathrm{Ker}(f_2))$.

Hence, if f_3 is surjective, so is f_2 and if f_2 is injective, so is f_3.

Proof. To prove 1), let $n_2 \in Im(f_2)$. Thus there is an element $m_2 \in M_2$ such that $f_2(m_2) = n_2$. By the commutativity of the middle square

$$\beta_2(n_2) = \beta_2(f_2(m_2)) = f_3(\alpha_2(m_2)) \in Im(f_3).$$

This implies $n_2 \in \beta_2^{-1}(Im(f_3))$, and this for any $n_2 \in N_2$. Hence

$$Im(f_2) = \beta_2^{-1}(Im(f_3)).$$

To prove the converse, pick an element $n_2 \in \beta_2^{-1}(Im(f_3))$. Then, $\beta_2(n_2) = n_3 \in Im(f_3)$. Hence, there is $m_3 \in M_3$ such that $f_3(m_3) = n_3$, and since the bottom row is exact $\beta_3(n_3) = 0$. Using the commutativity of the right square

$$f_4(\alpha_3(m_3)) = \beta_3(f_3(m_3)) = 0,$$

and since f_4 is $1:1$, $\alpha_3(m_3) = 0$. Therefore, using the exactness of the first row, $m_3 \in \text{Ker}(\alpha_3) = \text{Im}(\alpha_2)$, which means that there is an $m_2 \in M_2$ such that $\alpha_2(m_2) = m_3$. Now, consider the element $n_2 - f_2(m_2)$ in N_2. Then

$$\beta_2(n_2 - f_2(m_2)) = \beta_2(n_2) - \beta_2(f_2(m_2)) = n_3 - n_3 = 0.$$

Hence, $(n_2 - f_2(m_2)) \in \text{Ker}(\beta_2) = \text{Im}(\beta_1)$. It follows that there is an $n_1 \in N_1$ such that $\beta_1(n_1) = n_2 - f_2(m_2)$, and since f_1 is onto, there is an element $m_1 \in M_1$ such that $f_1(m_1) = n_1$.

Now, consider $\alpha_1(m_1) + m_2$ in M_2. Using the commutativity of the left square we get

$$f_2(\alpha_1(m_1) + m_2) = f_2(\alpha_1(m_1)) + f_2(m_2) = \beta_1(f_1(m_1)) + f_2(m_2)$$
$$= \beta_1(n_1) + f_2(m_2) = n_2 - f_2(m_2) + f_2(m_2) = n_2,$$

which shows that $n_2 \in \text{Im}(f_2)$, and since this occurs for any element of $\beta_2^{-1}(\text{Im}(f_3))$ we get

$$\beta_2^{-1}(\text{Im}(f_3)) \subset \text{Im}(f_2).$$

To prove 2), let $m_3 \in M_3$. Then, $f_3(m_3) = 0$ and using the commutativity of the right square

$$f_4(\alpha_3(m_3)) = \beta_3(f_3(m_3)) = \beta_3(0) = 0,$$

and since f_4 is $1:1$, we obtain $\alpha_3(m_3) = 0$, i.e. $m_3 \in \text{Ker}(\alpha_3) = \text{Im}(\alpha_2)$ by the exactness of the first row. This means there is $m_2 \in M_2$ such that $\alpha_2(m_2) = m_3$.

Now consider $n_2 = f_2(m_2) \in N_2$. Using the commutativity of the middle square again we get

$$\beta_2(n_2) = \beta_2(f_2(m_2)) = f_3(\alpha_2(m_2)) = f_3(m_3) = 0,$$

and so $n_2 \in \text{Ker}(\beta_2) = \text{Im}(\beta_1)$ by the exactness of the second row. Hence, there is an $n_1 \in N_1$ such that $\beta_1(n_1) = n_2$, and since f_1 is onto, there is an $m_1 \in M_1$ with $f_1(m_1) = n_1$. Now take the element $m_2 - \alpha_1(m_1)$ in M_2. The commutativity of the left square yields,

$$f_2(m_2 - \alpha_2(m_1)) = f_2(m_2) - f_2(\alpha_1(m_1))$$
$$= f_2(m_2) - \beta_1(f_1(m_1)) = n_2 - n_2 = 0,$$

which shows that $m_2 - \alpha_1(m_1)$ is in $\text{Ker}(f_2)$. Also,

$$\alpha_2(m_2 - \alpha_1(m_1)) = \alpha_2(m_2) - \alpha_2(\alpha_1(m_1)) = \alpha_2(m_2) = m_3.$$

Hence $m_3 \in \alpha_2(\text{Ker}(f_2))$. From this it follows that

$$\text{Ker}(f_3) \subset \alpha_2(\text{Ker}(f_2)).$$

For the converse, take an element $m_3 \in \text{Ker}(f_2)$. Then, there is $m_2 \in \text{Ker}(f_2)$ such that $\alpha_2(m_2) = m_3$. Again, using the commutativity of the middle square

$$f_3(m_3) = f_3(\alpha_2(m_2)) = \beta_2(f_2(m_2)) = \beta_2(0) = 0,$$

showing $m_3 \in \text{Ker}(f_3)$ and therefore

$$\alpha_2(\text{Ker}(f_2)) \subset \text{Ker}(f_2).$$

Finally, the last assertion is a direct consequence of 1) and 2). $\square$

Lemma 6.2.5. (*The Five Lemma*) *Suppose that in the diagram*

$$
\begin{array}{ccccccccc}
M_1 & \xrightarrow{\alpha_1} & M_2 & \xrightarrow{\alpha_2} & M_3 & \xrightarrow{\alpha_3} & M_4 & \xrightarrow{\alpha_4} & M_5 \\
\downarrow{f_1} & & \downarrow{f_2} & & \downarrow{f_3} & & \downarrow{f_4} & & \downarrow{f_5} \\
N_1 & \xrightarrow[\beta_1]{} & N_2 & \xrightarrow[\beta_2]{} & N_3 & \xrightarrow[\beta_3]{} & N_4 & \xrightarrow[\beta_4]{} & N_5
\end{array}
$$

the rows are exact, and the four vertical maps f_1, f_2, f_4 and f_5 are isomorphisms. Then, so is f_3.

Proof. The Four Lemma tells us it is enough to show f_3 is surjective. To see this, let $n_3 \in N_3$. Since f_4 is surjective, there is an $m_4 \in M_4$ such that $f_4(m_4) = \beta_3(n_3)$ in N_3. The commutativity of the rightmost square tells us

$$\beta_4(\beta_3(n_3)) = \beta_4(f_4(m_4)) = f_5(a_4(m_4)),$$

and since the second row is exact at N_3, i.e. $\text{Im}(\beta_3) = \text{Ker}(\beta_4)$,

$$\beta_4(\beta_3(n_3)) = \beta_4(f_4(m_4)) = 0.$$

Since f_5 is injective $a_4(m_4) = 0$, and so $m_4 \in \text{Ker}(a_4) = \text{Im}(a_3)$. Therefore, there is some δ in M_3 such that $a_3(\delta) = m_4$.

Now, $f_4(a_3(\delta)) = f_4(m_4) = \beta_3(n_3)$ and the commutativity of the squares gives us $f_4(a_3(\delta)) = \beta_3(f_3(\delta))$. Hence $\beta_3(f_3(\delta)) = \beta_3(n_3)$, i.e. $\beta_3(f_3(\delta) - n_3) = 0$, which implies $(f_3(\delta) - n_3) \in \text{Ker}(\beta_3) = \text{Im}(\beta_2)$. This means there is an $n_2 \in N_2$ such that $\beta_2(n_2) = f_3(\delta) - n_3$. But f_2 is surjective, there is an $m_2 \in M_2$ such that $f_2(m_2) = n_2$, therefore $\beta_2(f_2(m_2)) = f_3(\delta) - n_3$. The commutativity of the square implies $\beta_2(f_2(m_2)) = f_3(a_2(m_2)) = f_3(\delta) - n_3$. Thus

$$n_3 = f_3(\delta) - f_3(a_2(m_2)) = f_3(\delta - a_2(m_2)).$$

Hence f_3 is surjective.   □

Exercise 6.2.6. If in the following diagram of R-module homomorphisms

$$M_1 \xrightarrow{\ \alpha_1\ } M_2 \xrightarrow{\ \alpha_2\ } M_3 \longrightarrow 0$$

$$\Big\downarrow f$$

$$N$$

the row is exact, and $f \circ \alpha_1 = 0$, show there is a unique homomorphism $\gamma : M_3 \to N$ such that $\gamma \circ \alpha_2 = f$.

Exercise 6.2.7. If in the following diagram of R-module homomorphisms

$$N$$

$$f \downarrow$$

$$0 \longrightarrow M_1 \xrightarrow{\alpha_1} M_2 \xrightarrow{\alpha_2} M_3$$

the row is exact and $\alpha_2 \circ f = 0$, show there is a unique homomorphism $\gamma : N \to M_1$ such that $\alpha_1 \circ \gamma = f$.

The following exercise is an important tool in homological algebra.

Exercise 6.2.8. (The Snake Lemma) Consider the following commutative diagram of R-modules:

$$
\begin{array}{ccccccc}
M_1 & \xrightarrow{\alpha_1} & M_2 & \xrightarrow{\alpha_2} & M_3 & \longrightarrow & 0 \\
\downarrow{f_1} & & \downarrow{f_2} & & \downarrow{f_3} & & \\
0 \longrightarrow & N_1 & \xrightarrow{\beta_1} & N_2 & \xrightarrow{\beta_2} & N_3 &
\end{array}
$$

Show there is an exact sequence relating the kernels and the co-kernels of f_1, f_2 and f_3

$$\mathrm{Ker}(f_1) \longrightarrow \mathrm{Ker}(f_2) \longrightarrow \mathrm{Ker}(f_3) \xrightarrow{\partial} \mathrm{Coker}(f_1) \longrightarrow$$

$$\longrightarrow \mathrm{Coker}(f_2) \longrightarrow \mathrm{Coker}(f_3)$$

where ∂ is an homomorphism known as the *connecting homomorphism*.

Exercise 6.2.9. (**The** 3×3 **Lemma**) Consider the following diagram of R-module homomorphisms

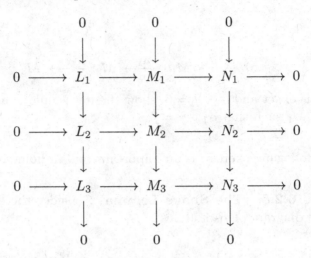

If all three rows and either the first two columns or the last two columns are short exact sequences, then so is the remaining column.

6.2.2 · Free Modules

Here M will be an R-module, with R a commutative ring with identity.

Definition 6.2.10. We say the subset S of M *generates* M if for each $m \in M$

$$m = r_1 s_1 + \cdots + r_n s_n,$$

where $s_i \in S$, $i = 1, ..., n$, and $r_i \in R$, $i = 1, ..., n$. If S is a finite subset, then M is *finitely generated* and the elements of S are called the *generators* of M. In the same vein, we shall say the subset S of M is *linearly independent* if whenever s_1, ..., s_n are distinct non-zero elements of S, and r_1, ..., r_n are in R, then,

$$r_1 s_1 + r_2 s_2 + \cdots + r_n s_n = 0, \quad \Rightarrow \quad r_1 = \cdots = r_n = 0.$$

Definition 6.2.11. A subset, S, of M is called a *basis of M over R* if it is linearly independent and generates M. A *free R-module* is an R-module that has a basis.

The point here is that in a vector space (i.e. when R is a field) this happens rather easily, but when R is merely a commutative ring with identity and perhaps has a more complicated ideal theory (such as $\mathbb{Z}$!) this is more difficult to achieve.

It is straightforward to see that if S is a basis of M, then every $m \in M$ can be written uniquely as

$$m = \sum_{i=1}^{n} r_i s_i,$$

where $s_1, ..., s_n$ are distinct elements of S, and $r_1, ..., r_n$ are in R.

Theorem 6.2.12. *Let M be an R-module. Then the following statements are equivalent:*

1. *$M = \bigoplus_{i \in I} R_i$ where $R_i = R$ for each $i \in I$.*

2. *M has a basis.*

Proof. $1 \implies 2$ It is clear that $(1, 0, 0, ...)$, $(0, 1, 0, ...)$, ... form a basis for $\bigoplus_{i \in I} R_i$, with $R_i = R$ for all i, and since isomorphisms send bases to bases, M has a basis.

$2 \implies 1$ Let $(m_i)_{i \in I}$ be a basis for M. Then

$$M = \bigoplus_{i \in I} R m_i.$$

To show that $R m_i = R$ for any i, consider the map

$$\psi : R \longrightarrow R m_i \quad : \quad \psi(r) = r m_i.$$

This is a homomorphism, and obviously surjective. In addition, if $r m_i = 0$, then $r = 0$ since m_i is a basis. Hence ψ is an isomorphism. $\square$

We remark that if M is a free R-module with bases S and T, then it is possible that $|S| \neq |T|$, even when both bases are finite. For an example, see Hungerford, [60], exercise 13, p. 190. This leads us to the following definition.

Definition 6.2.13. We say that R has the *invariant dimension property* if whenever M is a free R-module with bases S and T, then $|S| = |T|$.

If R has the invariant dimension property and M is a free module with basis S, then the *dimension* or *rank* of M is $\dim(M) = |S| = \mathrm{rank}(M)$.

As we know R is a field, or even a division ring, it has the invariant dimension property. More generally, as we shall see, if R is a Euclidean ring (or even a PID) and M has a *finite basis* then it also has the invariant dimension property. However, for commutative ring with identity this is generally not the case (see [60] Theorem 2.7, p. 186 and Corollary 2.12, pg. 186).

Theorem 6.2.14. *(Universal Property of a Basis) Suppose M is a free R-module, S is a basis of M, and N is any R-module. If $f : S \to N$ is any map, then there exists a unique R-module homomorphism $\phi : M \to N$ such that the diagram*

commutes, where $i : S \to M$ is the inclusion map.

Proof. First we prove existence. Suppose $S = \{s_i \mid i \in I\}$. Then for $m \in M$, $m = \sum_{i \in I} r_i s_i$, where the $r_i \in R$ and only finitely many r_i are different from 0. Therefore, this is really a finite sum. Define

$$\phi(m) = \sum_{i \in I} r_i f(s_i).$$

Since the coefficients, r_i, for $i \in I$ are uniquely determined by m, ϕ is well defined.

To show that ϕ is R-linear, suppose m_1 and m_2 are in M and r is in R. Say $m_1 = \sum_{i \in I} r_i s_i$ and $m_2 = \sum_{i \in I} \lambda_i s_i$. Then, since the set of all i in I with $r_i \neq 0$ or $\lambda_i \neq 0$ is finite,

$$\phi(rm_1 + m_2) = \phi\left(r \sum_{i \in I} r_i s_i + \sum_{i \in I} \lambda_i s_i\right)$$

$$= \phi\left(\sum_{i \in I}(rr_i + \lambda_i)s_i\right)$$

$$= \sum_{i \in I}(rr_i + \lambda_i)f(s_i)$$

$$= r \sum_{i \in I} r_i f(s_i) + \sum_{i \in I} \lambda_i f(s_i)$$

$$= r\phi(m_1) + \phi(m_2).$$

It follows from the definition of ϕ that $\phi(s) = f(s)$ for s in S since $s = 1 \cdot s$. To prove uniqueness, suppose $\psi : S \to M$ is an R-module homomorphism and $\psi(s) = f(s)$ for all s in S. If $m = \sum_{i \in I} r_i s_i$ is in M, then

$$\psi(m) = \psi\left(\sum_{i \in I} r_i s_i\right)$$

$$= \sum_{i \in I} r_i \psi(s_i) \quad (\psi \text{ is } R\text{-linear})$$

$$= \sum_{i \in I} r_i f(s_i) \quad (\psi(s) = f(s) \, \forall s \in S)$$

$$= \phi(m).$$

Therefore $\psi = \phi$. $\qquad\square$

Exercise 6.2.15. Suppose R is a ring and M and N are R-modules. Show that composition of functions defines an $\mathrm{End}_R(N)$-module structure on $\mathrm{Hom}_R(M, N)$: $\mathrm{End}_R(N) \times \mathrm{Hom}_R(M, N) \to \mathrm{Hom}_R(M, N)$ by $(f, \phi) \mapsto f \circ \phi$. Show that $\mathrm{Hom}_R(M, N)$ is unital if R has an identity and M and N are unital.

Exercise 6.2.16. Suppose R is a ring with identity and M and N are free, unital R-modules with finite bases, $S = \{s_1, \ldots, s_m\}$ and $B = \{b_1, \ldots, b_n\}$ respectively.

1. Find an R-module isomorphism, $\gamma \colon M \to R^m$.

2. Find an R-module isomorphism, $\psi_{M,N} : \mathrm{Hom}_R(M, N) \to M_{n,m}(R)$.

3. Show that when $M = N$, the isomorphism $\psi_{N,N} : \mathrm{End}_R(N) \to M_n(R)$ is an R-algebra isomorphism.

4. Show that the diagram

$$
\begin{array}{ccc}
\mathrm{End}_R(N) \times \mathrm{Hom}_R(M, N) & \longrightarrow & \mathrm{Hom}_R(M, N) \\
\downarrow & & \downarrow \\
M_n(R) \times M_{n,m}(R) & \longrightarrow & M_{n,m}(R)
\end{array}
$$

commutes, where the top horizontal arrow is given by composition of functions, the left vertical arrow is the product, $\psi_{N,N} \times \psi_{M,N}$, of the two maps in 2) and 3), the right vertical arrow is the map, $\psi_{M,N}$, in (2), and the bottom horizontal map is matrix multiplication.

6.2.3   The Tensor Product of R-modules

Here we extend the notion of a tensor product to modules. The first step in doing so is due to H. Whitney [119], who in 1938 defined the product $A \otimes_{\mathbb{Z}} B$ for Abelian groups A and B. A few years later Bourbaki's volume on algebra contained a definition of tensor products of modules in the current form. When R is a commutative ring with identity, the tensor product of R-modules is similar to that of k-vector spaces (as in Chapter 3). However, when R is non-commutative the situation is very complicated and we will not deal with this.

Let M, N, and P be R-modules, where R is a commutative ring with identity.

Definition 6.2.17. A map $\phi : M \times N \longrightarrow P$ is called *R-bilinear* if $\phi(\cdot, n) : M \longrightarrow P$ and $\phi(m, \cdot) : N \longrightarrow P$ are R-linear for all $m \in M$ and $n \in N$.

Definition 6.2.18. (Universal Property) The tensor product of M and N over R is an R-module $M \otimes_R N$ together with a bilinear map $\otimes : M \times N \longrightarrow M \otimes_R N$ such that the following universal property holds: for any R-module P and for every bilinear map $\phi : M \times N \longrightarrow P$ there is a unique linear map $\widehat{\phi} : M \otimes_R N \longrightarrow P$ such that $\phi = \widehat{\phi} \circ \otimes$, i.e. such that the following diagram commutes:

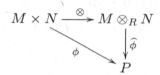

Theorem 6.2.19. *(Uniqueness of Tensor Products)* *The tensor product is unique up to an isomorphism in the following sense: if T is another tensor product for M and N over R, then, there is a unique R-module isomorphism $\widehat{\phi} : M \otimes_R N \longrightarrow T$ such that $\phi = \widehat{\phi} \circ \otimes$.*

Theorem 6.2.20. *(Existence of Tensor Products)* *Any two R-modules have a tensor product.*

Some properties of the tensor product of R-modules are the following:

Proposition 6.2.21. *For any R-modules M, N, and P there are natural isomorphisms as follows:*

1. *$M \otimes N = N \otimes M$.*

2. *$M \otimes R = M$.*

3. *$(M \oplus N) \otimes P \cong (M \otimes P) \oplus (N \otimes P)$.*

Proof. 1) The map $M \times N \to N \otimes M$ defined by $(m, n) \mapsto n \otimes m$ is bilinear so by the universal property of the tensor product, it induces a (unique) linear map $\phi : M \otimes N \to N \otimes M$ defined by $\phi(m \otimes n) = n \otimes m$.

Similarly, we get a unique linear map $\psi : N \otimes M \to M \otimes N$ such that $\psi(n \otimes m) = m \otimes n$. Therefore,

$$(\psi \circ \phi)(m \otimes n) = m \otimes n \quad \text{so } \psi \circ \phi = 1 \text{ on pure tensors.}$$

But the pure tensors generate $M \otimes N$ over R. Therefore $\psi \circ \phi = 1_{M \otimes N}$. Similarly we get $\phi \circ \psi = 1_{N \otimes M}$, i.e. ϕ is an isomorphism.

2) Again using the universal property, one sees that the bilinear map $M \times R \to M$ such that $(m, r) \mapsto rm$ induces a linear map

$$\phi : M \otimes R \longrightarrow M \quad \phi(m \otimes r) = rm.$$

Moreover, there is a linear map

$$\psi : M \longrightarrow M \otimes R \quad \psi(m) = m \otimes 1.$$

One gets

$$(\psi \circ \phi)(m \otimes r) = rm \otimes 1 = m \otimes r \quad \text{and} \quad (\phi \circ \psi)(m) = m$$

which shows that ϕ is an isomorphism.

3) Again, the bilinear map

$$(M \oplus N) \times P \longrightarrow (M \otimes P) \oplus (N \otimes P) \quad : \quad \Big((m, n), p\Big) \mapsto (m \otimes p, n \otimes p),$$

induces a linear map

$$\phi : (M \oplus N) \otimes P \longrightarrow (M \otimes P) \oplus (N \otimes P) \quad : \quad \phi\Big((m, n), p\Big) \mapsto (m \otimes p, n \otimes p).$$

Similarly, we get maps

$$M \otimes P \longrightarrow (M \oplus N) \otimes P \quad : \quad m \otimes p \mapsto (m, 0) \otimes p,$$

and

$$N \otimes P \longrightarrow (M \oplus N) \otimes P \quad : \quad n \otimes p \mapsto (0, n) \otimes p.$$

Adding, one gets the linear map

$$\psi : (M \otimes P) \oplus (N \otimes P) \longrightarrow (M \oplus N) \otimes P$$

defined as $\psi(m \otimes p, n \otimes p) = (m, 0) \otimes p + (0, n) \otimes p$.

One verifies easily that ϕ and ψ are inverses of one another on the pure tensors and hence also on the tensor product. $\square$

Let I and J be co-prime ideals in a ring R. Then, there are elements $a \in I$ and $b \in J$ with $a + b = 1$. Hence in the tensor product $R/I \otimes R/J$ for all monomial tensors $\bar{r} \otimes \bar{s}$ with r, $s \in R$ we obtain,

$$\bar{r} \otimes \bar{s} = (a + b)(\bar{r} \otimes s) = \overline{ar} \otimes s + \bar{r} \otimes \overline{bs} = 0.$$

This is because $\overline{ar} = 0 \in R/I$ and $\overline{bs} = 0 \in R/J$. As these monomial tensors generate the tensor product, we conclude that $R/I \otimes R/J = 0$.

As a concrete example,

$$\mathbb{Z}_p \otimes_{\mathbb{Z}} \mathbb{Z}_q = 0$$

for any two distinct primes p and q.

This shows that a tensor product space need not be bigger than its factors, and could even be 0 when none of its factors are.

(Extension of Scalars). Let M be an R-module, and $\overline{R}$ an R-algebra (so that $\overline{R}$ is a ring as well as an R-module). In addition, for any $a \in \overline{R}$ we denote by

$$\kappa_a : \overline{R} \longrightarrow \overline{R}, \quad r \mapsto \kappa_a(r) := ar$$

the multiplication map, which is obviously a homomorphism of R-modules. Then, setting

$$M_{\overline{R}} := M \otimes_R \overline{R},$$

we obtain a scalar multiplication by $\overline{R}$ on $M_{\overline{R}}$ given by

$$\overline{R} \times M_{\overline{R}} \longrightarrow M_{\overline{R}}$$

such that

$$(a, m \otimes r) \mapsto a \cdot (m \otimes r) := (1 \otimes \kappa_a)(m \otimes r) = m \otimes (ar)$$

which makes $M_{\overline{R}}$ into an $\overline{R}$-module. We say that $M_{\overline{R}}$ is obtained from M by an *extension of scalars* from R to $\overline{R}$. Note that any R-module homomorphism $\phi : M \longrightarrow N$ then gives rise to an "*extended*" $\overline{R}$-module homomorphism

$$\phi_{\overline{R}} := \phi \otimes \mathrm{id} : M_{\overline{R}} \longrightarrow N_{\overline{R}}.$$

(Multivariate Polynomial Rings as Tensor Products). Let R be a ring. We claim that the polynomial ring in 2 variables $R[x, y]$ is given by

$$R[x, y] \cong R[x] \otimes_R R[y]$$

as R-algebras, i.e. that polynomial rings in several variables can be seen as tensor products of polynomial rings in one variable.

Indeed, by the universal property of the tensor product, there are R-module homomorphisms

$$\phi : R[x] \otimes_R R[y] \longrightarrow R[x; y], \quad f \otimes g \mapsto fg,$$

and

$$\chi : R[x, y] \longrightarrow R[x] \otimes R[y], \quad \text{such that} \quad \sum_{i,j} a_{ij} x^i y^j \mapsto \sum_{i,j} a_{ij} x^i \otimes y^j.$$

Since

$$(\chi \circ \phi)(x^i \otimes y^j) = \chi(x^i \otimes y^j) = x^i \otimes y^j$$

and

$$(\phi \circ \chi)(x^i \otimes y^j) = \phi(x^i \otimes y^j) = x^i y^j$$

for all $i, j \in \mathbb{N}$ and since $x^i \otimes y^j$ generate $R[x] \otimes_R R[y]$, and $x^i y^j$ generate $R[x, y]$ as R-modules, we see that ϕ and χ are inverse to each other. Moreover, ϕ is also a ring homomorphism with the multiplication in $R[x] \otimes_R R[y]$, since

$$\phi((f \otimes g) \cdot (f' \otimes g')) = \phi((ff') \otimes (gg')) = ff' gg' = \phi(f \otimes g) \cdot \phi(f' \otimes g').$$

Hence

$$R[x, y] \cong R[x] \otimes_R R[y]$$

as R-algebras.

Finally, we conclude this section by studying how tensor products behave when applied to exact sequences. The following shows that the tensor products are right exact. That is, tensoring an exact sequence on the right results in an exact sequence.

Theorem 6.2.22. *(Tensor Products are Right Exact) Let N be an R-module and let*

$$M_1 \xrightarrow{\alpha_1} M_2 \xrightarrow{\alpha_2} M_3 \longrightarrow 0$$

be an exact sequence of R-modules. Then, the sequence

$$M_1 \otimes N \xrightarrow{\alpha_1 \otimes i} M_2 \otimes N \xrightarrow{\alpha_2 \otimes i} M_3 \otimes N \longrightarrow 0,$$

where $i : N \to N$ is the identity endomorphism, is also exact.

Proof. Since both α_2 and i are surjective, so is $\alpha_2 \otimes i$. Now, $\operatorname{Ker}(\alpha_2 \otimes i)$ is the submodule of $M_2 \otimes N$ generated by pure tensors $m_2 \otimes n$, with $m_2 \in \operatorname{Ker}(\alpha_2)$ and $n \in N$. Because of the exactness $\operatorname{Im}(\alpha_1) = \operatorname{Ker}(\alpha_2)$, so there is $m_1 \in M_1$ such that $\alpha_1(m_1) = m_2$, which implies

$$m_2 \otimes n = \alpha_1(m_1) \otimes n = \alpha_1(m_1) \otimes i(n) = (\alpha_1 \otimes i)(m_1 \otimes n) \in \operatorname{Im}(\alpha_1 \otimes i).$$

Because the pure tensors generate $\operatorname{Ker}(\alpha_2 \otimes i))$ it follows that

$$\operatorname{Ker}(\alpha_2 \otimes i) \subset \operatorname{Im}(\alpha_1 \otimes i). \tag{1}$$

On the other hand,

$$(\alpha_2 \otimes i) \circ (\alpha_1 \otimes i) = (\alpha_2 \circ \alpha_1) \otimes (i \circ i) = 0 \otimes i,$$

which implies

$$\operatorname{Im}(\alpha_1 \otimes i) \subset \operatorname{Ker}(\alpha_2 \otimes i). \tag{2}$$

From (1) and (2) we get our conclusion. $\square$

Proposition 6.2.23. *There is a natural isomorphism*

$$\operatorname{Hom}(M \otimes N, P) \cong \operatorname{Hom}(M, \operatorname{Hom}(N, P))$$

gotten by sending any $\alpha : M \otimes N \to P$ to $\hat{\alpha}$, where

$$\hat{\alpha}(m) = \alpha(m \otimes -) : N \to P,$$

for any $m \in M$.

Proof. The map

$$\mathrm{Hom}(M, \mathrm{Hom}(N, P)) \longrightarrow \mathrm{Hom}(M \otimes N, P)$$

sends the map $f : M \to \mathrm{Hom}(N, P)$ to the map $\hat{f} : M \otimes N \to P$ which is defined by $\hat{f}(m \otimes n) = f(m)(n)$. This map is bilinear and so it induces a map on the tensor product. It is easily checked that these two maps are inverses of one another. $\square$

Exercise 6.2.24. Let a, b be positive integers with $d = \gcd(a, b)$ then

$$\mathbb{Z}/a\mathbb{Z} \otimes_{\mathbb{Z}} \mathbb{Z}/b\mathbb{Z} \cong \mathbb{Z}/d\mathbb{Z}$$

as Abelian groups. In particular,

$$\mathbb{Z}/a\mathbb{Z} \otimes_{\mathbb{Z}} \mathbb{Z}/b\mathbb{Z} = 0$$

if and only if $d = 1$.

Exercise 6.2.25. If the sequence of R-modules

$$0 \longrightarrow M_1 \xrightarrow{\alpha_1} M_2 \xrightarrow{\alpha_2} M_3 \longrightarrow 0$$

is a short exact sequence which splits, then, so does the sequence

$$0 \longrightarrow M_1 \otimes N \xrightarrow{\alpha_1 \otimes i} M_2 \otimes N \xrightarrow{\alpha_2 \otimes i} M_3 \otimes N \longrightarrow 0$$

where $i : N \to N$ is the identity endomorphism.

Exercise 6.2.26. Prove that for any R-modules M, N, N_1, N_2 and P, there exist natural isomorphisms as follows:

1. $(M \otimes N) \otimes P = M \otimes (N \otimes P)$.

2. $\otimes(N_1 \times N_2) \cong (M \otimes N_1) \times (N \otimes N_2)$.

6.3   *R*-Modules vs. Vector Spaces

Although the definitions of a vector space and of a module appear to be completely similar, except that the field k of scalars has been replaced by the ring R, because of this there are many *significant differences* between these two structures. For example, in a module rx can be 0 with neither $r \in R$ nor $x \in M$ being 0. For example, this occurs when the module is a ring with zero divisors operating on itself by left translation. As we have observed, vector spaces, even over a division ring, have the invariant dimension property. However, as mentioned above this fails for modules over a commutative ring with 1 (see [60]).

In the case of vector spaces, as we saw in Chapter 3, every subspace has a complement. However, even over $\mathbb{Z}$, a submodule need not have a complement. For example $\mathbb{Z}$ is a $\mathbb{Z}$-module (a module over itself). Its submodules are precisely the ideals, i.e. the sets $n\mathbb{Z}$. Hence, any two non-zero submodules of $\mathbb{Z}$ have non-zero intersections. For if m, $n > 0$, then $0 \neq mn \in m\mathbb{Z} \cap n\mathbb{Z}$. Thus, none of the submodules $n\mathbb{Z}$, with $n \neq 0, \ 1$, have complements. Alternatively, since $\mathbb{Z}$ is torsion free and all elements of $\mathbb{Z}/n\mathbb{Z} = \mathbb{Z}_n$, have finite order, $n\mathbb{Z}$ cannot be a direct summand.

A subspace of a finite dimensional vector space is itself finitely generated and as we shall see in this chapter, the same is true over $\mathbb{Z}$. More generally, we will prove in Vol. II, Chapter 9 that in a finitely generated module over a Noetherian ring every submodule is finitely generated. However, in general, this is not so.

In any vector space the set $S = \{v\}$, consisting of a single non-zero vector v, is linearly independent. However, in a module, this need not be the case. For example, in the $\mathbb{Z}$-module $\mathbb{Z}_n$ any point $k \in \mathbb{Z}_n - \{0\}$, satisfies $nk = 0$. Therefore we see that no singleton set $\{k\}$ is linearly independent. Similarly, every vector space has a basis. But as above there are modules which do not even have linearly independent elements, and hence have no basis. Since the entire module is a spanning set, a minimal spanning set need not be a basis.

In a vector space, a set S of vectors is linearly dependent if and only if one of the vectors in S is a linear combination of the others. For arbitrary modules this is not true. For example, in the $\mathbb{Z}$-module $\mathbb{Z}^2$, consisting of all ordered pairs of integers, the ordered pairs $(2,0)$ and $(3,0)$ are linearly dependent, since $3(2,0) - 2(3,0) = (0,0)$, but neither one of these ordered pairs is a linear combination (i.e. a scalar multiple) of the other.

In a vector space, a set of vectors is a basis if and only if it is a minimal spanning set, or equivalently, a maximal linearly independent set. For the modules the following is the best we can do in general:

Let $\mathcal{B}$ be a basis for an R-module M. Then,

1. $\mathcal{B}$ is a minimal spanning set.

2. $\mathcal{B}$ is a maximal linearly independent set.

That is, in a module, neither a minimal spanning set nor a maximal linearly independent set is necessarily a basis. Such sets may not exist.

The following exercises show further differences between R-modules and vector spaces.

Exercise 6.3.1. Let $\mathbb{Q}_+$ be the additive group of rational numbers. Clearly, it is a $\mathbb{Z}$-module. Prove that any two elements of $\mathbb{Q}_+$ are linearly dependent over $\mathbb{Z}$. Conclude that $\mathbb{Q}_+$ cannot have a basis over $\mathbb{Z}$.

Exercise 6.3.2. Here we give some additional examples of R-modules. We ask the reader to verify the following statements.

1. Let A be some index set and R be a ring with identity. Then $\prod_A R = \{\phi : A \to R\}$ with pointwise operations is an R-module called the direct product.

2. Let A be some index set and R be a ring with identity. Then $\sum_A R = \{\phi : A \to R : \phi = 0 \quad \text{for all but finitely many } A\}$ with pointwise operations is an R-submodule of the direct product called the weak direct product.

3. Of course, if A is a finite set these two coincide and we write R^n, where $n = \text{card}(A)$. If $n = 1$ we just get R itself. Thus any ring R with identity is an R-module. Its submodules are the left ideals.

6.4 Finitely Generated Modules over a Euclidean Ring

Here we shall be concerned with a crucial generalization of a vector space, namely a module M over a Euclidean ring R (or more generally one could work over a principal ideal domain). The structure theorem we will prove can then be turned back on itself to give important results in linear algebra, namely the Jordan canonical form.

As we shall see, the structure of a finitely generated R-module M will ultimately depend on the complexity of the ideal theory in R, the case of a field being trivial, while that of $\mathbb{Z}$ being only slightly less so (for all this see Chapter 5). Notice that in the case of a finitely generated vector space the structure has been completely determined in Chapter 3. This is because when R is a field there are essentially no ideals, but when R is a Euclidean ring then there are many ideals, but not so many since they are all generated by a single element. However, the fact that here we cannot simply divide by non-zero elements (or equivalently that there are no non-trivial ideals) will change things in important ways. Thus, the ideal theoretic properties of R will determine much about the structure of M.

The two most significant cases of R being a Euclidean ring are $R = \mathbb{Z}$, or $R = k[x]$, polynomials with coefficients from a field k. The first of these mean that $M(R)$ is an *Abelian group*. Another, but less important example is the Gaussian integers $\mathcal{G}$. In each of these the group of units is rather small.

Recall the group of units of $\mathbb{Z}$, $k[x]$, $\mathcal{G}$ are, respectively, ± 1, $k^{\times}$, $\{\pm 1, \pm i\}$.

Exercise 6.4.1. The reader should also check that Gaussian elimination does not work over $\mathbb{Z}$. What is the problem here?

Our main purpose here is to prove a finitely generated module M over a Euclidean ring is the direct sum of a finite number of cyclic submodules. In the case of a finite dimensional vector space this is of course true where the cyclic submodules there all have dimension 1. In the present situation we shall see that these submodules can be more complicated. So we first ask what is a cyclic module of M?

Definition 6.4.2. A cyclic module M is one that is generated by a single element, say x. In other words $M = \{rx : r \in R\}$.

Let M be cyclic and consider the map $r \mapsto rx$. This is a surjective R-module homomorphism where R is regarded as an R acting on itself by ring multiplication. Calling the map ϕ, what is Ker (ϕ)? This is $\{r \in R : rx = 0\}$ which is evidently an *ideal* in R. If x has "infinite order", then this is (0) and $M \cong R$. Otherwise, since R is a PID (theorem 5.5.4), $M \cong R/(r)$, where $r \in R$. This is the additional level of complication. The cyclic modules may not just be R itself. Some of them could also be of the form $R/(r)$, where r is a non-zero element and a non-unit in R.

Theorem 6.4.3. *Every finitely generated module $M(R)$ over a Euclidean ring is the direct sum of a finite number of cyclic submodules, $M = \bigoplus M_i$. Here each $M_i \cong R$, or $R/(r_i)$, where $r_i \in R$ is non-zero and not a unit. Further we can arrange that each $r_i | r_{i+1}$. These r_i are called the elementary divisors of M and r, the number of R factors, is called the free rank of M. The r and the r_i are a complete set of invariants for M (meaning that they uniquely determine M up to a module isomorphism).*

Proof. Consider all finite generating sets for M and choose one of *minimal length* $\{x_1, \ldots x_k\}$ of M. We will assume by induction that the theorem is true for *all* finitely generated R-modules of length $\leq k - 1$. Now there are two possibilities for M. Either

    1. *Every* minimal generating set satisfies *some* non-trivial relation $\sum_{i=1}^{k} r_i x_i = 0$ (where r_i not all 0). Or,

    2. There is *some* minimal generating set satisfying *no* non-trivial relation.

In case 2, each $x \in M$ can be written uniquely as $x = \sum_{i=1}^{k} r_i x_i$. For if there were another representation $x = \sum_{i=1}^{k} r_i' x_i$, then

$$\sum_{i=1}^{k} (r_i - r_i') x_i = 0$$

would be a non-trivial relation. Hence the map $x \mapsto (r_1, \ldots r_k)$ from $M \to \sum_{i=1}^{k} \bigoplus R$ is an R-module map which is clearly bijective and so M is the direct sum of cyclic R-modules $\{r x_i : r \in R\} \cong R$. This is the torsion free case which corresponds exactly to the situation for vector spaces where V is the direct sum of a finite number of copies of the field.

The reason M is torsion free here is if $x = \sum_{i=1}^{k} r_i x_i \in M$ and for some $r \in R$ we have $rx = \sum_{i=1}^{k} r r_i x_i = 0$, where $r \neq 0$, then by uniqueness $r r_i = 0$ for all i. Since $r \neq 0$ and R is an integral domain, each $r_i = 0$ and hence x itself is zero.

We now turn to case 1 where every minimal generating set satisfies some non-trivial relation. In particular, this is true of $\{x_1, \ldots x_k\}$ hence $\sum_{i=1}^{k} s_i x_i = 0$, s_i not all zero. Among all such relations choose one which has an s_i of *smallest possible grade*, calling this coefficient again s_i and if necessary reorder the generators so that it is s_1 with smallest possible grade. Suppose $\sum_{i=1}^{k} r_i x_i = 0$ is any other non-trivial relation among the $\{x_1, \ldots x_k\}$. Then $s_1 | r_1$. For $r_1 = q s_1 + r$, where $0 < g(r) < g(s_1)$ and since $\sum_{i=1}^{k} s_i x_i = 0$, we know $\sum_{i=1}^{k} q s_i x_i = 0$. But also $\sum_{i=1}^{k} r_i x_i = 0$. Therefore $\sum_{i=1}^{k} (r_i - q s_i) x_i = 0$. Since this is a relation whose leading coefficient has a lower grade than s_1 we conclude $r = 0$ and $s_1 | r_1$. Moreover, $s_1 | s_i$ for all $i \geq 2$. Suppose for example, $s_2 = q s_1 + r$, where $0 < g(r) < g(s_1)$. Then $\{x_1 + q x_2, x_2 \ldots x_k\}$ would generate M since $1(x_1 + q x_2) - q x_2 = x_1$. This is another generating set of minimal length. Yet,

$$s_1 (x_1 + q x_2) + r x_2 + s_3 x_3 + \ldots + s_k x_k$$

$$= s_1 x_1 + (s_1 q + r) x_2 + s_3 x_3 + \ldots + s_k x_k = \sum_{i=1}^{k} s_i x_i = 0.$$

Hence $r = 0$ and $s_1 | s_2$. Similarly $s_i = q_i s_1$ for all $i \geq 2$.

Now consider $\{x_1 + q_2x_2 + \ldots + q_kx_k, x_2, \ldots, x_k\}$. This generates everything that $\{x_2, \ldots, x_k\}$ generates and also it generates x_1. Hence it generates M. Its length is k so we have here another minimal generating set. Let $\bar{x}_1 = x_1 + q_2x_2 + \ldots q_kx_k$. Then $s_1\bar{x}_1 = \sum_{i=1}^{k} s_ix_i = 0$. So we have a very simple, but non-trivial relation among these generators. Namely, $s_1\bar{x}_1 + 0x_2 + \ldots + 0x_k = 0$. Now if

$$r_1\bar{x}_1 + r_2x_2 + \ldots + r_kx_k = 0 \qquad\qquad (*)$$

is any other relation among these generators, then as above $s_1 | r_1$ and so $r_1\bar{x}_1 = 0$.

Let M' be the R-submodule generated by $\{x_2, \ldots, x_k\}$ and M_1 be the cyclic submodule generated by $\bar{x}_1$. Then M_1 and M' together generate M. We show actually $M = M_1 \oplus M'$. This is because if $r_1\bar{x}_1 + x' = \tilde{r}_1\bar{x}_1 + \tilde{x}$, where $r_1, \tilde{r}_1 \in R$ and $x', \tilde{x} \in M'$ and $r_1 \neq \tilde{r}_1$, then substituting we get a non-trivial relation $\nu_1\bar{x}_1 + \nu_2x_2 + \ldots + \nu_kx_k = 0$. By $(*)$ it follows that $\nu_1\bar{x}_1 = 0$. Hence $\nu_2x_2 + \ldots + \nu_kx_k = 0$. Therefore, $r_1\bar{x}_1 = \tilde{r}_1\bar{x}_1$, and $x' = \tilde{x}$. Thus $M = M_1 \oplus M'$.

By induction there exist $\bar{x}_2 \ldots \bar{x}_k$ which generate M' with $\mathrm{order}(\bar{x}_i) | \mathrm{order}(\bar{x}_{i+1})$ for $i \geq 2$ so that M' is the direct sum of these cyclic submodules (of both M' and M), with dividing orders. Since $s_1 | \mathrm{order}(\bar{x}_2)$ we get divided orders throughout and $M = R/(r_1) \oplus R/(r_j) \oplus R \ldots \oplus R$.   □

Exercise 6.4.4. Show any submodule N of a finitely generated module M over a Euclidean ring R is itself finitely generated. (We shall see a generalization of this to Noetherian rings in Vol. II, Chapter 9.) If M has m generators where $m = l(M)$, then N has $\leq m$ generators.

We remark that the following corollary has, among many important applications, one in algebraic topology since the homology groups of a finite simplicial complex are finitely generated Abelian groups.

Corollary 6.4.5. *(**Fundamental Theorem of Abelian Groups**)*
Every finitely generated Abelian group $(A, +)$ is the direct sum of a finite number of cyclic subgroups:

$$A = \mathbb{Z}^r \oplus \mathbb{Z}_{n_1} \oplus \mathbb{Z}_{n_2} \oplus \cdots \oplus \mathbb{Z}_{n_t}$$

where $n_i | n_{i+1}$.

The r is called the *free rank* and the n_i's the *torsion numbers* of A. From this follows a number of important properties of finitely generated Abelian groups.

Corollary 6.4.6. *Any finitely generated Abelian group is a direct sum of a finite group and a finitely generated free Abelian group. Any finite Abelian group is a direct sum of cyclic groups.*

Corollary 6.4.7. *The torsion subgroup of a finitely generated Abelian group is finite. In particular, any finitely generated Abelian torsion group is finite.*

Corollary 6.4.8. *Any finitely generated torsion free Abelian group is free.*

Corollary 6.4.9. *Every finitely generated Abelian group $(G, +)$ is the direct sum of a finite number of cyclic subgroups the finite ones being of prime power order.*

6.4.0.1 Exercises

Exercise 6.4.10. Show that actually Theorem 6.4.3 is valid when R is a principal ideal domain rather than just a Euclidean ring.

Exercise 6.4.11. Classify all finitely generated R-modules, where R is the ring $\mathbb{Q}[x]/(x^2 + 1)^2$.

6.5 Applications to Linear Transformations

The main interest in Theorem 6.4.3 from the point of view of linear algebra is when $R = k[x]$ and k is an arbitrary field. Later we will specialize to an algebraically closed field.

Let V be a finite dimensional vector space over a field k and $T \in \text{End}_k(V)$. From this data we can make a $k[x]$ module of V in much the same way as when we considered the group algebra (see example 5.9.13). We let $k[x]$ act on V through T. Thus, if $p(x) \in k[x]$ and $v \in V$

we define $p(x) \cdot v = p(T)v$. It is an easy check, which we leave to the reader, that this is a module which we call $V_T(k)$. Thus, in addition to the given action by k on V we now also have all polynomials in T acting on V. We also leave the easy proof of the following proposition to the reader as an exercise.

Proposition 6.5.1. *The following are important module theoretic properties.*

1. *A submodule W of V_T is exactly a T-invariant subspace.*

2. *A module homomorphism between two such modules $f : V_T(k) \to U_S(k)$ is a k-linear map $f : V \to U$ which is also an intertwining operator, that is $Sf = fT$.*

3. *In particular, $V_T(k)$ and $U_S(k)$ are isomorphic if and only if S and T are similar.*

Now we define the *minimal polynomial of T*.

Proposition 6.5.2. *For each $T \in \mathrm{End}_k(V)$ there exists a unique monic polynomial $m_T(x)$ which annihilates V and has lowest degree among all such annihilating polynomials. It is called the minimal polynomial of T.*

Notice that by Proposition 6.5.2 the minimal polynomial depends only on T and not on any matrix representation. Hence by the third part of Proposition 6.5.1 similar transformations, or matrices, have the same minimum polynomial.

Proof. Let $x \in V$ and consider $\{x, T(x), T^2(x), \ldots, T^n(x)\}$. Since V has dimension n, these $n + 1$ elements can not be linearly independent. Hence, there must be constants $c_0, \ldots c_n$, not all zero so that $\sum_{i=0}^{n} c_i T^i(x) = 0$. Let $p = \sum_{i=0}^{n} c_i T^i$. Thus each $x \in V$ has an annihilating polynomial p where $p(T)x = 0$. Now let $\{x_1 \ldots x_n\}$ be a basis of V and consider the corresponding annihilating polynomials $p_i(T)x_i = 0$. Let $p(x) = \prod_{i=1}^{n} p_i(x)$. Then $p \in k[x]$ and because of commutativity of

$k[x]$ it follows that $p(T)(x_i) = 0$ for all i. Hence for $x = \sum_{i=1}^{n} c_i x_i$ we get

$$p(T)(x) = p(T)(\sum_{i=1}^{n} c_i x_i) = \sum_{i=1}^{n} p(T)(x_i) = 0.$$

Thus, $p(T)(V) = (0)$, i.e. there is some polynomial which annihilates all of V. Choose one of lowest possible degree, normalize it so that it is monic and call it m_T. Let p be any other polynomial such that $p(T)(V) = (0)$. Write $p = q m_T + r$, where $\deg(r) < \deg(m_T)$. Then $r(T)(V) = (0)$, a contradiction. Therefore $r = 0$ and $p = q m_T$. Thus m_T is the *unique monic generator* of the ideal of all annihilators of V. $\qquad\square$

The previous proposition tells us that $V_T(k)$ is a torsion module. Therefore, there is no free part in Theorem 6.4.3. That is,

$$V_T \cong k[x]/(f_1(x)) \oplus \ldots \oplus k[x]/(f_j(x)), \qquad (6.1)$$

where the polynomials f_i all have degree at least 1. Another way to say this is that V_T actually has *"bounded order"*.

Example 6.5.3. Let T be the (nilpotent) operator whose matrix has zeros everywhere except on the super diagonal where it has ones. Then $T^n = 0$ and so $m_T | x^n$. By unique factorization this means $m_T = x^i$ where $i \leq n$. But no lower power of T annihilates V. Hence $m_T(x) = x^n$.

This example can be generalized as follows.

Example 6.5.4. Let T be just as above except that it also has λ on the diagonal. Then $T - \lambda I = N$, where N is nilpotent. Hence, as above, $(T - \lambda I)^n = 0$ and no lower exponent will do this. Thus $m_T(x) = (x - \lambda)^n$.

Next let T be a block triangular matrix where each $n_i \times n_i$ block is of the form $T_i - \lambda_i I = N_i$, the $\lambda_i \in k$ be *distinct*, and there are j blocks. Then $m_T(x) = \prod_{i=1}^{j} (x - \lambda_i)^{n_i}$.

Exercise 6.5.5. Notice that in all these cases $\deg(m_T) = \dim V$. Show here that $m_T = \pm \chi_T$, where χ_T is the characteristic polynomial of T.

A similarity invariant is a property shared by all operators STS^{-1} similar to a given operator T.

Example 6.5.6. Now let T be a block triangular matrix where each $n_i \times n_i$ block is of the form $T_i - \lambda_i I = N_i$, $\lambda_i \in k$, and there are j blocks, but the λ_i are not all distinct. Since the minimum polynomial is a similarity invariant. For simplicity we may arrange the blocks so that in each group the λ_i are the same. Then the minimal polynomial of such a block is $(x - \lambda_i)^{\bar{n}_i}$, where $\bar{n}_i$ is the largest of these blocks. Let D be the set of distinct λ_i. Then $m_T(x) = \prod_D (x - \lambda_i)^{\bar{n}_i}$. Thus here $\deg m_T(x) < \dim V$.

We now come to the *first Jordan form* and the *companion matrices*. For this we need a key preliminary result.

Proposition 6.5.7. *Let T be a linear operator on V and v be a cyclic generator of V with monic annihilating polynomial*

$$f(x) = x^n + c_{n-1}x^{n-1} + \ldots + c_1 x + c_0.$$

Then $f = m_T$, the minimal polynomial, $\dim V = n$ and a basis of T is given by $B = \{x, T(x), \ldots, T^{n-1}(x)\}$. Finally, the matrix of T with respect to this basis is the companion matrix. See just below.

Proof. First consider the $k[x]$-module $k[x]/(f(x))$ (which is also a k-vector space). We show

$$\{1 + (f(x)), x + (f(x)), \ldots, x^{n-1} + (f(x))\}$$

is a basis. First we check these are linearly independent. Suppose

$$c_0(1 + (f(x))) + c_1(x + (f(x))) + \ldots + c_{n-1}(x^{n-1}(f(x))) = 0,$$

that is $= (f(x))$. Then

$$c_0 + c_1 x + \ldots + c_{n-1}x^{n-1} \in (f(x)).$$

But this is a polynomial of degree $\leq n - 1$ and everything on this ideal is a multiple of $f(x)$ and so has degree $\geq n$. It follows that all c_i must be 0, so these elements are linearly independent. On the other hand, anything in the $k[x]$-module is of the form $g(x) + (f(x))$, where g is some polynomial. But $g(x) = q(x)f(x) + r(x)$ where $\deg r < n$. Therefore in the residue classes we may assume $\deg g < n$. As a result

$$\{1 + (f(x)), x + (f(x)), \ldots, x^{n-1} + (f(x))\}$$

also generates the module and so is a basis. This means $\dim V = n$. Also, $f = m_T$ since it annihilates V, is monic and has lowest degree among all annihilating polynomials.

Now consider the isomorphism between $k[x]/(f(x))$ and $V_T(k)$. This comes from a surjective $k[x]$-module homomorphism $k[x] \to V$ which sends 1 to v and $(f(x))$ to 0. Since

$$\{1 + (f(x)), x + (f(x)), \ldots, x^{n-1} + (f(x))\}$$

is a basis of $k[x]/(f(x))$, its image must be a basis of V. But its image is $\{v, Tv, \ldots, T^{n-1}v\}$.

What is the matrix of T with respect to the basis,

$$\{v, Tv, \ldots, T^{n-1}v\}?$$

Obviously, for each of the first $n-1$ elements, T acts as a shift operator, while $T(T^{n-1}v) = T^n(v)$. But since $f(T) = T^n + c_{n-1}T^{n-1} + \ldots + c_1T + c_0 = 0$ it follows that $T^n(v) = -c_0 v - c_1 T(v) - \ldots - c_{n-1}T^{n-1}(v)$. Therefore the matrix has for its first $n-1$ columns,

$$T = \begin{pmatrix} 0 & 0 & 0 & \ldots & 0 & -c_0 \\ 1 & 0 & 0 & \ldots & 0 & -c_1 \\ 0 & 1 & 0 & \ldots & 0 & -c_2 \\ 0 & 0 & 1 & \ldots & 0 & \vdots \\ \vdots & \vdots & \vdots & \vdots & \vdots & \vdots \\ 0 & 0 & 0 & \ldots & 0 & -c_{n-1} \end{pmatrix}.$$

This is called the *companion matrix*.   □

This last result leads to the estimate of $\deg m_T$ in general.

Corollary 6.5.8. *Let $T \in \text{End}_k(V)$. Then $\deg m_T \leq \dim V$.*

Proof. In equation 6.1 we have written V as a direct sum of the T-invariant subspaces $V_1, \ldots, V_j$, where $V_i = k[x]/(f_i(x))$ and each f_i divides f_{i+1}. Since we know $\prod_{i=1}^{j} f_i$ annihilates V_T and f_j is a multiple of each of the f_i, where $i < j$. Actually, f_j itself annihilates V. Hence $m_T | f_j$. In particular, $\deg m_T \leq \deg f_j$. On the other hand, $\dim V \geq \dim V_j = \deg f_j$. Thus $\deg m_T \leq \dim V$.   □

We have also just proved the *First Jordan Form*.

Theorem 6.5.9. *(First Jordan Form) Let T be a linear transformation on the finite dimensional vector space V. Then V is the direct sum of a finite number of T-invariant subspaces V_i on which T acts cyclicly with annihilating polynomials each successively dividing the next. Thus T, in an appropriate basis, is block diagonal with blocks of the form, the companion matrices.*

Exercise 6.5.10. Observe that the minimal polynomial of a companion matrix C_f is f. What is the characteristic polynomial of C_f?

This observation gives the Cayley-Hamilton theorem which in turn gives Corollary 6.5.8 in a stronger form. The Cayley-Hamilton theorem also gives us another way to get at the minimum polynomial m_T, for it must divide χ_T (and therefore have degree $\leq n$).

Theorem 6.5.11. *(Cayley-Hamilton Theorem) Any operator T satisfies its characteristic equation $\chi_T(T) = 0$.*

Proof. Since on each C_f block $m_T(x) = \chi_T(x)$, when we take account of possible repetitions we get $m_T(x)|\chi_T(x)$. Hence $\chi_T(T) = 0$.   □

By using the Chinese Remainder theorem (5.5.14), we can make a refinement of the conclusion of our Theorem 6.4.3. Since R has unique factorization (see 5.6.3), for each summand $R/(r)$ write $r = p_1^{e_1} \ldots p_k^{e_k}$. Then $R/(r) = \bigoplus_{i=1}^{k} R/(p_i^{e_i})$. Hence each of the cyclic summands of the torsion part of M is of the form $R/(p^e)$, where p is a prime of R. Combining this with the first Jordan form we have,

Corollary 6.5.12. *(Second Jordan Form) Let T be a linear transformation on the finite dimensional vector space V. Then V is the direct sum of a finite number of T-invariant subspaces V_i on which T acts cyclically with annihilating polynomials each a prime power, $p(x)^e$. Thus T is conjugate to a block diagonal matrix with blocks that are of the form, the companion matrices of type $p_i(x)_i^e$.*

6.6   The Jordan Canonical Form and Jordan Decomposition

We now consider the situation of the previous section, but with the additional assumption that the field K is *algebraically closed*. In this case, irreducible polynomials are of degree 1. Hence monic irreducible polynomials are of the form $x - a$. For $T \in \mathrm{End}_K(V)$ we know from the second Jordan form that $V_T(K)$ can be decomposed further into the direct sum of sub $K[x]$-modules of the form $K[x]/(p(x)^e)$, where p is irreducible and monic. Hence here submodules are of the form $K[x]/((x-a)^e)$. Thus on each block we have an cyclic operator leaving the appropriate subspace W invariant with a minimal polynomial $m_T(x) = (x-a)^e$. Therefore, $(T - aI)^e = 0$, so that $T - aI$ is a nilpotent operator of index e where $e = \dim W$. Since aI also leaves W invariant, $T - aI$ also leaves W invariant. It follows that by choosing an appropriate basis for W, $T - aI|W$ has matrix form all zeros except for 1 on the super diagonal. This basis is $\{w_0, (T - aI)(w_0), \ldots, (T - aI)^{e-1}(w_0)\}$, where $w_0 \in W$ is a cyclic generator. Since aI has the same matrix with respect to any basis, we see that $T|W$ has matrix form all zeros except a on the diagonal and 1's on the super diagonal,

$$
\begin{pmatrix}
a & 1 & \ldots\ldots & 0 \\
0 & a & 1 & \ldots 0 \\
\multicolumn{4}{c}{\cdots\cdots\cdots\cdots} \\
0 & \ldots\ldots & a & 1 \\
0 & 0 & \ldots & 0 \quad a
\end{pmatrix}.
$$

Since we can decompose V into a direct sum of T-invariant subspaces, finally we get

$$
T = \begin{pmatrix}
\begin{pmatrix}
a_1 & 1 & \cdots & \cdots & 0 \\
0 & a_1 & 1 & \cdots & 0 \\
\multicolumn{5}{c}{\cdots\cdots\cdots\cdots\cdots} \\
0 & \cdots & \cdots & a_1 & 1 \\
0 & 0 & \cdots & 0 & a_1
\end{pmatrix} & & \text{\Large 0} \\[2ex]
& \ddots & \\[2ex]
\text{\Large 0} & & \begin{pmatrix}
a_n & 1 & \cdots & \cdots & 0 \\
0 & a_n & 1 & \cdots & 0 \\
\multicolumn{5}{c}{\cdots\cdots\cdots\cdots\cdots} \\
0 & \cdots & \cdots & a_n & 1 \\
0 & 0 & \cdots & 0 & a_n
\end{pmatrix}
\end{pmatrix}. \qquad (*)
$$

Exercise 6.6.1. Use the companion matrices to prove that any monic polynomial in $K[x]$ of degree n is the characteristic polynomial of some $T \in \text{End}(V)$, where $\dim V = n$.

Corollary 6.6.2. *(**Third Jordan Form**) Let T be a linear transformation on the finite dimensional vector space V over an algebraically closed field. Then V is the direct sum of a finite number of T-invariant subspaces V_i on which T acts as above. Thus T is conjugate in $\text{GL}(n,k)$ to a block triangular matrix of the form $(*)$. The a_i are the full set of eigenvalues of T.*

Note, however, that some of these a_i may be equal thus making $m_T \neq \chi_T$ and some of the n_i may be 1 (such as when the operator is diagonalizable). Also, $V = \oplus V_{a_i}$, where $V_{a_i} = \{v \in V : (T - a_iI)^{n_i}(v) = 0\}$. These are called the algebraic eigenspaces, or generalized eigenspaces, in contrast to the geometric eigenspace associated with the eigenvalue a in $\{v \in V : Tv = av\}$.

We make one final remark concerning Corollary 6.6.2. The reason for assuming that K is algebraically closed is to be sure that the eigenvalues of T are in the field. But it may be that for *some operators* this happens even if the field is not algebraically closed. For example, if T is nilpotent, all eigenvalues are zero and so lie in any field. Similarly, if T is unipotent all eigenvalues are 1. Another important case is when we have a real matrix which happens to have all its eigenvalues real. In particular, under all these circumstances, T is similar to a triangular matrix.

Corollary 6.6.3. *Any linear transformation on the finite dimensional vector space over an algebraically closed field can be put in triangular form. That is, it similar to a triangular matrix. In particular, the a_i of the third Jordan form are precisely the eigenvalues of the operator.*

Corollary 6.6.4. *Let k be an arbitrary field (not necessarily algebraically closed) and $T \in \text{End}_k(V)$. Over the splitting field of χ_T the operator T is similar to a triangular matrix.*

Proof. Over the ground field, $m_T | \chi_T$ by the Cayley-Hamilton theorem. Hence the roots of m_T all lie in the splitting field[2] of χ_T. $\qquad \square$

Now extending the ground field does not affect the characteristic polynomial and hence does not affect its coefficients. Moreover because the coefficients are similarity invariants, we can calculate them from the triangular form. In particular this applies to $\text{tr}\, T$ or $\det T$.

Corollary 6.6.5. *The sum of the eigenvalues of T is $tr(T)$, while $\det(T)$ is the product of these eigenvalues.*

Corollary 6.6.6. *Let $T \in \text{End}_{\mathbb{R}}(V)$. Then, T is similar to a block triangular matrix, where the blocks corresponding to the various complex conjugate eigenvalues are 2×2 blocks of the form,*

$$\begin{pmatrix} a_j & b_j \\ -b_j & a_j \end{pmatrix}$$

together with (any) real 1×1 blocks.

This follows immediately from the third Jordan form upon considering the eigenvalues of T. These will be either real or pairs of complex conjugates. The former will give real triangular blocks and the later will

---

[2] The splitting field is extensively dealt with in Chapter 11.

give complex triangular blocks which make real 2×2 block triangular matrices of the following form:

$$\begin{pmatrix} a & b \\ -b & a \end{pmatrix}.$$

Proposition 6.6.7. *Let K be algebraically closed. T is diagonalizable if and only if $m_T(x)$ is the product of distinct linear factors.*

Proof. If T is diagonalizable, then, $\chi_T(x) = (x - a_1)^{n_1} \ldots (x - a_r)^{n_r}$, where the a_i are distinct and the n_i are the multiplicities. Thus, to annihilate the geometric eigenspace V_{a_i}, we need only $(T - a_iI)^1$. So, $m_{T|V_{a_i}} = x - a_i$. Hence, $m_T(x) = \prod_{i=1}^{r}(x - a_i)$. Conversely, if $m_T(x)$ is the product of distinct linear factors the third Jordan form immediately shows that T is diagonalizable since the blocks are 1×1. $\qquad\square$

Proposition 6.6.7 leads to a useful, necessary and sufficient condition for diagonalizability as follows.

The next corollary and the exercises below it give characterizations of diagonalizable operators.

Corollary 6.6.8. *Let K be algebraically closed. Then T is similar to a diagonal matrix if and only if all its Jordan blocks are 1×1.*

Exercise 6.6.9. Prove that over any field, T is diagonalizable if and only if V has a basis of eigenvectors.

Exercise 6.6.10. Prove that over an algebraically closed field, for each eigenvalue of T, the geometric multiplicity is $\leq$ to the algebraic multiplicity, and they are equal for all eigenvalues if and only if T is diagonalizable.

Exercise 6.6.11. Recall Exp is the map of real or complex matrices given by the everywhere convergent power series, $\mathrm{Exp}\, X = \sum_{i=0}^{\infty} \frac{X^i}{i!}$. Show that

$$\mathrm{Exp} : M_n(\mathbb{C}) \to \mathrm{GL}(n, \mathbb{C})$$

is surjective.

6.6.1   The Minimal Polynomial (continued...)

The following result gives us a characterization of the minimal polynomial.

Proposition 6.6.12. *For any polynomial p, $p(T) = 0$, if and only if $m_T \mid p$.*

Proof. $\Longrightarrow$ By the Euclidean algorithm, there are polynomials q and r such that $p = q m_T + r$ with $0 \leq \deg(r) < \deg(m_T)$. Now,

$$r(T) = p(T) - q(T) m_T(T) = 0 - 0 = 0,$$

so by definition of the minimal polynomial, $r = 0$. Hence $m_T \mid p$.
$\Longleftarrow$ Obvious. $\square$

An implication of the proposition above is: the Cayley-Hamilton Theorem says that $\chi_T(T) = 0$, where χ_T is the characteristic polynomial of T, and so by the above proposition, $m_T \mid \chi_T$.

From now on, we will assume that the field is algebraically closed. Therefore, we can write

$$\chi_T(t) = \pm (t - \lambda_1)^{r_1} \cdots (t - \lambda_k)^{r_k}$$

where $\lambda_1, \ldots, \lambda_k$ are the distinct eigenvalues of T and $r_i \geq 1$ for each i.

We now give an alternative proof of Proposition 6.6.7 which avoids the use of the Jordan canonical form.

Theorem 6.6.13. *The minimal polynomial has the form*

$$m_T(t) = (t - \lambda_1)^{s_1} \cdots (t - \lambda_k)^{s_k} \qquad (1)$$

for some numbers s_i with $1 \leq s_i \leq r_i$. Moreover, T is diagonalizable if and only if each $s_i = 1$.

Proof. First, the Cayley-Hamilton Theorem tells us that m_T has the form (1) with $s_i \leq r_i$. Secondly, we need to see that each $s_i \geq 1$, i.e. that every eigenvalue is a root of the minimal polynomial. So, let

$i \in \{1, \ldots, k\}$ and let v be a λ_i-eigenvector. For any polynomial $p(x)$, we have $p(T)v = p(\lambda_i)v$, and in particular this holds for $p = m_T$. Hence $m_T(\lambda_i)v = 0$. But $v \neq 0$ (being an eigenvector), so $m_T(\lambda_i) = 0$, as required.

Third, suppose we know that T is diagonalizable. As we know this means it has a basis of eigenvectors, whose eigenvalues are $\lambda_1, \ldots, \lambda_k$, and it is easy to calculate that

$$(t - \lambda_1) \cdots (t - \lambda_k)$$

is an annihilating polynomial for T. So this is the minimal polynomial. Finally, we have to show that if all the s_i's are 1 (i.e. the minimal polynomial splits into distinct linear factors) then T is diagonalizable. To prove this we need the following lemma:

Lemma 6.6.14. *If*

$$U \xrightarrow{\phi} V \xrightarrow{\psi} W$$

are finite-dimensional vector spaces and linear maps, then

$$\dim Ker(\psi \circ \phi) \leq \dim Ker(\psi) + \dim Ker(\phi).$$

Proof. First we remark that $\mathrm{Ker}((\psi \circ \phi) = \phi^{-1}(\mathrm{Ker}(\psi))$. Now, consider the function

$$\xi : \phi^{-1}\big(\mathrm{Ker}(\psi)\big) \longrightarrow \mathrm{Ker}(\psi) \ : \ u \mapsto \phi(u).$$

Applying the rank-nullity formula we get

$$\dim \phi^{-1}\big(\mathrm{Ker}(\psi)\big) = \dim\big(\mathrm{Im}(\xi)\big) + \dim\big(\mathrm{Ker}(\xi)\big).$$

This, with the facts that $\mathrm{Im}(\xi) \subseteq \mathrm{Ker}(\psi)$ and $\mathrm{Ker}(\xi) \subseteq Ker(\phi)$, prove the result. $\qquad\square$

Now, suppose that all the $s_i = 1$. Then, the composition of the maps

$$V \xrightarrow{T - \lambda_k I} V \xrightarrow{T - \lambda_{k-1} I} \cdots \xrightarrow{T - \lambda_1 I} V$$

is the 0 map. So,

$$\dim(V) = \dim \operatorname{Ker}\Big((T - \lambda_1 I) \cdots (T - \lambda_k I)\Big)$$
$$\leq \dim \operatorname{Ker}(T - \lambda_1 I) + \cdots + \dim \operatorname{Ker}(T - \lambda_k I)$$
$$= \dim \operatorname{Ker}(T - \lambda_1 I) \oplus \cdots \oplus \operatorname{Ker}(T - \lambda_k I).$$

where the inequality comes from the Lemma 6.6.14 (and an easy induction), and the second equality is justified by the fact that the sum of the eigenspaces is a direct sum. Hence, the sum of the eigenspaces has the same dimension as V, i.e. this sum is V, and T is diagonalizable.   □

6.6.2   Families of Commuting Operators

We now give an important extension of the previous discussion where, instead of considering a single operator T, we look at a family of *commuting* operators, $\mathcal{T}$,   The first part is a generalization of Corollary 6.6.3.

Theorem 6.6.15. *Let K be an algebraically closed field and $\mathcal{T} \subseteq \operatorname{End}_K(V)$ consisting of commuting operators. Then,*

1. *$\mathcal{T}$ can be simultaneously put in triangular form. That is, there is a fixed $P \in \operatorname{GL}(n, K)$ so that PTP^{-1} is triangular for every $T \in \mathcal{T}$.*

2. *If all $T \in \mathcal{T}$ are diagonalizable then $\mathcal{T}$ can be simultaneously put in diagonal form. That is, there is a fixed $P \in \operatorname{GL}(n, k)$ so that PTP^{-1} is diagonal for every $T \in \mathcal{T}$.*

Proof. Because $\mathcal{T}$ is a commuting family, we first show that if $T \in \mathcal{T}$ and $a \in K$, then $W = \operatorname{Ker}(T - aI)$ is stable under any $S \in \mathcal{T}$. Let $w \in W$, then $(T - aI)(S(w)) = TS(w) - aS(w) = ST(w) - aS(w) = S(Tw - aw) = S(0) = 0$.

Secondly, choose some T and a so that $(0) < W < V$. How can we do this? Since K is algebraically closed, let $v \neq 0$ be an eigenvector of T with eigenvalue a, and $W = V_a$, the geometric eigenspace. Therefore, $W \neq (0)$. If $W = V$, then T is a scalar multiple of the identity. If that is so, for every $T \in \mathcal{T}$, then both 1) and 2) are trivially true. So we may assume there is such a W.

Thirdly, since $\dim W < \dim V$ and $\mathcal{T}$ stabilizes W, in proving the first statement we can assume by induction that there exists a $v_1 \neq 0 \in W$ so that the line $[v_1]$ is $\mathcal{T}$-stable and so $\mathcal{T}$ acts on $V/[v_1]$ as a commuting family of endomorphisms. Because $\dim V/[v_1] < \dim V$, induction tells us there exists $v_2, \ldots, v_n \in V$ such that $v_2 + k v_1, \ldots, v_n + k v_n$ is a basis of $V/[v_1]$ and $\mathcal{T}$ acts by triangular operators with respect to this basis. Thus, $v_1, \ldots, v_n$ is a basis for V and $\mathcal{T}$ stabilizes $\mathrm{lin.sp}_K \{v_1, \ldots, v_i\}$ for each $i = 1, \ldots, n$. Therefore, $\mathcal{T}$ is triangular on V.

We now turn to the second statement. Suppose $\mathcal{T}$ consists of diagonalizable operators. For each $T \in \mathcal{T}$ we can write $V = V_1 \oplus \ldots V_r$, where the V_i are the geometric eigenspaces associated with distinct eigenvalues $a_1, \ldots, a_r$ of T. (But of course all this depends on T!) But by 1, each V_i is $\mathcal{T}$-stable. Since $\mathcal{T}$ is diagonalizable on V and the V_i are $\mathcal{T}$-invariant $\mathcal{T}$ acts diagonalizably on each V_i by Lemma 6.6.21 of the following subsection. If for some T, $r(T) \geq 2$, then *all* $V_i(T) < V$. By inductive hypothesis $\mathcal{T}$ can be simultaneously diagonalized on each V_i and therefore $\mathcal{T}$ can be simultaneously diagonalized on V. The only case remaining is when $r(T) = 1$ for all T. That is, each $T \in \mathcal{T}$ is a scalar where the conclusion is trivially true. $\qquad\qquad\square$

Corollary 6.6.16. *If T and S commute and both are diagonalizable (respectively triangularizable), then so are $S \pm T$ and ST.*

Corollary 6.6.17. *Let A be an $n \times n$ matrix over an algebraically closed field K, or more generally if we are over k and the eigenvalues of A all lie in k. Then, the transpose, A^t, is conjugate over k to A.*

Proof. By our hypothesis, using the first Jordan canonical form, A can be put in lower triangular form and hence A^t can be put in upper triangular form all over k. It follows that A and A^t have the same spectrum. Now, apply the third Jordan form to A. Thus, A is similar to a block diagonal matrix with blocks of the form A_λ. Therefore, it is sufficient to consider an individual block, e.g., of order m. Let P be the $m \times m$ matrix of the form

$$P = \begin{pmatrix} 0 & \cdots & 1 \\ \vdots & 1 & \vdots \\ 1 & \cdots & 0 \end{pmatrix}.$$

Then, P has its entries in k, $P = P^{-1}$ and one checks easily that $PA_\lambda P = A\lambda^t$.   □

6.6.3   Additive & Multiplicative Jordan Decompositions

The final topic in this section is the *additive and multiplicative* Jordan decompositions.

Definition 6.6.18. An operator T on a finite dimensional vector space V over k is called *semi-simple* if T is diagonalizable over the algebraic closure of k.

Theorem 6.6.19. *(Jordan-Chevalley Decomposition)*[3] *Let V be a finite dimensional space over an algebraically closed field K and $T \in \text{End}_K(V)$. Then $T = S + N$ where S is diagonalizable (semi-simple), N is nilpotent and they commute. These conditions uniquely characterize S and N. Moreover, there exist polynomials p and q without constant term in $K[x]$ such that $S = p(T)$ and $N = q(T)$. Hence, not only do S and N commute with T, but they commute with any operator which commutes with T. If $A \subset B$ are subspaces of V and $T(B) \subset A$, then, $S(B) \subset A$ and $N(B) \subset A$. In particular, if A is a T-invariant subspace then it is S and N invariant. If $T(B) = 0$, then $S(B) = 0$ and $N(B) = 0$.*

The Jordan form of T consists of blocks each of which is the sum of a scalar and a nilpotent operator. This proves the first statement. Regarding the uniqueness, we first note that

Lemma 6.6.20. *An operator S is semi-simple if and only if every S-invariant W space has an S-invariant complement.*

---

[3]M. Jordan (1838-1922) French mathematician, known for his important work in group theory and linear algebra as well as the Jordan curve theorem and Jordan measure.

Proof. Suppose that S is semi-simple and k is algebraically closed. Then we can write $V = \bigoplus_{\alpha \in A} V_\alpha$ where V_α are 1-dimensional S-invariant vector spaces and A is a finite set. Consider the family of sets of the form $\bigcup_{\beta \in B} \{V_\beta\} \cup \{W\}$ where $B \subset A$ and the V_β's and W are linearly independent. This family is non-empty as it contains $\{N\}$. Since it is finite, it has a maximal $K = \bigcup_{\beta \in B} \{V_\beta\} \cup \{W\}$. Let

$$M' = \bigoplus_{\beta \in B} V_\beta \oplus W.$$

We prove that $M' = M$. Otherwise there exists an $\alpha \in A$ such that $V_\alpha \nsubseteq M'$. Since V_α is 1-dimensional $V_\alpha \cap M' = 0$. Hence,

$$L = \bigcup_{\beta \in B} \{V_\beta\} \cup \{W\} \cup \{V_\alpha\}$$

in the family and $K \subset L$. This contradicts the maximality of K. Thus, $M' = M$. Now, take $W' = \bigoplus_{\alpha \in A} V_\alpha$. Then, $M = W \oplus W'$ and W' is S invariant.

The converse can be proved by induction, since there is always an eigenvector for S over an algebraically closed field. $\square$

Lemma 6.6.21. *The restriction of a semi-simple operator S to an invariant subspace W is still a semi-simple operator.*

Proof. Let U be an S-invariant subspace of W. Then U is an S-invariant subspace of V. By previous lemma there is an S-invariant subspace U' of V which complements U in V. Then, $U' \cap W$ is an S-invariant subspace of W which complements U in W. $\square$

Lemma 6.6.22. *If S and S' are diagonalizable and commute, then $S + S'$ is diagonalizable. If N and N' are nilpotent and commute, then $N + N'$ is nilpotent.*

Proof. For $\alpha \in \mathrm{Spec}(S)$ let V_α be the eigenspace of α. Then, V is the direct sum of the V_α. If $v \in V_\alpha$, then, $S(v) = \alpha v$ and so $S'S(v) = \alpha S'(v) = SS'(v)$, so that $S'(V_\alpha) \subset V_\alpha$. Now, the restriction of S' to each V_α is still semi-simple. Choose a basis in each V_α in which the restriction is diagonal and in this way get a basis of V. Since on each V_α, $S = \alpha I$ it follows that $S + S'$ is diagonal. Then, $N^n = 0 = (N')^m$, so by the binomial theorem 0.4.2 $(N + N')^{n+m} = 0$.   □

Now, if $T = S + N = S' + N'$, then $S' - S = N - N'$. Since S' commutes with N' and itself it commutes with T and hence with S and N. Similarly, N' commutes with S and N. In particular $S' - S = N - N'$ is both diagonalizable and nilpotent. Such an operator is clearly 0. This proves the uniqueness part of Theorem 6.6.19.

Completion of the proof of theorem 6.6.19:

Proof. Let $\chi_T(x) = \prod (x - \alpha_i)^{n_i}$ be the factorization of the characteristic polynomial of T into distinct linear factors over k. Since the α_i are distinct the $(x - \alpha_i)^{n_i}$ are pairwise relatively prime. Consider the following

$$p(x) \equiv \alpha_i \quad (\mathrm{mod}\ (x - \alpha_i)^{n_i})$$
$$p(x) \equiv 0 \qquad (\mathrm{mod}\ x).$$

If no $\alpha_i = 0$ then x together with the $(x - \alpha_i)^{n_i}$ are also relatively prime. If some $\alpha_i = 0$ then the last congruence follows from the others. In either case by the Chinese Remainder theorem there is a polynomial p satisfying them. As a result,

$$p(T) - \alpha_i I = \phi_i(T)(T - \alpha_i I)^{n_i}.$$

Therefore, on each V_{α_i}, $p(T) = \alpha_i I$. This is equal to S on V_{α_i} and so $S = p(T)$. Taking $q(x) = x - p(x)$ we see that $q(0) = 0$ and $q(T) = T - p(T) = T - S = N$. Suppose $A \subset B$ are subspaces of V and $T(B) \subset A$. Since $S = \sum \alpha_i T^i$, we get $S(B) \subset A$ if we can show that $T^i(B) \subset A$ for $i \geq 1$. Now, $T^2(B) = T(T(B)) \subset T(A) \subset T(B) \subset A$ and proceed by induction.   □

We now turn to the *multiplicative Jordan-Chevalley decomposition.*

Corollary 6.6.23. *Let $g \in \mathrm{GL}(n, \mathbb{C})$. Then $g = g_u g_s$, where g_u is unipotent g_s is semi-simple, and g_u and g_s commute. Finally g_u and g_s are uniquely determined by these properties.*

Proof. First write the additive Jordan decomposition $g = g_n + g_s$. As g is invertible so is g_s. Let $g_u = I + g_s^{-1} g_n$. Then since g_n and g_s commute so do g_n and g_s^{-1}. Therefore, $(g_s^{-1} g_n)^\nu = (g_s^{-1})^\nu (g_n)^\nu$ for every positive integer ν. It follows that since g_n is nilpotent so is $g_s^{-1} g_n$. Therefore g_u is unipotent. Moreover,

$$g_u g_s = g_s + g_s^{-1} g_n g_s$$

and since g_s and g_n commute, this is $g_s + g_n = g$. Also,

$$g_s g_u = g_s + g_n = g.$$

Thus $g = g_u g_s$ and these Jordan factors commute. Turning to uniqueness, suppose $g = su$ where s is semi-simple, u is unipotent and they commute. Then $u = I + n$, where n is nilpotent. Therefore, $g = s(I + n) = s + sn$. Since s commutes with u, it also commutes with $I + n$ and therefore commutes with n. Hence, as above, sn is nilpotent and because s and sn commute $g = s + sn$ is the additive Jordan decomposition of g. By its uniqueness $s = g_s$ and $sn = g_n$. But then,

$$g_u = I + g_s^{-1} g_n = I + s^{-1} sn = I + n = u.$$

□

The following is another approach to getting the multiplicative Jordan-Chevalley decomposition from the additive one: Let $g \in GL(n, \mathbb{C})$. Then by 6.6.11, there is some $X \in M_n(\mathbb{C})$ so that $\mathrm{Exp}(X) = g$. Let $X = X_n + X_s$ be its additive Jordan decomposition. Since X_n and X_s commute, $\mathrm{Exp}(X) = \mathrm{Exp}(X_n) \mathrm{Exp}(X_s)$. Let $g_u = \mathrm{Exp}(X_n)$ and $g_s = \mathrm{Exp}(X_s)$. Then g_u is unipotent, g_s is semi-simple and $g = g_u g_s$. Moreover since X_n and X_s commute and Exp is given by a convergent power series so do $\mathrm{Exp}(X_n) = g_u$ and $\mathrm{Exp}(X_s) = g_s$. The uniqueness of the mutliplicative Jordan components must be proved as above, but this tells us that even though X does not uniquely depend on g, nonetheless its additive Jordan components exponentiate uniquely onto the multiplicative Jordan components of g.

6.7 The Jordan-Hölder Theorem for R-Modules

Here M is an R-module, where R is a ring with identity.

Definition 6.7.1. A *composition series* for the module M is a finite sequence

$$(0) = M_0 \subset M_1 \subset \ldots \subset M_n = M$$

of submodules of M such that for every i, M_i/M_{i-1} is a simple R-module.

We remind the reader that an R-module M is *simple* if M does not have any proper submodules.

To be sure that composition series exist we require the following:

Definition 6.7.2. M is said to obey the *ascending chain condition*, written as ACC, if for every increasing sequence of submodules

$$M_1 \subset M_2 \subset \cdots \subset M_i \subset \cdots$$

there is an integer n such that $M_n = M_{n+1} = \ldots$, that is, if the chain stabilizes after n steps. Similarly, M is said to obey the *descending chain condition*, written as DCC, if for every decreasing sequence of submodules

$$M_1 \supset M_2 \supset \cdots \supset M_i \supset \cdots$$

there is an integer n such that $M_n = M_{n+1} = \ldots$, that is the chain stabilizes after n steps.

Definition 6.7.3. An R-module M is said to be of *finite length* if it satisfies one of the following equivalent conditions:

1. M satisfies both ACC and DCC conditions.

2. There exists a series $0 = M_0 \subset M_1 \subset \cdots \subset M_n = M$ of submodules of M such that for every i the quotient module M_i/M_{i-1} is a simple R-module. In other words, M has a compositions series.

Such a series is called a Jordan-Hölder series for M. The number n is called the *length* of this series and the quotient submodules M_i/M_{i-1} are called the *quotient factors* of this series. Thus, the length $l(M)$ of a module M which satisfies both the ACC and DCC is well defined, and if N is a proper submodule of M, then $l(N) < l(M)$.

We leave as an exercise for the reader to show these conditions are indeed equivalent.

Theorem 6.7.4. *(**Jordan-Hölder Theorem**) Let M be an R-module of finite length and let*

$$0 = M_0 \subset M_1 \subset \cdots \subset M_n = M \tag{1}$$

$$0 = N_0 \subset N_1 \subset \cdots \subset N_m = N. \tag{2}$$

be two Jordan-Hölder series for M. Then $m = n$ and, after rearrangement, the quotient factors of these series are the same.

Proof. We prove the result by induction on k, where k is the length of a Jordan-Hölder series of M of minimum length. Without loss of generality suppose that the series (1) is a series of M with minimum length. In particular, we have $m \geq n$. If $n = 1$, then M is a simple module and the length of every other Jordan-Hölder series of M is also 1, the only quotient factor is M and the result is proved.

Now, suppose that $n > 1$. Consider two submodules M_{n-1} and N_{m-1}, and set $K = M_{n-1} \cap N_{m-1}$. There are two possibilities:

1. $M_{n-1} = N_{m-1}$

2. $M_{n-1} = N_{m-1}$.

In the first case we set $K = M_{n-1} = N_{m-1}$ and we consider the following two Jordan-Hölder series:

$$0 = M_0 \subset M_1 \subset \cdots \subset M_{n-2} \subset K,$$

$$0 = N_0 \subset N_1 \subset \cdots \subset N_{m-2} \subset K.$$

Now, the R-module K has a Jordan-Hölder series of length $\leq n-1$, so the induction hypothesis implies that $n-1 = m-1$ and the quotient factors of the above series are the same. Consequently, the Jordan-Hölder series in (1) and (2) have the same length and the same quotient factors after rearrngement.

In the second case, $K \subsetneq M_{n-1}$ and $K \subsetneq N_{m-1}$. As $M_{n-1} \neq N_{m-1}$ and M_{n-1} and N_{m-1} are maximal in M, we get $M_{n-1} + N_{m-1} = M$. Therefore,

$$M_{n-1}/K = M_{n-1}/(M_{n-1} \cap N_{m-1}) \cong (M_{n-1}+N_{m-1})/N_{m-1} = M/N_{m-1}.$$

In other words,

$$M_{n-1}/K = M/N_{m-1}. \tag{3}$$

Similarly,

$$N_{m-1}/K = M/M_{n-1}. \tag{4}$$

In particular, the two quotient modules M_{n-1}/K and N_{m-1}/K are simple modules. As M satisfies both the ACC and DCC, K has the following Jordan-Hölder series:

$$0 = K_0 \subset K_1 \subset \cdots \subset K_r = K.$$

Hence, for M we get two new Jordan-Hölder series:

$$0 = K_0 \subset K_1 \subset \cdots \subset K_r = K \subset M_{n-1} \subset M_n = M \tag{5}$$

and

$$0 = K_0 \subset K_1 \subset \cdots \subset K_r = K \subset N_{m-1} \subset N_m = M. \tag{6}$$

Now, by (1), M_{n-1} has a Jordan-Hölder series of length $\leq n - 1$, and so we can apply the inductive hypothesis to M_{n-1}, and get a Jordan-Hölder series for it giving two Jordan-Hölder series of M of the same length. By (5), M_{n-1} has a Jordan-Hölder series of length $r+1$. Hence, $r + 1 = n - 1$ and the two Jordan-Hölder series

$$0 = K_0 \subset K_1 \subset \cdots \subset K_r = K \subset M_{n-1}$$

and

$$0 = M_0 \subset M_1 \subset \cdots \subset M_{n-1}$$

have the same quotient factors. Therefore, the length and the quotient factors of two series (1) and (5) are the same. Also, by (6), N_{m-1} has a series of length $r + 1 = n - 1$. By induction, the length and the quotient factors of the following Jordan-Hölder series of N_{m-1} are the same:

$$0 = K_0 \subset K_1 \subset \cdots \subset K_r = K \subset N_{n-1}$$

and

$$0 = N_0 \subset N_1 \subset \cdots \subset N_{m-1}.$$

This implies the length and the quotient factors of two series, (2) and (6) are the same. By (3) and (4), the length and the quotient factors of two series, (5) and (6) are the same. So, the lengths and the quotient factors of series (1) and (2) are the same.   □

We remark that the proof of the theorem also works just as well for two Jordan-Hölder series of a group G. With this we have also completing the proof of theorem 2.1.5.

Exercise 6.7.5. Given an Abelian group G show that it satisfies the ACC and DCC if and only if it is finite. Prove that an infinite simple (non-Abelian) group must satisfy the two chain conditions. Show that if G is non-Abelian this is not necessarily so. For example if $S(X)$ is the group of all permutations of the infinite set X leaving all but a finite number of elements fixed and $A(X)$ are those with signature -1. Then $A(X)$ is simple and infinite.

6.8   The Fitting Decomposition and Krull-Schmidt Theorem

Here we deal with the *Fitting decomposition* of a module over a ring satisfying the ACC and DCC. When applied to linear transformations on a finite dimensional vector space, it is related to some of our other decomposition theorems above. Here we also call an endomorphism φ of an R-module M nilpotent if $\varphi^k = 0$ for some positive integer k.

Theorem 6.8.1. *Let $M(R)$ be an R-module satisfying the ACC and DCC and $T \in \mathrm{End}_R(M)$. Then for a positive integer i,*

1. $\mathrm{Ker}(T^i) \subseteq \mathrm{Ker}(T^{i+1})$ *and* $Im(T^i) \supseteq Im(T^{i+1})$ *everything being* T-*invariant*.

2. *Since these are all submodules and the ACC and DCD hold, there must be a positive integer n so that* $\mathrm{Ker}(T^n) = \mathrm{Ker}(T^{n+1})$ *and* $Im(T^n) = Im(T^{n+1})$.

3. $M = \mathrm{Ker}(T^n) \oplus Im(T^n)$.

4. *This means $M = M_1 \oplus M_2$, the direct sum of T-invariant submodules with $T \mid M_1$ nilpotent and $T \mid M_2$ an isomorphism.*

Proof. The first two statements are obvious.

Suppose $T^n(x) = 0$ and $x = T^n(y)$, where $x, y \in M$. Then $T^n(x) = T^{2n}(y)$. Thus $y \in \mathrm{Ker}(T^{2n})$. But $\mathrm{Ker}(T^{2n}) = \mathrm{Ker}(T^n)$, so that $T^n(y) = 0$. Hence $x = 0$. Thus $\mathrm{Ker}(T^n) \cap Im(T^n) = (0)$. That $\mathrm{Ker}(T^n) + Im(T^n) = M$ follows from the first isomorphism for modules. Namely, $M/\mathrm{Ker}(T^n) = Im(T^n)$. This proves the third statement.

As to the fourth, Let $M_1 = \mathrm{Ker}(T^n)$ and $M_2 = Im(T^n)$. Then these are T-invariant and so T^n invariant submodules. Then $T^n(x) = 0$ for every $x \in M_1$. Thus $T \mid M_1$ is nilpotent. Now let $x \in M_2$ and suppose $T(x) = 0$. Then $x \in M_2 \cap M_1 = (0)$ so $x = 0$. Thus $T \mid M_2$ is injective. Now let $y \in M_2 = Im(T^n)$. Then $y = T^n(x)$ for some $x \in M$. But $x = x - T^n(x) + T^n(x) = x - y + y$ and, by the above, $x - y \in \mathrm{Ker}(T^n)$. That is, $T^n(x - y) = 0$ so that $T^n(x) = T^n(y)$. Thus $y = T^n(y)$. and so $T^n \mid M_2$ is surjective. □

This of course applies when V is a finite dimensional vector space over k and M is the $k[x]$-module V_T over $k[x]$ as above. Then we get the following,

Corollary 6.8.2. *Any linear transformation T decomposes V into the direct sum of T-invariant subspaces $V_\oplus V_2$ in which $T \mid V_1$ is nilpotent and $T \mid V_2$ is an isomorphism.*

We now turn to the Krull-Schmidt theorem.

Definition 6.8.3. A module M is called *indecomposable* if $M = M_1 \oplus M_2$ implies $M_1 = 0$ or $M_2 = 0$.

A special case of Fitting's lemma is when M is indecomposable.

Corollary 6.8.4. *Let M be indecomposable module of finite length and $\varphi \in \mathrm{End}_R(M)$, then either φ is an isomorphism or it is nilpotent.*

Lemma 6.8.5. *Let M be as in Corollary 6.8.4 and φ, φ_1, $\varphi_2 \in \mathrm{End}_R(M)$, $\varphi = \varphi_1 + \varphi_2$. If φ is an isomorphism, then at least one of φ_1, φ_2 is also an isomorphism.*

Proof. Without loss of generality we may assume that $\varphi = Id$. But in this case φ_1 and φ_2 commute. If both φ_1 and φ_2 are nilpotent, then $\varphi_1 + \varphi_2$ is nilpotent, but this is impossible as $\varphi_1 + \varphi_2 = Id$.   □

Corollary 6.8.6. *Let M be as in Lemma 6.8.5. Let $\varphi = \varphi_1 + \cdots + \varphi_k \in \mathrm{End}_R(M)$. If φ is an isomorphism, then for some i, φ_i is an isomorphism.*

Evidently, if M satisfies ACC and DCC then M has a decomposition

$$M = M_1 \oplus \cdots \oplus M_k,$$

where all M_i are indecomposable.

Theorem 6.8.7. *(Krull-Schmidt) Let M be a module of finite length and*

$$M = M_1 \oplus \cdots \oplus M_k = N_1 \oplus \cdots \oplus N_l$$

for some indecomposable M_i and N_j. Then $k = l$ and there exists a permutation σ such that each $M_i \cong N_{\sigma(j)}$ for some j.

Proof. Let $\pi_i : M_1 \to N_i$ be the restriction to M_1 of the natural projection $\pi : M \to N_i$, and $\rho_j : N_j \to M_1$ be the restriction to N_j of the natural projection $\rho : M \to M_1$. Then $\rho_1 \pi_1 + \cdots + \rho_l \pi_l = Id$, and by the Corollary above, there is some i so that $\rho_i \pi_i$ is an isomorphism. Then Lemma 6.8.5 above implies that $M_1 \cong N_i$. The proof then follows by induction on k.   □

6.9   Schur's Lemma and Simple Modules

We now turn to another aspect of modules which will be pursued in Vol. II, Chapter 12. To give some idea of what we really have in mind here we will, per force, take R to be non-commutative. Here we shall be interested in *simple modules.*

We now turn to Schur's lemma.[4]

Most of the statements in this section are variants of what is known as Schur's lemma. We remark that some of these ideas work even when the space being acted on is not finite dimensional.

Theorem 6.9.1. *(Schur's Lemma) Let M and M' be simple R-modules and $f : M \to M'$ an R-module homomorphism. Then either f is an isomorphism, or f is identically zero.*

Proof. Since Ker f is a submodule of M it is either (0) in which case f is injective or its all of M, that is, f is identically zero. So assuming f is not identically zero, we know its injective. Now since M' is also simple and $f(M)$ is a submodule of M', then $f(M)$ is either (0) or all of M'. But the latter cannot be since f is not identically zero. Hence f is also surjective.   □

We now give several variants of Schur's lemma.

Corollary 6.9.2. *(Variant 1) Let M be a simple R-module and $f : M \to M$ an R-module homomorphism. Suppose T is an R-module map of M to itself and $fT = Tf$. Then T is an isomorphism, or $T = 0$.*

---

[4]Issai Schur (1875-1941) a Russian (German, Latvian) Jewish mathematician was a student of Frobenius, one of the two main founders of the theory of group representations, particularly of finite groups (the other being Burnside). Schur very much advanced this work. Schur's lemma and the Schur orthogonality relations are a crucial part of the subject. He also worked in number theory, analysis and theoretical physics. As a full professor in Berlin he had a large number of students, with many diverse interests, many of whom became quite famous in their own right. His lectures were very popular with often 500 in attendance until he was dismissed by the Nazis in 1935. He died in Jerusalem.

Proof. Both $\mathrm{Ker}\,T$ and $T(M)$ are submodules of M. They are also f-invariant because if $T(x) = 0$, then $f(T(x)) = T(f(x)) = 0$ so $f(x) \in \mathrm{Ker}\,T$. Similarly, if $y = T(x)$, then $f(y) = f(T(x)) = T(f(x))$ so $f(y) \in T(M)$. Since M is simple $\mathrm{Ker}\,T = 0$ or M. In the latter case $T = 0$, so we can assume T is injective. But $T(M)$ is also a submodule so it is also either 0 or all of M. The case $T = 0$ having already been dealt with we conclude that T is an isomorphism.   □

Now let V be a finite dimensional vector space over a field k and ρ be a *representation* of a group G on V over k. We can make (G, V, ρ), sometimes just called (V, ρ) into an R-module as follows. Let $R = k(G)$, the group algebra. This is the ring (actually k-algebra with identity). We make V into a module over this ring by letting G act through ρ. Thus

$$\left(\sum_i c_i \rho_{g_i}, v \right) \mapsto \sum_i c_i \rho_{g_i}(v),$$

and leave the easy formal verification that $(V, k(G), \rho)$ is an R-module to the reader.[5]

The following definition is crucial to studying group representations.

Definition 6.9.3.   1. Given a representation ρ, a subspace W of V_ρ is called *invariant* if $\rho(g)(W) \subseteq W$ for each $g \in G$.

2. Given two representations ρ and σ of G, a linear operator $T : V_\rho \to V_\sigma$ is called an *intertwining operator* for these representations if

$$\sigma(g) \circ T = T \circ \rho(g)$$

for each $g \in G$.

Observe that the module structure $V_{k(G)}$ reflects exactly that of the representation ρ. The simple proof of the following proposition is left to the reader as an exercise.

---

[5]That is, $\rho : G \to \mathrm{GL}(V_\rho)$ is a homomorphism with values in the general linear group of the vector space V_ρ, over the field k. We then call ρ a (*linear*) *representation* of G. We shall deal with this subject extensively in Vol. II, Chapter 12. In case that G has more structure, ρ is would be required to respect that structure.

Proposition 6.9.4.

1. *A subspace W of V is a submodule if and only if W is ρ invariant.*

2. *If ρ' is another representation of G on V' over the same field k and $f : V \rightarrow V'$ is a k linear map, then f is a module homomorphism if and only if for all $g \in G$, $f\rho_g = \rho'_g f$. That is, f is an intertwining operator for the representations ρ and ρ'.*

In view of Proposition 6.9.4 just above we have the following corollary.

Corollary 6.9.5.   *(Variant 2)*

1. *(V, ρ) is a simple R-module if and only if ρ is irreducible.*

2. *The representations ρ and ρ' are equivalent if and only if (V, ρ) and (V', ρ') are isomorphic as $k(G)$ modules.*

These follow from 1 and 2, respectively, of Proposition 6.9.4. Notice that a representation of degree 1 is always irreducible.

In particular, when M is a $k(G)$ module as above, suppose $Tf = fT$. Then since λI, for $\lambda \in k$ also commutes with f so does $T - \lambda I$. If λ is chosen so that $T - \lambda I$ is singular, then $T = \lambda I$.

From this we get as an immediate corollary,

Corollary 6.9.6.   *(Variant 3) Let (V, ρ) and (V', ρ') be finite dimensional irreducible representations and $f : V \rightarrow V'$ an intertwining operator. Then either f is an equivalence of representations, or f is identically zero.*

If the field is algebraically closed we can say more.

Corollary 6.9.7.   *(Variant 4) Suppose K is algebraically closed and (V, ρ) and (V', rho') are irreducible representations over K. If an intertwining operator $T : V \rightarrow V'$ is not identically zero it must be a scalar multiple of the identity.*

Proof. Since $T\rho_g = \rho_g T$ for every $g \in G$ it follows that also $(T - \lambda I)\rho_g = \rho_g(T - \lambda I)$ for all $g \in G$. Let λ be an eigenvalue of T. Since $T - \lambda I$ is singular $T - \lambda I = 0$. $\square$

Let S be a subset of $\mathrm{End}_k(V)$. We shall say S acts irreducibly on V if the only S-invariant subspaces of V are V and (0). In particular, if ρ is a representation of G on V we say ρ is irreducible when $S = \rho(G)$. In Vol. II, Chapter 13 we shall use a more specialized notion of irreducibility.

Corollary 6.9.8. *(Variant 5) If K is algebraically closed and (V, ρ) is an irreducible representation of G, then $\rho_z = \lambda(z)I$ for every $z \in \mathcal{Z}(G)$, its center. In particular, if G is Abelian every irreducible representation is one dimensional.*

Proof. For $z \in \mathcal{Z}(G)$ and $g \in G$, $\rho_z\rho_g = \rho_{zg} = \rho_g\rho_z$. It follows that each ρ_z is an intertwining operator. Hence by irreducibility $\rho_z = \lambda(z)I$ for every $z \in \mathcal{Z}(G)$. If G is Abelian, $\rho_g = \lambda(g)I$ for every $g \in G$. But then every subspace is invariant. Hence ρ can not be irreducible unless it has degree 1. $\square$

We remark that Corollaries 6.9.7 and 6.9.8 only hold when K is algebraically closed. For example, consider the representation where $\mathbb{R}$ acts on $\mathbb{R}^2$ by rotations. This is a representation of G over the field $\mathbb{R}$ of degree 2 which is clearly irreducible. But the operators are not scalar multiples of the identity. Far from it, their eigenvalues are distinct complex conjugates of one another $\cos\theta \pm i\sin\theta$.

Similarly, $\mathbb{R}$ can act irreducibly on $\mathbb{R}^2$ by unipotent operators. Here ρ is injective. These operators are also not scalar multiples of the identity either since their eigenvalues are all 1. If they were ρ would be identically I which cannot be since ρ is injective.

Appendix

A. Pell's Equation

The Diophantine equation

$$x^2 - dy^2 = 1 \qquad (1)$$

where d is an integer is one in which we ask if there are infinitely many integer solutions in x and y. This has been mistakenly called *Pell's equation* by Leonhard Euler, after the English mathematician John Pell (1611-1685). However, Pell's only contibution to the subject was the publication of some partial results of Wallis and Brouncker. Much earlier it had been investigated by Indian mathematicians in special cases: Brahmagupta (628) solved the case $d = 92$; then Bhaskara II (1150) did so for $d = 61$; and Narayana (1340-1400) for $d = 103$. The first to deal systematically with it was Fermat who said that he had proved the result in the key case where $d > 0$ and not a perfect square (but as usual he didn't write down the proof). Euler had shown that there are infinitely many solutions if there is one, but it was Lagrange who later published the first proof, using continued fractions.

To solve the equation (1), we first observe that if $d = -1$, then the only integral solutions are $(\pm 1, 0)$ and $(0, \pm 1)$, while if $d < -1$, the only solutions are $(\pm 1, 0)$. Therefore, henceforth we will consider only the case when $d > 0$.

Clearly, an obvious solution of (1) is the *trivial solution* $(\pm 1, 0)$. In addition, if d is a perfect square, say $d = d_1^2$, then (1) becomes

$$x^2 - (d_1 y)^2 = 1$$

and since the only two squares which differ by 1 are the integers 0 and 1, the only solutions of (1) are $(\pm 1, 0)$. Hence, we are looking for non-trivial solutions of (1), with $d > 0$ and not a perfect square. In addition, since, if (x, y) is a solution of (1) so are $(-x, y)$, $(x, -y)$, and $(-x, -y)$, we will concentrate on solutions (x, y) with x, y positive integers.

Our strategy will be to prove the following:

1. Each solution produces an infinite number of different solutions.

2. A solution, if it exists, called a minimal solution, produces all the others.

3. There exist a minimal solution, and the theory of continued fractions gives a way to find it.

To do this we shall need the following theorem:

Theorem A.1. (The Approximation Theorem) *If ξ is a real number and k is a positive integer, then there are integers n and m such that*

$$\left| \xi - \frac{n}{m} \right| \leq \frac{1}{m(k+1)}, \quad 1 \leq m \leq k.$$

Proof. Consider the following $k + 1$ numbers:

$$0\xi - [0\xi], \ 1\xi - [1\xi], \ \dots, \ k\xi - [k\xi].$$

All these numbers lie in the interval $[0, 1]$. We label this finite number in increasing order $P_0, P_1, \dots, P_k$. Writing a telescoping sum we see that

$$(P_1 - P_0) + (P_2 - P_1) + \cdots + (P_k - P_{k-1}) + (1 - P_k) = 1.$$

Thus we have $k + 1$ positive numbers whose sum is 1, and therefore, at least one of them must be smaller than $\frac{1}{k+1}$. Let this be the number $P_l - P_{l-1}$. This number is of the form $p\xi - q\xi - [(p-q)\xi]$, with $p, q \leq k$. Now, let $n = \pm[(p-q)\xi]$ and $m = p - q$. These satisfy our inequality. $\square$

Theorem A.2. *Let ξ be an irrational number. Then, if n, m are positive integers, the inequality*

$$|n - m\xi| < \frac{1}{m} \tag{2}$$

has an infinite number of solutions.

Proof. From Theorem A.1, we know that $\left|\xi - \frac{n}{m}\right| \le \frac{1}{m(k+1)}$, $1 \le m \le k$, which implies

$$\left|\frac{n}{m} - \xi\right| < \frac{1}{m^2}$$

since $m \le k$. The above inequality exhibits one solution of (2). To see that there are infinitely many, we remark that since ξ is assumed to be an irrational number, $|\xi - \frac{n}{m}|$ is always strictly positive. Therefore, by choosing m large enough and applying the same theorem once again, we can find another pair (n_1, m_1) such that

$$\left|\xi - \frac{n_1}{m_1}\right| < \left|\xi - \frac{n}{m}\right|.$$

Hence there are infinitely many solutions. $\qquad\square$

Proposition A.3. *There are infinitely many solutions of the equation*

$$x^2 - dy^2 = k$$

with x, y positive integers, for some k such that $|k| < 1 + 2\sqrt{d}$.

Proof. By Theorem A.2 for the irrational number $\xi = \sqrt{d}$, we know that there are infinitely many integral solutions of (2). Let (x, y) be one of them. Then

$$|x + y\sqrt{d}| = |x - y\sqrt{d} + 2y\sqrt{d}| < \frac{1}{y} + 2y\sqrt{d} \le (1 + 2\sqrt{d})y.$$

Thus

$$|x^2 - dy^2| = |x - y\sqrt{d}| \cdot |x + y\sqrt{d}| < \frac{1}{y}(1 + 2\sqrt{d})y = 1 + 2\sqrt{d}.$$

Now, since there is an infinite number of distinct pairs of integers (x, y) satisfying the above inequality, but only finitely many integers smaller than $1 + 2\sqrt{d}$, there must exist some integer k such that $|k| < 1 + 2\sqrt{d}$ for which there are infinitely many solutions to the equation $x^2 - dy^2 = k$. $\qquad\square$

With this, we can prove the following theorem,[6]

Theorem A.4. *For any positive non-square integer d, the equation*

$$x^2 - dy^2 = 1$$

has a non-trivial integral solution.

Proof. Let $g = [1 + \sqrt{d}\,]$. Then, the number of integers $\neq 0$ between $-1 - \sqrt{d}$ and $1 + \sqrt{d}$ is $2g$. Now, we choose $n = 2g^4$ pairs of integers $(x_1, y_1), (x_2, y_2), \ldots, (x_n, y_n)$, satisfying the inequalities

$$|x_i^2 - dy_i^2| < 1 + 2\sqrt{d}, \quad i = 1, 2, \ldots, n,$$

and in such a way that

$$|x_1 - y_1\sqrt{d}| > |x_2 - y_2\sqrt{d}| > \cdots.$$

The differences

$$x_1^2 - dy_1^2, \quad x_2^2 - dy_2^2, \quad \cdots, \quad x_n^2 - dy_n^2, \tag{1}$$

represent integers between $-1 - 2\sqrt{d}$ and $1 + 2\sqrt{d}$. Let us denote these integers by $N_1, N_2, \ldots, N_{2g}$ and let $M_1, M_2, \ldots, M_{2g}$ denote the number of times the integers N_i occur in series (1). Obviously,

$$M_1 + M_2 + \cdots + M_{2g} = 2g^4$$

and the largest of M_j, $j = 1, \ldots, 2g$ must be $\geq g^3$. This means for some $k = \pm 1, \pm 2, \ldots, \pm g$, the equation

$$x^2 - dy^2 = k, \tag{2}$$

---

[6] J. V. Uspensky and M. A. Heaslet, *Elementary Number Theory*, McGraw-Hill Book Company, New York and London, 1939.

is satisfied by at least $g^3 > g^2 \geq k^2$ pairs of integers x, y. Now, for the pairs of integers (x, y) and (z, w), we define the relation $(x, y) \sim (z, w)$ if and only if $z \equiv x \mod k$ and $w \equiv y \mod k$. Then, the number of non-congruent pairs is k^2 and since (2) is satisfied by more than k^2 pairs, at least two of them must be congruent $\mod k$, say (x, y) and (z, w). Hence

$$x^2 - dy^2 = z^2 - dw^2 = k,$$

and

$$x = z, \text{ and } w \equiv y \mod k.$$

In addition we can assume that

$$|z - w\sqrt{d}| < |x - y\sqrt{d}|. \tag{3}$$

Now consider the quotient

$$\frac{x - y\sqrt{d}}{z - w\sqrt{d}} = \frac{(x - y\sqrt{d})(z + w\sqrt{d})}{z^2 - dw^2} = \frac{xz - dyw + (xw - zy)\sqrt{d}}{k}. \tag{4}$$

We claim that $xz - dyw$ and $xw - zy$ are divisible by k. Indeed, since $(x, y) \sim (z, w)$,

$$xz - dyw \equiv (x^2 - dy^2) \mod k \equiv 0 \mod k, \quad xw - zy \equiv 0 \mod k.$$

Therefore,

$$X = \frac{xz - dyw}{k}, \quad Y = \frac{xw - zy}{k}$$

are integers, and from (4) we get

$$x - y\sqrt{d} = (z - w\sqrt{d})(X + Y\sqrt{d}).$$

Therefore,

$$x^2 - dy^2 = (z^2 - dw^2)(X^2 - dY^2)$$

and, since $x^2 - dy^2 = z^2 - dw^2 = k \neq 0$, we get

$$X^2 - dY^2 = 1.$$

From (3) we see that the left side of (4) is > 1. Hence $|X + Y\sqrt{d}| > 1$, $Y \neq 0$. In other words, we proved that the equation $X^2 - dY^2 = 1$ has an integral solution (X, Y), different from the trivial solution $X = \pm 1$, $Y = 0$. $\qquad\qquad\qquad\qquad\qquad\qquad\qquad\qquad\qquad\qquad\qquad\qquad\square$

It now follows that any two solutions of equation (1) produce a third one. Indeed, we have

Proposition A.5. *If* (x_1, y_1) *and* (x_2, y_2) *are two solutions of Pell's equation, then* (x, y), *defined by the equation*

$$x + y\sqrt{d} = (x_1 + y_1\sqrt{d})(x_2 + y_2\sqrt{d}), \qquad (\star)$$

is also a solution.

Proof. Since x_i, y_i, for $i = 1, 2$ are integers, equation $(\star)$, tells us that

$$x = x_1 x_2 + d y_1 y_2$$

$$y = x_1 y_2 + x_2 y_1$$

and so

$$x - y\sqrt{d} = (x_1 - y_1\sqrt{d})(x_2 - y_2\sqrt{d}).$$

Therefore,

$$x^2 - dy^2 = (x + y\sqrt{d})(x - y\sqrt{d}) = (x_1^2 - dy_1^2)(x_2^2 - dy_2^2) = 1,$$

which means that (x, y) is also a solution. $\qquad\qquad\qquad\qquad\square$

Proposition A.6. *Let* (x_1, y_1) *be a solution of* (1). *Then, there are infinitely many solutions* (x_n, y_n), *where* x_n *and* y_n *are given by*

$$x_n = \frac{\left(x_1 + y_1\sqrt{d}\right)^n + \left(x_1 - y_1\sqrt{d}\right)^n}{2}$$

$$y_n = \frac{\left(x_1 + y_1\sqrt{d}\right)^n - \left(x_1 - y_1\sqrt{d}\right)^n}{2\sqrt{d}}.$$

Proof. By assumption, $x_1^2 - dy_1^2 = 1$. In other words,

$$\left(x_1 - y_1\sqrt{d}\right)\left(x_1 + y_1\sqrt{d}\right) = 1$$

and by taking n-th powers we get

$$\left(x_1 - y_1\sqrt{d}\right)^n \left(x_1 + y_1\sqrt{d}\right)^n = 1.$$

The binomial formula 0.4.2 tells us that the two factors on the left side of the above equation are of the form

$$\left(x_1 - y_1\sqrt{d}\right)^n = x_n + y_n\sqrt{d}$$

$$\left(x_1 + y_1\sqrt{d}\right)^n = x_n + y_n\sqrt{d},$$

where x_n and y_n are polynomials on x_1 and y_1, with x_n is the integral part and y_n the coefficient of $\sqrt{d}$ in the binomial expansion $\left(x_1 + y_1\sqrt{d}\right)^n$.

Hence,

$$x_1^2 - dy_1^2 = 1 \implies x_n^2 - dy_n^2 = 1,$$

for any $n \in \mathbb{N}$. That x_n and y_n are integers is easily seen since, by the binomial theorem, both numerators in the two expressions have all their terms powers of $\sqrt{d}$. □

Remark A.7. All solutions (x_n, y_n) are distinct because all real numbers $\left(x_1 + y_1\sqrt{d}\right)^n$ are distinct, since the only complex numbers z for which $z^n = z^m$, $n < m$, are the roots of unity, and the only real roots of unity are ± 1. Indeed, by taking the 0-th power of $x_1 + y_1\sqrt{d}$ we obtain the trivial solution, $(1, 0)$. In addition, $(x_1 + y_1\sqrt{d})^{-1} = x_1 - y_1\sqrt{d}$ and taking negative integral powers we get more such solutions. These solutions are actually *all* the integral solutions. The proof of this depends on the following lemma whose proof we shall omit.

Lemma A.8. *Let (x, y) be a non-trivial integral solution of Pell's equation. Then, x and y are both positive if and only if $x + y\sqrt{d} > 1$.*

B. The Kronecker Approximation Theorem

Consider n-tuples $\{\alpha_1, \ldots, \alpha_n\}$, where $\alpha_i \in \mathbb{R}$ and n is an integer $n \geq 1$. We shall say $\{\alpha_1, \ldots, \alpha_n\}$ is *generic* if whenever $\sum_{i=1}^{n} k_i \alpha_i \in \mathbb{Z}$ for $k_i \in \mathbb{Z}$, then all $k_i = 0$.

Here is an example of a generic set. Let θ be a transcendental real number (such as π or e) and consider its powers, $\alpha_i = \theta^i$. Then for *any* positive integer n, $\{\theta^1 \ldots, \theta^n\}$ is generic. For if $k_1\theta^1 + \ldots k_n\theta^n = k$, where the k_i and k are integers, then since $\mathbb{Z} \subseteq \mathbb{Q}$, θ satisfies a polynomial equation over $\mathbb{Q}$, a contradiction.

Proposition B.1. *The set* $\{\alpha_1, \ldots, \alpha_n\}$ *is generic if and only if* $\{1, \alpha_1, \ldots, \alpha_n\}$ *is linearly independent over* $\mathbb{Q}$.

Proof. Suppose $\{1, \alpha_1, \ldots, \alpha_n\}$ is linearly independent over $\mathbb{Q}$. Let $\sum_{i=1}^{n} k_i \alpha_i = k$, where $k \in \mathbb{Z}$. We may assume $k \neq 0$. For if $k = 0$ then since the subset $\{\alpha_1, \ldots, \alpha_n\}$ is linearly independent over $\mathbb{Q}$ and $k_i \in \mathbb{Q}$ for each i it follows that $k_i = 0$. On the other hand, if $k \neq 0$ we divide and get $\sum_{i=1}^{n} \frac{k_i}{k} \alpha_i = 1$. So 1 is a $\mathbb{Q}$-linear combination of α_i's. This contradicts our hypothesis regarding linear independence.

Conversely, suppose $\{\alpha_1, \ldots, \alpha_n\}$ is generic and $q1 + q_1\alpha_1 + \ldots + q_n\alpha_n = 0$, where q and all the $q_i \in \mathbb{Q}$. If $q = 0$, then clearing denominators gives a relation $k_1\alpha_1 + \ldots + k_n\alpha_n = 0$, where $k_i \in \mathbb{Z}$. Since the α_i are generic and $0 \in \mathbb{Z}$ we get each $k_i = 0$. Hence each q_i is also $= 0$. Thus $\{1, \alpha_1, \ldots, \alpha_n\}$ is linearly independent over $\mathbb{Q}$. On the other hand, if $q \neq 0$, by dividing by q we get, $1 + s_1\alpha_1 + \ldots + s_n\alpha_n = 0$, where $s_i \in \mathbb{Q}$. Again clearing denominators yields $k + k_1\alpha_1 + \ldots + k_n\alpha_n = 0$, where k and $k_i \in \mathbb{Z}$. Since $k_1\alpha_1 + \ldots + k_n\alpha_n = -k$ and the original α_i are generic, each $k_i = 0$. Therefore each s_i is also 0 and thus $1 = 0$, a contradiction. $\square$

Here is another way to "find" generic sets. We consider $\mathbb{R}$ to be a vector space over $\mathbb{Q}$. Since $1 \neq 0$ we see 1 is linearly independent over $\mathbb{Q}$. Extend this to a basis of $\mathbb{R}$ over $\mathbb{Q}$. Then any finite subset of this basis gives a generic set after removing 1.

We now prove *Kronecker's approximation theorem*. We first need to define the character group. For a locally compact Abelian topological group G its *character group*, $\widehat{G}$, is the *continuous* homomorphisms $G \to \mathbb{T}$. Here characters are multiplied pointwise and convergence is uniform on compact subsets of G. Then $\widehat{G}$ is also a locally compact Abelian topological group.

Proposition B.2. *Let G and H be locally compact Abelian groups (written additively) and $\beta : G \times H \to \mathbb{T}$ be a non-degenerate, jointly continuous bilinear function. Consider the induced map $\omega_G : G \to \widehat{H}$ given by $\omega_G(g)(h) = \beta(g, h)$. Then ω_G is a continuous injective homomorphism with dense range. Similarly, $\omega_H : H \to \widehat{G}$ given by $\omega_H(h)(g) = \beta(g, h)$. Now ω_H is also a continuous injective homomorphism with dense range.*

Proof. By symmetry we need only consider the case of ω_G. Clearly $\omega_G : G \to \widehat{H}$ is a continuous homomorphism. If $\omega_G(g) = 0$ then for all $h \in H$, $\beta(g, h) = 0$. Hence $g = 0$ so ω_G is injective. To prove that $\omega_G(G)$ is a dense subgroup of $\widehat{H}$ we show that its annihilator in $\widehat{\widehat{H}}$ is trivial. Identifying H with its second dual $\widehat{\widehat{H}}$, its annihilator consists of all $h \in H$ so that $\beta(g, h) = 0$ for all $g \in G$. By non-degeneracy (this time on the other side) the annihilator of $\omega_G(G)$ is trivial. Hence $\omega_G(G)$ is dense in $\widehat{H}$ (see [100]). $\qquad \square$

We now come to the Kronecker theorem itself. What it says is that one can simultaneously approximate $(x_1, \ldots, x_n) \bmod (1)$ by $k(\alpha_1, \ldots, \alpha_n)$. If we denote by $\pi : \mathbb{R} \to \mathbb{T}$ the canonical projection with $\operatorname{Ker} \pi = \mathbb{Z}$, the Kronecker theorem says that any point, $(\pi(x_1), \ldots, \pi(x_n))$ on the n-torus, $\mathbb{T}^n$, can be approximated to any required degree of accuracy by integer multiples of $(\pi(\alpha_1), \ldots, \pi(\alpha_n))$.

In particular, any point on the torus can be approximated to any degree of accuracy by real multiples of $(\pi(\alpha_1), \ldots \pi(\alpha_n))$. The image under π of such a line, namely the real multiples of $(\alpha_1, \ldots, \alpha_n)$, called the *generic set*, always exists.

Theorem B.3. *Let $\{\alpha_1, \ldots, \alpha_n\}$ be a generic set, $\{x_1, \ldots, x_n\} \in \mathbb{R}$ and $\epsilon > 0$. Then there is a $k \in \mathbb{Z}$ and $k_i \in \mathbb{Z}$ such that $|k\alpha_i - x_i - k_i| < \epsilon$.*

Proof. Consider the bilinear form $\beta : \mathbb{Z} \times \mathbb{Z}^n \to \mathbb{T}$ given by $\beta(k, (k_1, \ldots k_n)) = \pi(k \sum_{i=1}^n k_i \alpha_i)$. Then β is additive in each variable separately and of course is jointly continuous since here the groups are discrete. The statement is equivalent to saying that image of the map $\beta_G : \mathbb{Z} \to \widehat{\mathbb{Z}^n} \simeq \mathbb{T}^n$ is dense.

We prove that β is non-degenerate. That is, if $\beta(k, (k_1, \ldots k_n)) = 0$ for all k, then $(k_1, \ldots k_n) = 0$ and if $\beta(k, (k_1, \ldots k_n)) = 0$ for all $(k_1, \ldots k_n)$ then $k = 0$.

Suppose $\beta(k, (k_1, \ldots k_n)) = 0$ for all k. That is $\pi(k \sum_{i=1}^{n} k_i \alpha_i) = 0$, or $k \sum_{i=1}^{n} k_i \alpha_i$ is an integer. Choose any $k \neq 0$. Then $\sum_{i=1}^{n} k k_i \alpha_i$ is an integer. Because of our hypothesis regarding the α_i we conclude all $k k_i = 0$ therefore $k_i = 0$. On the other hand, suppose $\beta(k, (k_1, \ldots k_n)) = 0$ for all $(k_1, \ldots k_n)$. Then we show $k = 0$. Hence we have $k \sum_{i=1}^{n} k_i \alpha_i$ is an integer for all choices of $(k_1, \ldots k_n)$. Arguing as before, suppose $k \neq 0$. Choose k_i not all zero. Then as the α_i are a generic set, for some i, $k k_i = 0$ and therefore $k = 0$.

Hence by Proposition B.2 we get an injective homomorphism $\omega : \mathbb{Z} \to \mathbb{Z}^n = \mathbb{T}^n$ with dense range. Thus the cyclic subgroup $\omega(\mathbb{Z})$ in dense in $\mathbb{T}^n$. □

Exercise B.4.

(1) Show that in $\mathbb{R}^2$ a line is winding if and only if it has irrational slope.

(2) Find the generic sets when $n = 1$. What does this say about dense subgroups of $\mathbb{T}$?

C. Some Groups of Automorphisms

Definition C.1. Let G be a topological group. An *automorphism* of G is a bijective, bicontinuous map $\alpha : G \longrightarrow G$, which is a group homomorphism. We denote by $\mathrm{Aut}(G)$ the group of automorphisms of G.

The Automorphisms of $(\mathbb{R}, +)$

Consider the group $(\mathbb{R}, +)$. We want to find all the continuous linear functions

$$f : \mathbb{R} \longrightarrow \mathbb{R} \quad : \quad f(x + y) = f(x) + f(y).$$

Proposition C.2. *The continuous homomorphisms of $(\mathbb{R}, +)$ are of the form*

$$f(x) = \lambda x, \quad \lambda \in \mathbb{R}.$$

Proof. Obviously, $f(0) = f(0 + 0) = f(0) + f(0) = 2f(0)$. Therefore, $f(0) = 0$. Also, $0 = f(0) = f(x + (-x)) = f(x) + f(-x)$, and thus $f(-x) = -f(x)$.

First let x be an integer. Then $f(2) = f(1 + 1) = 2f(1)$. Using induction, suppose that $f(k - 1) = (k - 1)f(1)$. Then $f(k) = f(k - 1 + 1) = f(k - 1) + f(1) = (k - 1)f(1) + f(1) = kf(1)$.

Now, suppose x is a rational number. If $n \neq 0$, then

$$f(1) = f\left(\frac{n}{n}\right) = f\left(\underbrace{\frac{1}{n} + \ldots + \frac{1}{n}}_{n}\right) = nf\left(\frac{1}{n}\right) \quad \Rightarrow \quad f\left(\frac{1}{n}\right) = \frac{1}{n}f(1).$$

Therefore, if $x = \frac{m}{n}$,

$$f(x) = f\left(\frac{m}{n}\right) = f\left(\underbrace{\frac{1}{n} + \cdots + \frac{1}{n}}_{m}\right) = mf\left(\frac{1}{n}\right) = \frac{m}{n}f(1).$$

Finally, suppose x is irrational. Then $x = \lim_{i \mapsto \infty} p_i$ where p_i are all rational numbers. Since f is continuous we see

$$f(x) = f(\lim_{i \mapsto \infty} p_i) = \lim_{i \mapsto \infty} f(p_i) = \lim_{i \mapsto \infty} p_i f(1) = xf(1).$$

Setting $f(1) = \lambda$, the conclusion follows. $\qquad \square$

Remark C.3. Here the assumption of the continuity of the function f is crucial.[7]

---

[7] For discontinuous solutions see B. R. Gelbaum and J. M. H. Olmsted [41] or [42].

The Automorphisms of the Torus $\mathbb{T}^n$

Definition C.4. The *n-Torus* $\mathbb{T}^n$ is the Abelian group,

$$\mathbb{T}^n = \mathbb{R}^n / \mathbb{Z}^n.$$

Proposition C.5. *We have* $\operatorname{Aut}(\mathbb{T}^n) = \operatorname{SL}^{\pm}(n, \mathbb{Z})$.

Proof. Let $a \in \operatorname{Aut}(\mathbb{T}^n)$. Consider the natural projection map $\pi : \mathbb{R}^n \longrightarrow \mathbb{R}^n / \mathbb{Z}^n$. This is a covering map, and since $\mathbb{R}^n$ is simply connected we can lift a to $\tilde{a}$, such that the following diagram

$$
\begin{array}{ccc}
\mathbb{R}^n & \xrightarrow{\;\;\tilde{a}\;\;} & \mathbb{R}^n \\
\downarrow{\scriptstyle \pi} & & \downarrow{\scriptstyle \pi} \\
\mathbb{T}^n = \mathbb{R}^n / \mathbb{Z}^n & \xrightarrow{\;\;a\;\;} & \mathbb{T}^n = \mathbb{R}^n / \mathbb{Z}^n
\end{array}
$$

is commutative. Obviously $\tilde{a}$ is also an automorphism i.e. $\tilde{a} \in \operatorname{GL}(n, \mathbb{R})$. But not all automorphisms of $\mathbb{R}^n$ project to automorphisms of $\mathbb{T}^n$. For this to happen, $\tilde{a}$ has to leave the subgroup $\mathbb{Z}^n$ invariant. In other words we must have

$$\tilde{a}(\mathbb{Z}^n) \subseteq \mathbb{Z}^n.$$

Since the same must happen for $\tilde{a}^{-1}$, this implies $\det(\tilde{a}) = \pm 1$, i.e. $\tilde{a} \in \operatorname{SL}^{\pm}(n, \mathbb{Z})$. $\square$

When $n \geq 2$ these are pretty big infinite groups. However when $n = 1$ the group is finite.

Corollary C.6. *For* $n = 1$, *we obtain the* circle group $\mathbb{T} = \mathbb{R}/\mathbb{Z} \simeq S^1$, *therefore*

$$\operatorname{Aut}(\mathbb{T}) = \operatorname{SL}^{\pm}(1, \mathbb{Z}) = \{\pm I\}.$$

This means the circle group has only two automorphisms, the identity and the inversion.

Bibliography

[1] H. Abbaspour, M. Moskowitz, *Basic Lie Theory*, World Scientific Publishing Co., River Edge, NJ, 2007.

[2] A. Abbondandolo. and R. Matveyev, *How large is the shadow of a symplectic ball?*, arXiv: 1202.3614v3 [math.SG], 2013.

[3] I. T. Adamson, *Introduction to Field Theory*, Cambridge University Press, Cambridge, 2nd ed. 1982.

[4] M. Aigner, G. Ziegler, *Proofs from the book*, 4th ed., Springer-Verlag, Berlin, New York, 2010.

[5] Y. Akizuki, *Teilerkettensatz und Vielfachensatz*, Proc. Phys.-Math. Soc. Japan, vol. **17**, 1935, pp. 337-345.

[6] D. B. Ames, *An Introduction to Abstract Algebra*, Inter. Textbook Company, Scranton, Pennsylvania, 1969.

[7] V. I Arnold, *Mathematical Methods of Classical Mechanics*, Springer-Verlag, 2nd edition, 1978.

[8] Michael Aristidou, Andy Demetre *A Note on Quaternion Rings over $\mathbb{Z}_p$*, Inter. Journal of Algebra, vol. **3**, 2009, no. 15, pp. 725-728.

[9] Emil Artin, *Über einen Satz von Herrn J. H. Maclagan Wedderburn*, Abhandlungen aus dem Mathematischen Seminar der Hamburgischen Universität, **5** (1927), pp. 245-250.

[10] E. Artin, *Galois Theory*, University of Notre Dame, 1946.

[11] E. Artin, *Geometric Algebra*, Interscience Publishers, Inc., New York, NY., 1957.

[12] J. Baez, *Platonic Solids in all Dimensions*, math.ucr.edu/home/baez/platonic.html, 2006.

[13] J. C. Baez, *The Octonions*, Bull. Amer. Math. Soc., vol. **39**, 2002, pp. 145-205.

[14] J. Barnes, *Gems of Geometry*, Springer-Verlag Berlin Heidelberg, 2009.

[15] F. Beukers, E. Calabi, J. Kolk, *Sums of generalized harmonic series and volumes*, Nieuw Archief voor Wiskunde vol. **11** (1993), pp. 217-224.

[16] J. Barshey, *Topics in Ring Theory*, W.A. Benjamin, Inc., 1969.

[17] D. Birmajer, J. B. Gil, *Arithmetic in the ring of the formal power series with integer coefficients*, Amer. Math. Monthly, vol. **115**, 2008, pp. 541-549.

[18] D. Birmajer, J. B. Gil and M. D. Weiner, *Factoring polynomials in the ring of formal power series over* $\mathbb{Z}$, Intr. J. Number Theory, vol. **8**, no. 7, 2012, pp. 1763-1776.

[19] Borel A., *Compact Clifford-Klein forms of symmetric spaces*, Topology (2) 1963, 111–122.

[20] M. Boij and D. Laksov, *An Introduction to Algebra and Geometry via Matrix Groups*, Lecture Notes, 2008.

[21] P. Bürgisser, F. Cucker, *Condition. The Geometry of Numerical Algorithms*, Springer, 2013.

[22] H. Cartan and S. Eilenberg, *Homological Algebra*, Princeton University Press, 1956.

[23] K. Chandrasekharan, *Introduction to Analytic Number Theory*, Springer-Verlag, New York, 1968, p. 10.

[24] C. Chevalley, *On the theory of local rings*, Annals of Math., **44**, 1943, pp. 690-708.

[25] A. H. Clifford, *Representations Induced In An Invariant Subgroup* Ann. of Math. v. 38, no. 3 (1937).

[26] C. Crompton, *Some Geometry of the p-adic Rationals*, 2006.

[27] J. Dieudonné, *Topics in Local algebra.*, University of Notre Dame Press, 1967.

[28] J. Dieudonné, *Sur les générateurs des groupes classiques*, Summa Brad. Math., vol. **3**, 1955, pp. 149-180.

[29] L. E. Dickson, *On finite algebras*, Nachrichten der Gesellschaft der Wissenschaften zu Göttingen, (1905), pp. 358-393.

[30] L. E. Dickson, *Linear Algebras*, Cambridge University Press, Cambridge, 1930.

[31] J. Draisma, D. Gijswijt, *Invariant Theory with Applications*, Lecture Notes, 2009.

[32] S. Eilenberg, S. MacLane, *Group extensions and homology*, Annals of Math. vol **43**, 1942, pp. 757-831.

[33] D. Eisenbud, *Commutative Algebra, with a View Toward Algebraic Geometry*, Springer, 2004.

[34] P. Erdös, *Über die Reihe $\sum \frac{1}{p}$*, Mathematica, Zutphen B **7** (1938), pp. 1-2.

[35] L. Euler, *Introductio in Analysin Infinitorum*, Tomus Primus, Lausanne (1748), Opera Omnia, Ser. 1, vol. **90**.

[36] W. Feit and J. G. Thompson, *Solvability of groups of odd order*, Pacific J. Math., vol. **13** (1963), pp. 775-1029.

[37] L. Fuchs, *Infinite Abelian Groups.* vol. I, Academic Press, 1970.

[38] H. Fürstenberg, *On the infinitude of primes*, Amer. Math. Monthly **62** (1955), p. 353.

[39] J. Gallier, *Logarithms and Square Roots of Real Matrices*, in arXiv:0805.0245v1, 2008.

[40] C. F. Gauss, *Disquisitiones Aritmeticae*, (translated by Arthur A. Clarke), Yale University Press, 1965.

[41] B. R. Gelbaum. J. M. H. Olmsted, *Counterexamples in Analysis*, Dover Publications, Inc. (1964).

[42] B. R. Gelbaum. J. M. H. Olmsted, *Theorems and Counterexamples in Mathematics*, Springer Verlag, 1990.

[43] A. Gelfond, *Sur le septième Problème de Hilbert* Bulletin de l'Acadèmie des Sciences de l'URSS. Classe des sciences mathématiques et naturelles. VII (4): 1934, pp. 623-634.

[44] J. W. Gibbs, *Elements of Vector Analysis Arranged for the Use of Students in Physics*, Tuttle, Morehouse and Taylor, New Haven, 1884.

[45] J. W. Gibbs, *On Multiple Algebra*, Proceedings of the American Association for the Advancement of Science, vol. **35**, 1886.

[46] O. Goldman, *Hilbert rings and the Hilbert Nullstellensatz*, Math. Z., vol. **54**, 1951, pp. 136-140.

[47] D. M. Goldschmidt, *A group theoretic proof of the pq theorem for odd primes*, Math. Z., vol. **113**, 1970, pp. 373-375.

[48] R. Goodman, Nolan Wallach, *Symmetry, Representations, and Invariants*, Springer, 2009.

[49] M. de Gosson, *Introduction to Symplectic Mechanics: Lectures I-II-III*, Lecture Notes from a course at the University of St-Paulo, May-June 2006.

[50] M. A. de Gosson, *Symplectic Geometry, Wigner-Weyl-Moyal Calculus, and Quantum Mechanics in Phase Space.*

[51] M. Gray, *A radical approach to Algebra*, Addison-Wesley Publishing Company, 1970.

[52] M. Gromov, *Pseudo-holomorphic curves in symplectic manifolds*, Invent. Math., vol. **81**, 1985, pp. 307-347.

[53] Marshall Hall, *The Theory of Groups*, AMS, Second Edition, 1999.

[54] P. Halmos, *What Does the Spectral Theorem Say?*, American Mathematical Monthly, vol. **70**, number 3, 1963, pp. 241-247.

[55] A. Hatcher, *Algebraic Topology*, Cambridge University Press, 2002.

[56] T. Head, *Modules; A Primer of Structure Theorems.*, Brooks/Cole Publishing Company, 1974.

[57] I. N. Herstein, *Wedderburn's Theorem and a Theorem of Jacobson*, American Mathematical Monthly, vol. **68**, No. 3 (Mar., 1961), pp. 249-251.

[58] C. Hopkins, *Rings with minimum condition for left ideals*, Ann. Math. vol. **40**, 1039, pp. 712-730.

[59] L. Horowitz, *A proof of the Fundamental theorem of algebra by means of Galois theory and 2-Sylow groups*, Nieuw Arch. Wisk. (3) 14, 1966, p. 95-96.

[60] T. W. Hungerford, *Algebra.*, Springer, 1974.

[61] N. Jacobson, *Lie Algebras.* Wiley-Interscience, New York, 1962.

[62] N. Jacobson, *Structure theory of simple rings without finiteness assumptions*, Trans. Amer. Math. Soc., vol. **57**, 1945, pp. 228-245.

[63] N. Jacobson, *Basic algebra I, II*, p. 391 (transvections). Dover publications, NY, 1985.

[64] D. M. Kan, *Adjoint Functors*, Trans. Amer. Math. Soc., vol. **87**, 1958, pp. 294-329.

[65] I. Kaplansky, *Infinite Abelian groups*, University of Michigan Press, Ann Arbor, 1954.

[66] I. Kaplansky, *Commutative Rings*, Allyn and Bacon, Boston, MA, 1970.

[67] M. G. Katz, *Systolic Geometry and Topology*, Amer. Math. Soc., 2007.

[68] W. Krull, *Dimensionstheorie in Stellenringen*, J. reine angew. Math. **179**, 1938, pp. 204-226.

[69] W. Krull, *Jacobsonsches Radikal und Hilbertscher Nullstellensatz*, in Proceedings of the International Congress of Mathematicians, Cambridge, MA, 1950, vol. **2**, Amer. Math. Soc., Providence, 1952 pp. 56-64.

[70] S. Lang, *Algebra.*, Addison-Wesley Publishing Company, 1965.

[71] J. Levitzki, *On rings which satisfy the minimum condition for the right-hand ideals.*, Compositio Mathematica, vol. **7**, 1939, pp. 214-222.

[72] J. H. McKay, *Another proof of Cauchy's group theorem*, Amer. Math. Monthly **66**, 1959, p. 119.

[73] I. G. Macdonald, *Symmetric Functions and Hall Polynomials*, Oxford University Press, 2nd edition, 1995.

[74] G. W. Mackey, *The Scope and History of Commutative and Noncommutative Harmonic Analysis*, The History of Mathematics, AMS. vol. **5**, 1992.

[75] S. Mac Lane, *Homology*, Springer, 1995.

[76] P. Malcolmson, F. Okoh, *Rings Without Maximal Ideals*, Amer. Math. Monthly, vol. **107**, No. 1 (2000), pp. 60-61.

[77] H. Matsumura, *Commutative ring theory*, Cambridge University Press, Cambridge 1986.

[78] J. Reter May, *Munchi's Proof of the Nullstellenzatz*, The Amer. Math. Monthly, vol. **110**, No. 2, 2003, pp. 133-140.

[79] E. Meinrenken, *Symplectic Geometry*, Lecture Notes, University of Toronto, 2000.

[80] C. M. Mennen, *The Algebra and Geometry of Continued Fractions with Integer Quaternion Coefficients*, University of the Witwatersrand, Faculty of Science, School of Mathematics, 2015.

[81] P. Monsky, *On Dividing a Square into Trianles*, Amer. Math. Monthly, vol. **77** (2), 1970, pp. 161-164.

[82] L. J. Mordell, *A Statement by Fermat*, Proceedings of the London Math. Soc., vol. **18**, 1920, pp. v-vi.

[83] M. Moskowitz, *A Course in Complex Analysis in One Variable*, World Scientific Publishing Co., Inc., River Edge, NJ, 2002.

[84] M. Moskowitz, *Adventures in Mathematics*, World Scientific Publishing Co., Inc., River Edge, NJ, 2003.

[85] L. Nachbin, *The Haar integral*, D. Van Nostrand Co. Inc., Princeton, N.J.-Toronto-London 1965.

[86] M. Nagata, *Local Rings.*, Interscience Publishers, 1962.

[87] M. A. Naimark, *Normed Rings*, P. Noordoff N.V., The Netherlands, 1959.

[88] R. Narasimhan, *Analysis on Real and Complex Manifolds*, Advanced Studies in Pure Mathematics, Masson & Cie-Paris, 1973.

[89] M. Neusel, *Invariant Theory*, Am. Math. Soc., 2007.

[90] I. Niven, *A simple proof that* π *is irrational*, Bull. amer. Math. Soc., vol. **53**, 1947, p. 509.

[91] I. Niven, *Irrational numbers*, Carus Mathematical Monographs, no. **11**, A.M.S., Washington, D.C., 1985.

[92] E. Noether, *Ableitung der Elementarteilertheorie aus der Gruppentheorie*, Nachrichten der 27 Januar 1925, Jahresbericht Deutschen Math. Verein. (2. Abteilung) 34 (1926), 104.

[93] Yong-Geun Oh, *Uncertainty Principle, Non-Squeezing Theorem and the Symplectic Rigidity*, Lecture for the 1995 DAEWOO Workshop, Chungwon, Korea.

[94] T. O'Meara, *Introduction to Quadratic fields*, Springer 1973.

[95] O. Ore, *Number theory and its History*, Dover Publications, 1976.

[96] R. S. Palais, *The Classification of Real Division Algebras*, vol. **75**, No. 4, 1968, pp. 366-368.

[97] V. Perić and M. Vuković, *Some Examples of Principal Ideal Domain Which Are Not Euclidean and Some Counterexamples*, Novi Sad J. Math. vol. **38**, No. 1, 2008, pp. 137-154.

[98] H. Poincaré, *Analysis situs*, J. Éc. Polytech., ser. 21, 1895, pp. 1-123.

[99] H. Poincaré, *Second complément à l'analysis situs*, Proc. London Math. Soc., vol **32**, 1900, pp. 277-308.

[100] L. S. Pontryagin, *Topological groups*, Translated from the second Russian edition by Arlen Brown, Gordon and Breach Science Publishers Inc., New York-London-Paris 1966.

[101] M. Reid, *Undergraduate Algebraic Geometry*, Cambridge University Press, 1989.

[102] G. Ricci, T. Levi-Civita, *Méthodes de Calcul Différentiel Absolu et Leurs Applications*, Math. Annalen, vol. **54**, 1901, pp. 125-201.

[103] A. R. Schep, *A Simple Complex Analysis and an Advanced Calculus Proof of the Fundamental Theorem of Algebra*, Amer. Math. Monthly, vol. **116**, Number 1, 2009, pp. 67-68.

[104] I. Shafarevich, A. O. Remizov, *Linear Algebra and Geometry*, Springer, 2012.

[105] E. Sperner, *Neuer Beweis für die Invarianz der Dimensionszaht und des Gebietes*, Abh. Math. Sem. Hamburg, vol. **6**, 1928, pp. 265-272.

[106] D. Sullivan, *Geometric Topology. Localization, Periodicity, and Galois Symmetry*, The 1970 MIT notes, 2005.

[107] D. Suprunenko, *Matrix Groups* Translations of Mathematical Monographs, vol. **45**, AMS 1976.

[108] J. Tits, *Free subgroups in linear groups*, J. Algebra **20** (1972), 250-270.

[109] M. Tărnăuceanu, *A characterization of the quaternion group*, An. St. Univ. Ovidius Constantsa, vol. **21**(1), 2013, pp. 209-214.

[110] W. Vasconcelos, *Injective Endomorphisms of Finitely Generated Modules*, Proc. Amer. Math. Soc., **25**, 1970, pp. 900-901.

[111] V. S. Vladimirov, *P-adic Analysis and Mathematical Physics*, World Scientific Publishing Co. 1994.

[112] W. Voigt, *Die fundamentalen physikalischen Eigenschaften der Krystalle in elementarer Darstellung*, Verlag von Veit and Comp., Leipzig, 1898.

[113] E. Waring, *Meditationes Algebraicae*, Cambridge, England, 1770.

[114] J. H. Maclagan-Wedderburn, *A theorem on finite algebras*, Transactions of the American Mathematical Society, **6** (1905), pp. 349-352.

[115] C. A. Weibel, *An Introduction to Homological Algebra*, Cambridge University Press, 1994.

[116] A. Weinstein, *Symplectic Geometry*, Bulletin A.M.S. vol. **5**, no. 1, 1981.

[117] H. Weyl, *The Classical Groups, Their Invariants and Representations*, Princeton University Press, Princeton, N.J., 1939, 1946.

[118] H. Whitney, *Tensor products of abelian groups*, Duke Math. Journal, vol. **4**, 1938, pp. 495-528.

[119] H. Whitney, *Elementary structure of real algebraic varieties*, Ann. of Math. **66** (1957), 545-556.

[120] E. Witt, *Über die Kommutativität endlicher Schiefkörper*, Abhandlungen aus dem Maithematischen Seminar der Hamburgischen Universität, **8** (1931), p. 413, see also Collected Papers-Gesammelte Abhandlungen, ed. by Ina Kersten, Springer 1998.

[121] O. Zariski and P. Samuel, *Commutative algebra*, Vol. 1. With the cooperation of I. S. Cohen. Corrected reprinting of the 1958 edition. Graduate Texts in Mathematics, No. **28**. Springer-Verlag, New York-Heidelberg-Berlin, 1975.

[122] O. Zariski, *Generalized semi-local rings*, Summa Brasiliensis Math. **1**, fasc. 8, 1946, pp. 169-195.

[123] O. Zariski, *A new proof of Hilbert's Nullstellensatz*, Bull. Amer. Math. Soc., vol. **53**, 1947, pp. 362-368.

[124] H. J. Zassenhaus, *A group-theoretic proof of a theorem of Maclagan-Wedderburn*, Proceedings of the Glasgow Mathematical Association, **1** (1952), pp. 53-63.

[125] G. M. Ziegler, *Lectures on Polytopes*, Springer-Verlag New York, Inc., 1995.

Index

Printed in the United States
By Bookmasters